William Hamilton

Lectures on Metaphysics and Logic

Vol. II.

William Hamilton

Lectures on Metaphysics and Logic
Vol. II.

ISBN/EAN: 9783742803566

Manufactured in Europe, USA, Canada, Australia, Japa

Cover: Foto ©berggeist007 / pixelio.de

Manufactured and distributed by brebook publishing software
(www.brebook.com)

William Hamilton

Lectures on Metaphysics and Logic

LECTURES

ON

METAPHYSICS AND LOGIC

ON EARTH, THERE IS NOTHING GREAT BUT MAN;
IN MAN, THERE IS NOTHING GREAT BUT MIND.

LECTURES

ON

METAPHYSICS AND LOGIC

BY

SIR WILLIAM HAMILTON, BART.

PROFESSOR OF LOGIC AND METAPHYSICS IN THE
UNIVERSITY OF EDINBURGH

Advocate, A.M. (Oxon.) &c.; Corresponding Member of the Institute of France; Honorary Member of the
American Academy of Arts and Sciences; and of the Latin Society of Jena, &c.

EDITED BY THE

REV. H. L. MANSEL, B.D. LL.D.

WAYNFLETE PROFESSOR OF MORAL AND METAPHYSICAL PHILOSOPHY, OXFORD

AND

JOHN VEITCH, M.A.

PROFESSOR OF LOGIC, RHETORIC, AND METAPHYSICS, ST ANDREWS

IN FOUR VOLUMES

VOL. II.

SECOND EDITION, REVISED

WILLIAM BLACKWOOD AND SONS
EDINBURGH AND LONDON
MDCCCLXI

LECTURES

ON

METAPHYSICS

BY

SIR WILLIAM HAMILTON, BART.

EDITED BY THE

REV. H. L. MANSEL, B.D. LL.D.

[illegible]

AND

JOHN VEITCH, M.A.

[illegible]

VOL. II.

SECOND EDITION, REVISED

WILLIAM BLACKWOOD AND SONS
EDINBURGH AND LONDON
MDCCCLXI

CONTENTS OF VOL. II.

LECTURE XX.

LECTURE XXI.

LECTURE XXII.

LECTURE XXIII.

LECTURE XXIV.

LECTURE XXV.

LECTURE XXVI.

LECTURE XXVII.

LECTURE XXVIII.

LECTURE XXIX.

LECTURE XXX.

LECTURE XXXI.

LECTURE XXXII.

LECTURE XXXIII.

LECTURE XXXIV.

LECTURE XXXV.

LECTURE XXXVI.

LECTURE XXXVII.

LECTURE XXXVIII.

LECTURE XXXIX.

LECTURE XL.

LECTURE XLI.

LECTURE XLII.

LECTURE XLIII.

LECTURE XLIV.

LECTURE XLV.

LECTURE XLVI.

APPENDIX.

LECTURES ON METAPHYSICS.

LECTURE XX.

DISTRIBUTION OF THE SPECIAL COGNITIVE FACULTIES.

The Special Faculties of Knowledge.

GENTLEMEN,—We have now concluded the consideration of Consciousness, viewed in its more general relations, and shall proceed to analyse its more particular modifications, that is, to consider the various Special Faculties of Knowledge.

Three great classes of mental phænomena.

It is here proper to recall to your attention the division I gave you of the Mental Phænomena into three great classes,—viz., the phænomena of Knowledge, the phænomena of Feeling, and the phænomena of Conation. But as these various phænomena all suppose Consciousness as their condition,—those of the first class, the phænomena of knowledge, being, indeed, nothing but consciousness in various relations,—it was necessary, before descending to the consideration of the subordinate, first to exhaust the principal; and in doing this the discussion has been protracted to a greater length than I anticipated.

The first class,—Phænomena of Knowledge.

I now proceed to the particular investigation of the first class of the mental phænomena,—those of Knowledge or Cognition,—and shall commence by delineating to you the distribution of the cognitive faculties which I shall adopt;—a distribution different from

LECT. XX.

any other with which I am acquainted. But I would first premise an observation in regard to psychological powers, and to psychological divisions.

Mental powers.

As to mental powers,—under which term are included mental faculties and capacities,—you are not to suppose entities really distinguishable from the thinking principle, or really different from each other. Mental powers are not like bodily organs. It is the same simple substance which exerts every energy of every faculty, however various, and which is affected in every mode of every capacity, however opposite. This has frequently been wilfully or ignorantly misunderstood; and, among others, Dr Brown has made it a matter of reproach to philosophers in general, that they regarded the faculties into which they analysed the mind as so many distinct and independent existences.[a] No reproach, however, can be more unjust, no mistake more flagrant; and it can easily be shown that this is perhaps the charge of all others, to which the very smallest number of psychologists need plead guilty. On this point Dr Brown does not, however, stand alone as an accuser; and, both before and since his time, the same charge has been once and again preferred, and this, in particular, with singular infelicity, against Reid and Stewart. To speak only of the latter,—he sufficiently declares his opinion on the subject in a foot-note of the *Dissertation*:—"I quote," he says, "the following passage from Addison, *not* as a specimen of his metaphysical acumen, but as a proof of his good sense in divining and obviating a difficulty, which, I believe, most persons will acknowledge occurred to themselves when they first entered

Brown wrong as to the common philosophical opinion regarding these.

[a] *Philosophy of the Human Mind*, Lecture xvi., vol. i. p. 338, (second edition).—Ed.

on metaphysical studies:—'Although we divide the soul into several powers and faculties, there is no such division in the soul itself, since it is *the whole soul* that remembers, understands, wills, or imagines. Our manner of considering the memory, understanding, will, imagination, and the like faculties, is for the better enabling us to express ourselves in such abstracted subjects of speculation, not that there is any such division in the soul itself.' In another part of the same paper, Addison observes, 'that what we call the faculties of the soul are only the different ways or modes in which the soul can exert herself.'—*Spectator*, No. 600."[a]

What meant by mental power; and the relative opinion of philosophers.

I shall first state to you what is intended by the terms, *mental power*, *faculty*, or *capacity*; and then show you that no other opinion has been generally held by philosophers.

It is a fact too notorious to be denied, that the mind is capable of different modifications, that is, can exert different actions, and can be affected by different passions. This is admitted. But these actions and passions are not all dissimilar; every action and passion is not different from every other. On the contrary, they are like, and they are unlike. Those, therefore, that are like, we group or assort together in thought, and bestow on them a common name; nor are these groups or assortments manifold,—they are in fact few and simple. Again, every action is an effect; every action and passion a modification. But every effect supposes a cause; every modification supposes a subject. When we say that the mind exerts an energy, we virtually say that the mind is the cause of the energy; when we say that the mind acts or suffers,

a *Collected Works*, vol. I. p. 334.

LECT. XX.

we say in other words, that the mind is the subject of a modification. But the modifications, that is, the actions and passions, of the mind, as we stated, all fall into a few resembling groups, which we designate by a peculiar name; and as the mind is the common cause and subject of all these, we are surely entitled to say in general that the mind has the faculty of exerting such and such a class of energies, or has the capacity of being modified by such and such an order of affections. We here excogitate no new, no occult principle. We only generalise certain effects, and then infer that common effects must have a common cause; we only classify certain modes, and conclude that similar modes indicate the same capacity of being modified. There is nothing in all this contrary to the most rigid rules of philosophising; nay, it is the purest specimen of the inductive philosophy.

Faculty and Capacity distinguished.

On this doctrine, a *faculty* is nothing more than a general term for the causality the mind has of originating a certain class of energies; a *capacity* only a general term for the susceptibility the mind has of being affected by a particular class of emotions.[a] All mental powers are thus, in short, nothing more than names determined by various orders of mental phænomena. But as these phænomena differ from, and resemble, each other in various respects, various modes of classification may, therefore, be adopted, and, consequently, various faculties and capacities, in different views, may be the result.

Philosophical System,—its true place and importance.

And this is what we actually see to be the case in the different systems of philosophy; for each system of philosophy is a different view of the phænomena of mind. Now here I would observe that we might fall into one or other of two errors; by attributing either

a See above, vol. I. p. 177, *et seq.*—ED.

too great or too small importance to a systematic arrangement of the mental phænomena. It must be conceded to those who affect to undervalue psychological system, that system is neither the end first in the order of time, nor that paramount in the scale of importance. To attempt a definitive system or synthesis, before we have fully analysed and accumulated the facts to be arranged, would be preposterous, and necessarily futile; and system is only valuable when it is not arbitrarily devised, but arises naturally out of an observation of the facts, and of the whole facts, themselves; *τῆς πολλῆς πείρας τελευταῖον ἐπιγέννημα.*

On the other hand, to despise system is to despise philosophy; for the end of philosophy is the detection of unity. Even in the progress of a science, and long prior to its consummation, it is indeed better to assort the materials we have accumulated, even though the arrangement be only temporary, only provisional, than to leave them in confusion. For without such arrangement, we are unable to overlook our possessions; and as experiment results from the experiment it supersedes, so system is destined to generate system in a progress never attaining, but ever approximating to, perfection.

The opinion generally prevalent regarding mental powers.

Having stated what a psychological power in propriety is, I may add that this, and not the other, opinion, has been the one prevalent in the various schools and ages of philosophy. I could adduce to you passages in which the doctrine that the faculties and capacities are more than mere possible modes, in which the simple indivisible principle of thought may act and exist, is explicitly denied by Galen,[a] Lac-

[a] Galen, however, adopting Plato's threefold division of the faculties (*Ratio, Iracundia, Cupiditas*), expressly teaches that these have separate local seats, and that the mind is a whole composed of parts different

tantius,[α] Tertullian,[β] St Austin,[γ] Isidorus,[δ] Ireneus,[ε] Synesius,[ζ] and Gregory of Nyssa,[η] among the fathers of the Church; by Iamblichus,[θ] Plotinus,[ι]

both in kind and in nature (*genere et natura*). See his *De Hippocratis et Platonis Decretis*, lib. vi.; *Opera*, pp. 1003, 1004, *et seq.* (edit. Basle, 1549). Cf. lib. v. c. viii.—ED.

α [*De Opificio Dei*, c. 18.] [*Opera*, ii. 125 (edit. 1784); where, however, Lactantius merely pronounces the question in regard to the identity or difference of the *anima* and *animus*, insoluble, and gives the arguments on both sides.—ED.]

β [*De Anima*, c. 18.] [*Opera*, ii. 301, (edit. 1630): "Quid sensus, nisi ejus rei quæ sentitur intellectus? Quid intellectus, nisi ejus rei quæ intelligitur sensus? Unde ista tormenta cruciandæ simplicitatis, et suspendendæ veritatis? Quis mihi exhibebit sensum non intelligentem quod sentit? aut intellectum non sentientem quod intelligit? . . . Si corporalia quidem sentiuntur, incorporalia vero intelliguntur: verum genera diversa sunt non domicilia sensus et intellectus, id est, non anima et animus."—ED.]

γ See *De Trinitate*, lib. x. c. 8, § 18; *Opera*, viii. p. 898, (edit. Bened.): "Hæc tria, memoria, intelligentia, voluntas, quoniam non sunt tres vitæ, sed una vita; nec tres mentes, sed una mens: consequenter utique nec tres substantiæ sunt, sed una substantia. . . . . . . Quocirca tria hæc eo sunt unum, quo una vita, una mens, una essentia." Cf. *ibid.*, lib. xi. c. 3, §§ 5, 6, *Opera*, viii. p. 903, (edit. Bened.); lib. ix. c. 4, § 3, and c. 5, § 8. The doctrine of St Augustin on this point, however, divided the schoolmen. Henry of Ghent, and Gregory of Rimini, maintained that his opinion was Nominalistic, while others held that it might be identified with that of Aquinas. See Fromondus, *Philosophia Christiana de Anima*, lib. i. c. vi. art. iii. p. 160 *et seq.* (ed. 1649).—ED.

δ [*Originum*, lib. xi. c. 1.] [*Opera*, p. 91, (edit. 1617): "Hæc omnia adjuncta sunt animæ, ut una res sit. Pro efficientiis enim causarum diversa nomina sortita est anima. Nam et *memoria mens* est: dum ergo vivificat corpus, anima est; dum scit, *mens* est; dum vult, *animus* est; dum recolit, *memoria* est."—ED.]

ε [*Contra Hæreses*, lib. ii. c. 29.] [*Opera*, t. i. p. 392, (edit. Leipsic, 1848): "Sensus hominis, mens, et cogitatio, et intentio mentis, et ea quæ sunt hujusmodi, non aliud quid præter animam sunt; sed ipsius animæ motus et operationes, nullam sine anima habentes substantiam."—ED.]

ζ [*De Insomniis*,] [*Opera*, p. 103, (edit. 1553): "Ὅλῳ ἀκούει τῷ πνεύματι, καὶ ὅλῳ βλέπει, καὶ τὰ λοιπὰ πάντα δύναται. Δυνάμεις μία μὲν πᾶσαι κατὰ τὴν κοινὴν ῥίζαν· πολλαὶ δὲ κατὰ περίοδον.—ED.]

η [*De Hominis Opificio*, c. 6.] [*Opera*, t. i. p. 55.] [Οὐδὲ γὰρ ἡμῶν πολλαί τινες εἰσὶν αἱ ἀντιληπτικαὶ τῶν πραγμάτων δυνάμεις, εἰ καὶ πολυτρόπως διὰ τῶν αἰσθήσεων τῶν κατὰ ζωὴν ἐφαπτόμεθα. Μία γάρ τις ἐστὶ δύναμις, αὐτὸς ὁ ἐγκείμενος νοῦς, ὁ δι' ἑκάστου τῶν αἰσθητηρίων διεξιὼν, καὶ τῶν ὄντων ἐπιδρασσόμενος.—ED.]

θ "Anima quamvis videatur omnes rationes et totas in se species exhibere, tamen determinata semper est secundum aliquid unum, id est, unam speciem." *De Mysteriis*, as paraphrased by Marsilius Ficinus; *Opera*, p. 1879.—ED.

ι *Ennead*, iv. lib. iii. c. 3, p. 374, (ed. 1615): Τοῦτο δὲ οὐδέν' ἂν τὴν μὲν [ψυχὴν] ὅλην, τὴν δὲ μέρος ἂν εἶναι παραδέξοιτο, καὶ μάλιστα, οἷς τὸ αὐτὸ δυνάμεως πάρεστιν ἐπεὶ καὶ οἷς ἄλλο ἔργον, τῷ δὲ ἄλλο οἷον ὀφθαλμοῖς καὶ ὠσίν· οὐ μόριον ἄλλο ψυχῆς ὁράσει, ἄλλο δὲ ὠσὶ λεκτέον παρεῖναι, (ἄλλων δὲ τὸ μερίζειν οὕτως), ἀλλὰ τὸ αὐτό, κἂν

Proclus,[α] Olympiodorus,[β] and the pseudo Hermes Trismegistus,[γ] among the Platonists; by the Aphrodisian,[δ] Ammonius Hermiæ,[ε] and Philoponus[ζ] among the Aristotelians. Since the restoration of letters the same doctrine is explicitly avowed by the elder Scaliger,[η] Patricius,[θ] and Campanella;[ι] by Descartes,[κ] Male-

ἄλλῃ δύναμις ἐν ἑκατέρωι ἐνεργῇ. *Ibid.*, lib. ii. p. 363: Ψυχὴ μεριστὴ μέν, ὅτι ἐν πᾶσι μέρεσι τοῦ ἐν ᾧ ἐστιν· ἀμέριστος δὲ ὅτι ὅλη ἐν πᾶσι, καὶ ἐν ὁτῳοῦν αὐτοῦ ὅλη. Cf. lib. i. p. 361. —Ed.

α *In Platonis Theologiam*, lib. iv. c. xvi. p. 210, (edit. 1618): Διὰ γὰρ τὴν ἄκραν μετουσίαν τῆς συνοχῆς, ἀμέριστος ὁ νοῦς. Διὰ δὲ τὰ δεύτερα μέτρα τῆς μεθέξεως, ἡ ψυχὴ μεριστή, καὶ ἀμέριστός ἐστι, κατὰ μίαν σύγκρασιν. *Ibid.*, lib. i. c. xi. p. 25: Τὴν δὲ ψυχὴν ἓν καὶ πολλά;—thus rendered in the Latin version of Portus: "Animam unam et multa, [propter varias unius animæ facultates, et variarum rerum cognitionem, quam una anima habet."]—Ed.

β Olympiodorus adopts Plato's division of the soul into three principles. As regards the unity of the rational soul alone, something may perhaps be inferred from the Commentary on the *First Alcibiades*, where the rational soul is identified with the personal self. See especially pp. 203, 226, edit. Creuzer. Compare also a passage from his Commentary on the *Phædo*, cited by Cousin, *Fragments Philosophiques*, tom. i. p. 421, (ed. 1847). Neither passage, however, bears decisively on this question.—Ed.

γ [*De Intellectione et Sensu*, lib. xv. f. 42.] [Patricii, *Nova de Universis Philosophia*, (edit. 1593): Ἐν γὰρ τοῖς ἄλλοις ζώοις ἡ αἴσθησις τῇ φύσει ἥνωται, ἐν δ' ἀνθρώποις ἡ νόησις. Νοήσεως δὲ ὁ νοῦς διαφέρεται τοσοῦτον, ὅσον ὁ Θεὸς θειότητος. Ἡ μὲν γὰρ θειότης ὑπὸ τοῦ θεοῦ γίνεται, ἡ δὲ νόησις ὑπὸ τοῦ νοῦ, ἀδελφὴ οὖσα τοῦ λόγου, καὶ ὄργανα ἀλλήλων.—Ed.]

δ Πᾶσαι γὰρ αὗται (sc. ψυχὴ θρεπτική, αἰσθητική, φανταστική, ὁρμητική, ὀρεκτική) μία οὖσαι κατὰ τὸ ὑποκείμενον, ταῖς διαφοραῖς τῶν δυνάμεων αὐταῖς διαφέρονται. *In De Anima*, lib. i. f. 110a, (edit. Ven. 1534).—Ed.

ε Τῆς ἡμετέρας ψυχῆς διτταὶ αἱ ἐνέργειαι, αἱ μὲν γνωστικαί, οἷον νοῦς, δόξα, αἴσθησις, φαντασία, διάνοια, αἱ δὲ ζωτικαὶ καὶ ὀρεκτικαί, οἷον βούλησις, προαίρεσις, θυμός, καὶ ἐπιθυμία. *In Quinque Voces Porphyrii*, f. 7a, (edit. Aldine, 1546).—Ed.

ζ *In De Anima*, Proœm., f. 4a.: Οὐ γὰρ οἶδεν ἑαυτὴν ἡ ὄψις, ἢ ἡ ἀκοή, ἢ ἁπλῶς ἡ αἴσθησις· οὐδὲ ζητεῖ ποῖον ἐστι φύσιν· ἡ μέντοι ψυχὴ ἡ λογικὴ αὐτὴ ἑαυτὴν γινώσκει· αὐτὴ γοῦν ἐστιν ἡ ζητοῦσα, αὐτὴ ἡ ζητουμένη· αὐτὴ ἡ εὑρίσκουσα, αὐτὴ ἡ εὑρισκομένη· ἡ γινώσκουσα, καὶ γινωσκομένη. Cf. In lib. i. c. v., text 69, to end.—Ed.

η [*Exercitationes*, cccvii. § 1; cccvii. § 37.] [Cf. cccvii. § 15.—Ed.]

θ *Mystica Ægyptiorum et Chaldæorum*, lib. ii. c. iii. f. 4, col. 2: "Anima unica est et simplex; sed multiplicantur virtutes ejus, ultra substantiam, et si videtur operari plurima simul, ejus opera sunt multa ratione patientium. Si quidem corpora non recipiunt operationes animæ æqualiter, sed pro conditione sua; ergo pluralitas operationum inest rebus, non animæ." —Ed.

ι "Eandem animam sentientem et memorativam esse imaginativam et discursivam." See *De Sensu Rerum*, lib. ii. c. xxi. p. 77, (edit. 1637). Cf. cc. xix. xx.—Ed.

κ [*De Passionibus*, pars. ii. art. 68.]

LECT. XI.

branche,[α] Leibnitz,[β] and Wolf;[γ] by Condillac,[δ] Kant,[ε] and the whole host of recent philosophers. During the middle ages, the question was indeed one which divided the schools. St Thomas,[ζ] at the head of one party, held that the faculties were distinguished not only from each other, but from the essence of the mind; and this, as they phrased it, really and not formally. Henry of Ghent,[η] at the head of another party, maintained a modified opinion,—that the faculties were really distinguished from each other, but not from the essence of the soul. Scotus,[θ] again, followed by Occam[ι] and the whole sect of Nominalists, denied

α *Recherche de la Vérité*, lib. iii. c. i. § 1.—Ed.

β [*Nouveaux Essais*, liv. ii. c. xxi. § 6, p. 132—edit. Raspe.]

γ [*Psychologia Rationalis*, § 81.]

δ [*De l'Art de penser*, c. viii. *Cours*, t. iii. p. 304.]

ε *Kritik der reinen Vernunft*—Transc. Dial., B. ii. H. i. (p. 407, edit. 1799). Kant, however, while he admits this unity of the subject, as a conception involved in the fact of consciousness, denies that the conception can be legitimately transferred to the soul as a real substance.—Ed.

ζ *Summa*, pars i. qu. 77, art i. *et seq.* *Ibid.*, qu. 54, art. iii. Cf. *In Sent.*, lib. i. dist. iii. qu. 4, art. ii. St Thomas is followed by Capreolus, Cajetan, Ferrariensis, and Marsilius Ficinus. See Cottunius, *De Trip. Stat. Animæ Rationalis*, p. 281.—Ed.

η Henry of Ghent is, by Fromondus, classed with Gregory of Rimini and the Nominalists. See *De Anima*, lib. i. c. vi. art. 3. But see Genovesi, *Elementa Metaph.*, pars ii. p. 120.—Ed.

θ [See Zabarella, *De Rebus Naturalibus*, *Lib. De Facultatibus Animæ*, p. 685. Tennemann, *Gesch. der Philosophie*, viii. 2, p. 751.] ["Dico igitur," says Scotus, "quod potest sustineri, quod essentia animæ, indistincta re et ratione, est principium plurium actionum sine diversitate reali potentiarum, ita quod sint vel partes animæ, vel accidentia, vel respectus. . . . . Dices, quod erit ibi saltem differentia rationis. Concedo, sed hoc nihil faciet ad principium operationis realis." *In Sent.*, lib. ii. dist. xv. qu. 2, (quoted by Tennemann.) The Conimbricenses distinguish between the doctrine of Scotus, and that held in common by Gregory (Ariminensis), Occam, Gabriel Biel, Marsilius, and almost the whole sect of the Nominalists,—who, they say, concur in affirming,—"potentias [animæ] nec re ipsa, nec formaliter ex natura rei ab animæ essentia distingui, licet anima ex varietate actionum diversa nomina sortiatur;" whereas Scotus, according to them, is of opinion that, while the faculties cannot in reality (re ipsa) be distinguished from the mind, these may, however, be distinguished "formaliter, et ex natura rei." *In De Anima*, lib. ii. c. iii. qu. i, p. 150. Cottunius attributes the latter opinion to the Scotists universally. See his *De Triplici Statu Animæ Rationalis*, p. 280, (ed. 1628). Cf. Toletus, *In De Anima*, lib. ii. c. iv. f. 69.—Ed.]

ι *In Sent.*, lib. ii. dist. 16, qq. 21,

all real difference either between the several faculties, or between the faculties and the mind; allowing between them only a formal or logical distinction. This last is the doctrine that has subsequently prevailed in the latter ages of philosophy, and it is a proof of its universality, that few modern psychologists have ever thought it necessary to make an explicit profession of their faith in what they silently assumed. No accusation can, therefore, be more ungrounded than that which has been directed against philosophers,—that they have generally harboured the opinion that faculties are, like organs in the body, distinct constituents of mind. The Aristotelic principle, that in relation to the body "the soul is all in the whole and all in every part,"—that it is the same indivisible mind that operates in sense, in imagination, in memory, in reasoning, &c., differently indeed, but differently only because operating in different relations,[a]—this opinion is the one dominant among psychologists, and the one which, though not always formally proclaimed, must, if not positively disclaimed, be in justice presumptively attributed to every philosopher of mind. Those who employed the old and familiar language of philosophy, meant, in truth, exactly the same as those who would establish a new doctrine on a newfangled nomenclature.

The Aristotelic doctrine regarding the relation of the soul to the body.

From what I have now said, you will be better prepared for what I am about to state in regard to the classification of the first great order of mental phænomena, and the distribution of the faculties of Knowledge founded thereon. I formerly told you that the mental qualities,—the mental phænomena, are never

Psychological Division, what.

26. See Conimbricenses, *In De Anima*, p. 150. Cottunius, *De Trip. Stat. An. Rat.*, p. 290.—Ed.

a *De Anima*, lib. I. c. v. § 26 (ed. Trend.): 'Ἀλλ' οὐδὲν ἧττον ἐν ἑκατέρῳ τῶν μορίων ἅπαντ' ἐνυπάρχει τὰ μόρια τῆς ψυχῆς. κ.τ.λ. Cf. Plotinus, above, vol. ii. p. 6, note ι.—Ed.

presented to us separately; they are always in conjunction, and it is only by an ideal analysis and abstraction that, for the purposes of science, they can be discriminated and considered apart.[a] The problem proposed in such an analysis, is to find the primary threads which, in their composition, form the complex tissue of thought. In what ought to be accomplished by such an analysis, all philosophers are agreed, however different may have been the result of their attempts. I shall not state and criticise the various classifications propounded of the cognitive faculties, as I did not state and criticise the classifications propounded of the mental phænomena in general. The reasons are the same. You would be confused, not edified. I shall only delineate the distribution of the faculties of knowledge, which I have adopted, and endeavour to afford you some general insight into its principles. At present I limit my consideration to the phænomena of Knowledge; with the two other classes,—the phænomena of Feeling and the phænomena of Conation,—we have at present no concern.

The special faculties of knowledge, evolved out of Consciousness.

I again repeat that consciousness constitutes, or is coextensive with, all our faculties of knowledge,—these faculties being only special modifications under which consciousness is manifested. It being, therefore, understood that consciousness is not a special faculty of knowledge, but the general faculty out of which the special faculties of knowledge are evolved, I proceed to this evolution.

I. The Presentative Faculty.

In the first place, as we are endowed with a faculty of Cognition, or Consciousness in general, and since it cannot be maintained that we have always possessed the knowledge which we now possess, it will be admitted, that we must have a faculty of acquiring

a See above, vol. I. p. 183.—Ed.

knowledge. But this acquisition of knowledge can only be accomplished by the immediate presentation of a new object to consciousness, in other words, by the reception of a new object within the sphere of our cognition. We have thus a faculty which may be called the Acquisitive, or the Presentative, or the Receptive.

Subdivided, as External and Internal, into Perception and Self-Consciousness.

Now, new or adventitious knowledge may be either of things external, or of things internal, in other words, either of the phænomena of the non-ego or of the phænomena of the ego; and this distinction of object will determine a subdivision of this, the Acquisitive Faculty. If the object of knowledge be external, the faculty receptive or presentative of the qualities of such object, will be a consciousness of the non-ego. This has obtained the name of External Perception, or of Perception simply. If, on the other hand, the object be internal, the faculty receptive or presentative of the qualities of such subject-object, will be a consciousness of the ego. This faculty obtains the name of Internal or Reflex Perception, or of Self-Consciousness. By the foreign psychologists this faculty is termed also the Internal Sense.

Under the general faculty of cognition is thus, in the first place, distinguished an Acquisitive, or Presentative, or Receptive Faculty; and this acquisitive faculty is subdivided into the consciousness of the non-ego, or External Perception, or Perception simply, and into the Consciousness of the ego, or Self-Consciousness, or Internal Perception.

This acquisitive faculty is the faculty of Experience. External perception is the faculty of external, self-consciousness is the faculty of internal, experience. If we limit the term Reflection in conformity to its original employment and proper signification,—an

LECT. XX.

attention to the internal phænomena,—*reflection* will be an expression for self-consciousness concentrated.

II. The Conservative Faculty,—Memory Proper.

In the second place, inasmuch as we are capable of knowledge, we must be endowed not only with a faculty of acquiring, but with a faculty of retaining or conserving it when acquired. By this faculty, I mean merely, and in the most limited sense, the power of mental retention. We have thus, as a second necessary faculty, one that may be called the Conservative or Retentive. This is Memory, strictly so denominated, —that is, the power of retaining knowledge in the mind, but out of consciousness; I say retaining knowledge in the mind, but out of consciousness, for to bring the *retentum* out of memory into consciousness, is the function of a totally different faculty, of which we are immediately to speak. Under the general faculty of cognition is thus, in the second place, distinguished the Conservative or Retentive Faculty, or Memory Proper. Whether there be subdivisions of this faculty, we shall not here inquire.

III. The Reproductive Faculty.

But, in the third place, if we are capable of knowledge, it is not enough that we possess a faculty of acquiring, and a faculty of retaining it in the mind, but out of consciousness; we must further be endowed with a faculty of recalling it out of unconsciousness into consciousness, in short, a reproductive power. This Reproductive Faculty is governed by the laws which regulate the succession of our thoughts,—the laws, as they are called, of Mental Association. If these laws are allowed to operate without the intervention of the will, this faculty may be called Suggestion, or Spontaneous Suggestion; whereas, if applied under the influence of the will, it will properly obtain the name of Reminiscence or Recollection. By *reproduction*, it should be observed, that I strictly mean

Subdivided as without, or with Will, into Suggestion and Reminiscence.

the process of recovering the absent thought from unconsciousness, and not its representation in consciousness. This reproductive faculty is commonly confounded with the conversative, under the name of Memory; but most erroneously. These qualities of mind are totally unlike, and are possessed by different individuals in the most different degrees. Some have a strong faculty of conversation, and a feeble faculty of reproduction; others, again, a prompt and active reminiscence, but an evanescent retention. Under the general faculty of cognition, there is thus discriminated, in the third place, the Reproductive Faculty.

IV. The Representative Faculty,—Imagination.

In the fourth place, as capable of knowledge, we must not only be endowed with a presentative, a conservative, and a reproductive faculty; there is required for their consummation,—for the keystone of the arch,—a faculty of representing in consciousness, and of keeping before the mind the knowledge presented, retained, and reproduced. We have thus a Representative Faculty; and this obtains the name of Imagination or Phantasy.

The element of imagination is not to be confounded with the element of reproduction, though this is frequently, nay commonly, done; and this either by comprehending these two qualities under imagination, or by conjoining them with the quality of retention under memory. The distinction I make is valid. For the two faculties are possessed by different individuals in very different degrees. It is not, indeed, easy to see how, without a representative act, an object can be reproduced. But the fact is certain, that the two powers have no necessary proportion to each other. The representative faculty has, by philosophers, been distinguished into the Productive or Creative, and into

LECT. XX. the Reproductive, Imagination. I shall hereafter show you that this distinction is untenable.

Thus under the general cognitive faculty, we have a fourth special faculty discriminated,—the Representative Faculty,—Phantasy, or Imagination.

V. The Elaborative Faculty,—Comparison. In the fifth place, all the faculties we have considered are only subsidiary. They acquire, preserve, call out, and hold up, the materials, for the use of a higher faculty which operates upon these materials, and which we may call the Elaborative or Discursive Faculty. This faculty has only one operation, it only compares,—it is Comparison,—the faculty of Relations. It may startle you to hear that the highest function of mind is nothing higher than comparison, but, in the end, I am confident of convincing you of the paradox. Analysis and Synthesis. Under comparison, I include the conditions, and the result, of comparison. In order to compare, the mind must divide or separate, and conjoin or compose. Analysis and synthesis are, therefore, the conditions of comparison. Again, the result of comparison is either the affirmation of one thing of another, or the negation of one thing of another. If the mind affirm one thing of another, it conjoins them, and is thus again synthesis. If it deny one thing of another, it disjoins them, and is thus again analysis. Conception or Generalisation. Generalisation, which is the result of synthesis and analysis, is thus an act of comparison, and is properly denominated Conception. Judgment. Judgment is only the comparison Reasoning. of two terms or notions directly together; Reasoning, only the comparison of two terms or notions with each other through a third. Conception or Generalisation, Judgment and Reasoning, are thus only various applications of comparison, and not even entitled to the distinction of separate faculties.

Under the general cognitive faculty, there is thus discriminated a fifth special faculty in the Elaborative Faculty, or Comparison. This is Thought, strictly so called; it corresponds to the Διάνοια of the Greek, to the *Discursus* of the Latin, to the *Verstand* of the German philosophy; and its laws are the object of Logic.

VI. The Regulative Faculty,—Reason or Common Sense.

But in the sixth and last place, the mind is not altogether indebted to experience for the whole apparatus of its knowledge,—its knowledge is not all adventitious. What we know by experience, without experience we should not have known; and as all our experience is contingent, all the knowledge derived from experience is contingent also. But there are cognitions in the mind which are not contingent,—which are necessary,—which we cannot but think,—which thought supposes as its fundamental condition. These cognitions, therefore, are not mere generalisations from experience. But if not derived from experience, they must be native to the mind; unless, on an alternative that we need not at present contemplate, we suppose with Plato, St Austin, Cousin, and other philosophers, that Reason, or more properly Intellect, is impersonal, and that we are conscious of these necessary cognitions in the divine mind. These native,—these necessary cognitions, are the laws by which the mind is governed in its operations, and which afford the conditions of its capacity of knowledge. These necessary laws, or primary conditions, of intelligence, are phænomena of a similar character; and we must, therefore, generalise or collect them into a class; and on the power possessed by the mind of manifesting these phænomena, we may bestow the name of the Regulative Faculty. This faculty corresponds in some measure to what, in

LECT. XX.

the Aristotelic philosophy, was called Νοῦς,—νοῦς (*intellectus, mens*), when strictly employed, being a term, in that philosophy, for the place of principles,—the *locus principiorum*. It is analogous, likewise, to the term *Reason*, as occasionally used by some of the older English philosophers, and to the *Vernunft* (*reason*) in the philosophy of Kant, Jacobi, and others of the recent German metaphysicians, and from them adopted into France and England. It is also nearly convertible with what I conceive to be Reid's, and certainly Stewart's, notion of Common Sense. This, the last general faculty which I would distinguish under the Cognitive Faculty, is thus what I would call the Regulative or Legislative,—its synonyms being Νοῦς, Intellect, or Common Sense.

The term Faculty not properly applicable to Reason or Common Sense.

You will observe that the term *faculty* can be applied to the class of phænomena here collected under one name, only in a very different signification from what it bears when applied to the preceding powers. For νοῦς, intelligence or common sense, meaning merely the complement of the fundamental principles or laws of thought, is not properly a faculty, that is, it is not an active power at all. As it is, however, not a capacity, it is not easy to see by what other word it can be denoted.

These constitute the whole fundamental faculties of cognition.

Such are the six special Faculties of Cognition;—1°, The Acquisitive or Presentative or Receptive Faculty, divided into Perception and Self-Consciousness; 2°, The Conservative or Retentive Faculty, Memory; 3°, The Reproductive or Revocative Faculty, subdivided into Suggestion and Reminiscence; 4°, The Representative Faculty or Imagination; 5°, The Elaborative Faculty or Comparison, Faculty of Relations; and, 6°, The Regulative or Legislative Faculty, Intellect or

Intelligence Proper, Common Sense. Besides these faculties, there are, I conceive, no others; and, in the sequel, I shall endeavour to show you, that while these are attributes of mind not to be confounded,—not to be analysed into each other, the other faculties which have been devised by philosophers are either factitious and imaginary, or easily reducible to these.

Tabular view of the Faculties of Knowledge.

The following is a tabular view of the distribution of the Special Faculties of Knowledge:—

Cognitive Faculties		
	I. Presentative	External = Perception. Internal = Self-consciousness.
	II. Conservative	= Memory.
	III. Reproductive	Without will = Suggestion. With will = Reminiscence.
	IV. Representative	= Imagination.
	V. Elaborative	= Comparison,—Faculty of Relations.
	VI. Regulative	= Reason,—Common Sense.

LECTURE XXI.

THE PRESENTATIVE FACULTY.—I. PERCEPTION.—REID'S HISTORICAL VIEW OF THE THEORIES OF PERCEPTION.

LECT. XXI.

Recapitulation.

HAVING concluded the consideration of Consciousness as the common condition of the mental phænomena, and of those more general phænomena which pertain to consciousness as regarded in this universal relation; I proceeded, in our last Lecture, to the discussion of consciousness viewed in its more particular modifications,—that is, to the discussion of the Special Powers, —the Special Faculties and Capacities of Mind. And having called to your recollection the primary distribution of the mental phænomena into three great classes,—the phænomena included under our general faculty of Knowledge, or Thought, the phænomena included under our general capacity of Feeling, or of Pleasure and Pain, and the phænomena included under our general power of Conation, that is, of Will and Desire; I passed on to the consideration of the first of these classes,—that is, the phænomena of Knowledge. These phænomena are, in strictest propriety, mere modifications of consciousness, being consciousness only in different relations; and consciousness may, therefore, be regarded as the general faculty of knowledge: whereas the phænomena of the other classes, though they suppose consciousness as the condition of their manifestation, inasmuch as we cannot feel, nor will, nor desire, without knowing or being aware that we so do or suffer,—these phænomena are, however, something more than mere modifications of consciousness, seeing a new quality is superadded to that of cognition.

Employment of the term Cognition vindicated.

I may notice, parenthetically, the reason why I frequently employ *cognition* as a synonym of knowledge. This is not done merely for the sake of varying the expression. In the first place, it is necessary to have a word of this signification, which we can use in the plural. Now the term *knowledges* has waxed obsolete, though I think it ought to be revived. It is frequently employed by Bacon.[a] We must, therefore, have recourse to the term *cognition*, of which the plural is in common usage. But, in the second place, we must likewise have a term for knowledge, which we can employ adjectively. The word *knowledge* itself has no adjective, for the participle *knowing* is too vague and unemphatic to be employed, at least alone. But the substantive *cognition* has the adjective *cognitive*. Thus, in consequence of having a plural and an adjective, *cognition* is a word we cannot possibly dispense with in psychological discussion. It would also be convenient, in the third place, for psychological precision and emphasis, to use the word *to cognise* in connection with its noun *cognition*, as we use the decompound *to recognise* in connection with its noun *recognition*. But in this instance the necessity is not strong enough to warrant us doing what custom has not done. You will notice, such an innovation is always a question of circumstances; and though I would not subject Philosophy to Rhetoric more than Gregory

Condition under which the employment of new terms in philosophy is allowable.

a See above, vol. i. p. 57.—ED.

LECT. XXI.

the Great would Theology to Grammar, still, without an adequate necessity, I should always recommend you, in your English compositions, to prefer a word of Saxon to a word of Greek or Latin derivation. It would be absurd to sacrifice meaning to its mode of utterance,—to make thought subordinate to its expression; but still where no higher authority,—no imperious necessity, dispenses with philological precepts, these, as themselves the dictates of reason and philosophy, ought to be punctiliously obeyed. "It is not in language," says Leibnitz, "that we ought to play the puritan;"[α] but it is not either for the philosopher or the theologian to throw off all deference to the laws of language,—to proclaim of their doctrines,

> "Mysteria tanta
> Turpe est grammaticis submittere colla capistris."[β]

The general right must certainly be asserted to the philosopher of usurping a peculiar language, if requisite to express his peculiar analyses; but he ought to remember that the exercise of this right, as odious and suspected, is *strictissimi juris*, and that, to avoid the pains and penalties of grammatical recusancy, he must always be able to plead a manifest reason of philosophical necessity.[γ] But to return from this digression.

Having, I say, recalled to your observation the primary distribution of the mental phænomena into these three classes,—a distribution which, you will remember, I stated to you, was first promulgated by Kant,—I

α *Unvorgreiffliche Gedancken betreffend die Ausübung und Verbesserung der Teutschen Sprache*,—*Opera*, (edit. Dutens), vol. vi. pars ii. p. 13.—Ed.

β Buchanan, *Franciscanus*, l. 632.—Ed.

γ [Οὐχ ἡμεῖς οἱ ἐν τῷ τοιῷδε χορεύοντες, τῶν λόγων ὑπηρέται, ἀλλ' οἱ λόγοι οἱ ἡμέτεροι ὥσπερ οἰκέται.—Plato.] [*Theætetus*, p. 173.—Ed.] ["Hæc enim necessario extorquenda sunt a sapiente, quasi monstra monstris, absurda absurdis, inepta ineptis, ut inscitiæ minutissimas latebras vestigatas expugnemus." Scaliger, *In Arist. De Plant.*, lib. ii.] [f. 135ᵃ, ed. 1556.—Ed.]

proceeded to the subdivision of the first class of the general faculty of knowledge into its various special faculties,—a subdivision, I noticed, for the defects of which I am individually accountable. But before displaying to you a general view of my scheme of distribution, I first informed you what is meant by a power of mind, active or passive, in other words, what is meant by a mental faculty or a mental capacity; and this both in order to afford you a clear conception of the matter, and, likewise, to obviate some frivolous objections which have been made to such an analysis, or rather to such terms.

Phænomena of mind presented in composition.

The phænomena of mind are never presented to us undecomposed and simple, that is, we are never conscious of any modification of mind which is not made up of many elementary modes; but these simple modes we are able to distinguish, by abstraction, as separate forms or qualities of our internal life, since, in different states of mind, they are given in different proportions and combinations. We are thus able to distinguish as simple, by an ideal abstraction and analysis, what is never actually given except in composition; precisely as we distinguish colour from extension, though colour is never presented to us apart, nay, cannot even be conceived as actually separable, from extension. The aim of the psychologist is thus to analyse, by abstraction, the mental phænomena into those ultimate or primary qualities, which, in their combination, constitute the concrete complexities of actual thought. If the simple constituent phænomenon be a mental activity, we give to the active power thus possessed by the mind of eliciting such elementary energy the name of *faculty*; whereas if the simple or constituent phænomenon be a mental passivity, we give to the passive

LECT. XXI.

power thus possessed by the mind of receiving such an elementary affection, the name of *capacity*. Thus it is that there are just as many simple faculties as there are ultimate activities of mind, as many simple capacities as there are ultimate passivities of mind; and it is consequently manifest that a system of the mental powers can never be final and complete, until we have accomplished a full and accurate analysis of the various fundamental phænomena of our internal life. And what does such an analysis suppose? Manifestly three conditions:—1°, That no phænomenon be assumed as elementary which can be resolved into simpler principles; 2°, That no elementary phænomenon be overlooked; and, 3°, That no imaginary element be interpolated.

Three rules of psychological analysis.

These have not been observed by psychologists.

These are the rules which ought evidently to govern our psychological analyses. I could show, however, that these have been more or less violated in every attempt that has been made at a determination of the constituent elements of thought; for philosophers have either stopped short of the primary phænomenon, or they have neglected it, or they have substituted another in its room. I decline, however, at present an articulate criticism of the various systems of the human powers proposed by philosophers, as this would, in your present stage of advancement, tend rather to confuse than to inform you, and, moreover, would occupy a longer time than we are in a condition to afford: I therefore pass on to a summary recapitulation of the distribution of the cognitive faculties given in last Lecture. It is evident that such a distribution, as the result of an analysis, cannot be appreciated until the analysis itself be understood; and this can only be understood after the discussion of the several faculties and elementary phænomena has been carried through. You are, therefore, at present to look upon this scheme as little more than a table of contents to the various chapters, under which the phænomena of knowledge will be considered. I now only make a statement of what I shall subsequently attempt to prove. The principle of the distribution is, however, of such a nature that I flatter myself it can, in some measure, be comprehended even on its first enunciation: for the various elementary phænomena and the relative faculties which it assumes, are of so notorious and necessary a character, that they cannot possibly be refused; and, at the same time, they are discriminated from each other, both by obvious contrast, and by the fact that they are manifested in different individuals, each in very various proportions to each other.

Evolution of Special Faculties of Knowledge from Consciousness. I. The Acquisitive Faculty.

If man has a faculty of knowledge in general, and if the contents of his knowledge be not all innate, it is evident that he must have a special faculty of acquiring it,—an acquisitive faculty. But to acquire knowledge is to receive an object within the sphere of our consciousness; in other words, to present it, as existing, to the knowing mind. This Acquisitive Faculty may, therefore, be also called a Receptive or Presentative Faculty. The latter term, *Presentative Faculty*, I use, as you will see, in contrast and correlation to a *Representative Faculty*, of which I am immediately to speak. That the acquisition of knowledge is an ultimate phænomenon of mind, and an acquisitive faculty a necessary condition of the possession of knowledge, will not be denied. This faculty is the faculty of experience, and affords us exclusively all the knowledge we possess *a posteriori*, that is, our whole contingent knowledge, —our whole knowledge of fact. It is subdivided into two, according as its object is external or internal. In the former case it is called External Perception, or simply Perception; in the latter, Internal Perception, Reflex Perception, Internal Sense, or more properly, Self-Consciousness. Reflection, if limited to its original and correct signification, will be an expression for self-consciousness attentively applied to its objects, —that is, for self-consciousness concentrated on the mental phænomena.

II. The Conservative Faculty.

In the second place, the faculty of acquisition enables us to know,—to cognise an object, when actually presented within the sphere of external or of internal consciousness. But if our knowledge of that object terminated when it ceased to exist, or to exist within the sphere of consciousness, our knowledge would hardly deserve the name; for what we actually perceive by the faculties of external and of internal perception, is but an infinitesimal part of the knowledge which we actually possess. It is, therefore, necessary that we have not only a faculty to acquire, but a faculty to keep possession of knowledge, in short, a Conservative or Retentive Faculty. This is Memory strictly so denominated; that is, the simple power of retaining the knowledge we have once acquired. This conservation, it is evident, must be performed without an act of consciousness,—the immense proportion of our acquired and possessed riches must lie beyond the sphere of actual cognition. What at any moment we really know, or are really conscious of, forms an almost infinitesimal fraction of what at any moment we are capable of knowing.

III. The Reproductive Faculty.

Now this being the case, we must, in the third place, possess a faculty of calling out of unconsciousness into living consciousness the materials laid up by the conservative faculty, or memory. This act of calling out of memory into consciousness, is not identical with the act of conservation. They are not even similar or proportional; and yet, strange to say, they have always, or almost always, in the analyses of philosophers, been considered as inseparable. The faculty of which this act of revocation is the energy, I call the Reproductive. It is governed by the laws of Mental Association, or rather these laws are the conditions of this faculty itself. If it act spontaneously and without volition or deliberate intention, Suggestion is its most appropriate name; if, on the contrary, it act in subordination to the will, it should be called Reminiscence. The term Recollection, if not used as a synonym for reminiscence, may be employed indifferently for both.

IV. The Representative Faculty.

In the fourth place, the general capability of knowledge necessarily requires that, besides the power of evoking out of unconsciousness one portion of our retained knowledge in preference to another, we possess the faculty of representing in consciousness what is thus evoked. I will, hereafter, show you that the act of representation in the light of consciousness, is not to be confounded with the antecedent act of reproduction or revocation, though they severally, to a certain extent, infer each other. This Representative Faculty is Imagination or Phantasy. The word Fancy is an abbreviation of the latter; but with its change of form, its meaning has been somewhat modified. *Phantasy*, which latterly has been little used, was employed in the language of the older English philosophers as, like its Greek original, strictly synonymous with *Imagination*.

V. The Elaborative Faculty.

In the fifth place, these four acts of acquisition, conservation, reproduction, and representation, form a class of faculties, which we may call the Subsidiary,

LECT. XXI.

as furnishing the materials to a higher faculty, the function of which is to elaborate these materials. This elaborative or discursive faculty is Comparison; for under comparison may be comprised all the acts of Synthesis and Analysis, Generalisation and Abstraction, Judgment and Reasoning. Comparison, or the Elaborative or Discursive Faculty, corresponds to the Διάνοια of the Greeks, to the *Verstand* of the Germans. This faculty is Thought Proper; and Logic, as we shall see, is the science conversant about its laws.

VI. The Regulative Faculty.

In the sixth place, the previous faculties are all conversant about facts of experience,—acquired knowledge,—knowledge *a posteriori*. All such knowledge is contingent. But the mind not only possesses contingently a great apparatus of *a posteriori*, adventitious, knowledge; it possesses necessarily a small complement of *a priori*, native, cognitions. These *a priori* cognitions are the laws or conditions of thought in general; consequently, the laws and conditions under which our knowledge *a posteriori* is possible.

Knowledge *a priori* and *a posteriori*, explained.

By the way, you will please to recollect these two relative expressions. As used in a psychological sense, knowledge *a posteriori* is a synonym for knowledge empirical, or from experience; and, consequently, is adventitious to the mind, as subsequent to, and in consequence of, the exercise of its faculties of observation. Knowledge *a priori*, on the contrary, called likewise native, pure, or transcendental knowledge, embraces those principles which, as the conditions of the exercise of its faculties of observation and thought, are, consequently, not the result of that exercise. True it is that, chronologically considered, our *a priori* is not antecedent to our *a posteriori* knowledge; for the internal conditions of experience can only operate when an object of experience has been presented. In the order of time our knowledge, therefore, may be said to commence with experience, but to have its principle antecedently in the mind. Much as has been written on this matter by the greatest philosophers, this all-important doctrine has never been so well stated as in an unknown sentence of an old and now forgotten thinker. "Cognitio omnis a mente primam originem, a sensibus exordium habet primum."[a] These few words are worth many a modern volume of philosophy. You will observe the felicity of the expression. The whole sentence has not a superfluous word, and yet is absolute and complete. *Mens*, the Latin term for νοῦς, is the best possible word to express the intellectual source of our *a priori* principles, and is well opposed to *sensus*. But the happiest contrast is in the terms *origo* and *exordium;* the former denoting priority in the order of existence, the latter priority in the order of time.

Relation of our knowledge to experience,—how best expressed.

But to return whence I have diverged. These *a priori* principles form one of the most remarkable and peculiar of the mental phænomena; and we must class them under the head of a common power or principle of the mind. This power,—what I would call the Regulative Faculty,—corresponding to the Greek νοῦς when used as the *locus principiorum*, may be denominated Reason, using that word in the sense in which, as opposed to Reasoning, it was applied by some of the older English writers, and by Kant, Jacobi, and others of the more modern German philo-

a [Patricius, *Nova de Universis Philosophia*, p. 1.]

LECT. XXI. sophers. It may also be considered as equivalent to the term Common Sense, in the more correct acceptation of this expression.

The general faculty of knowledge is thus, according to this distribution, divided into six special faculties: first, the Acquisitive, Presentative, or Receptive; second, the Conservative; third, the Reproductive; fourth, the Representative; fifth, the Elaborative; and sixth, the Regulative. The first of these, the Acquisitive, is again subdivided into two faculties, —Perception and Self-Consciousness; the third into Suggestion and Reminiscence; and the fifth may likewise admit of subdivisions, into Conception, Judgment, and Reasoning, which, however, as merely applications of the same act in different degrees, hardly warrant a distinction into separate faculties.

The special faculties of Knowledge, considered in detail.

Having thus varied, amplified, and abridged the outline which I gave you in my last Lecture of the several constituents of the class of Cognitive Faculties, I now proceed to consider these faculties in detail.

1. The Presentative Faculty—Perception.

Perception, or the consciousness of external objects, is the first power in order. And in treating of this faculty,—the faculty on which turns the whole question of Idealism and Realism,—it is perhaps proper, in the first place, to take an historical survey of the hypotheses of philosophers in regard to Perception. In doing this, I shall particularly consider the views which Reid has given of these hypotheses: his authority on this the most important part of his philosophy is entitled to high respect; and it is requisite to point out to you, both in what respects he has misrepresented others, and in what been misrepresented himself.

Historical survey of hypotheses in regard to Perception, proposed.

The principal point in regard to Perception, on which opinions vary.

Before commencing this survey, it is proper to state in a few words, the one,—the principal, point in regard to which opinions vary. The grand distinction of philosophers is determined by the alternative they adopt on the question,—Is our perception or our consciousness of external objects, mediate or immediate?

As we have seen, those who maintain our knowledge of external objects to be immediate, accept implicitly the datum of consciousness which gives as an ultimate fact, in this act, an ego immediately known, and a non-ego immediately known. Those again who deny that an external object can be immediately known, do not accept one half of the fact of consciousness, but substitute some hypothesis in its place,—not, however, always the same. Consciousness declares that we have an immediate knowledge of a non-ego, and of an external non-ego. Now of the philosophers who reject this fact, some admit our immediate knowledge of a non-ego, but not of an external non-ego. They do not limit the consciousness or immediate knowledge of the mind to its own modes, but, conceiving it impossible for the external reality to be brought within the sphere of consciousness, they hold that it is represented by a vicarious image, numerically different from mind, but situated somewhere, either in the brain or mind, within the sphere of consciousness. Others, again, deny to the mind not only any consciousness of an external non-ego, but of a non-ego at all, and hold that what the mind immediately perceives, and mistakes for an external object, is only the ego itself peculiarly modified. These two are the only generic varieties possible of the representative hypothesis. And they have each their respective advan-

Two grand hypotheses of Mediate Perception.

LECT. XXI.

tages and disadvantages. They both equally afford a basis for idealism. On the former, Berkeley established his Theological, on the latter, Fichte his Anthropological Idealism. Both violate the testimony of consciousness, the one the more complex and the clumsier, in denying that we are conscious of an external non-ego, though admitting that we are conscious of a non-ego within the sphere of consciousness, either in the mind or brain. The other, the simpler and more philosophical, outrages, however, still more flagrantly the veracity of consciousness, in denying not only that we are conscious of an external non-ego, but that we are conscious of a non-ego at all.

Each of these admits of various subordinate hypotheses.

Each of these hypotheses of a representative perception admits of various subordinate hypotheses. Thus the former, which holds that the representative or immediate object is a *tertium quid*, different both from the mind and from the external reality, is subdivided according as the immediate object is viewed as material, as immaterial, or as neither, or as both, as something physical, or as something hyperphysical, as propagated from the external object, as generated in the medium, or as fabricated in the soul itself; and this latter either in the intelligent mind or in the animal life, as infused by God or by angels, or as identical with the divine substance, and so forth. In the latter, the representative modification has been regarded either as factitious, that is, a mere product of mind; or as innate, that is, as independent of any mental energy.[a]

Historical survey of opinions in regard to Perception.

I must return on this subject more articulately, when I have finished the historical survey. At present I only beg to call your attention to two facts which it is necessary to bear in mind: the first

a See *Reid's Works*, Note C, p. 816-819.—Ed.

regards a mistake of Reid, the second a mistake of Brown; and the proper understanding of these will enable you easily to apprehend how they have both wandered so widely from the truth.

Reid did not distinguish the two forms of the representative hypothesis.

Reid,[a] who, as I shall hereafter endeavour to show you, probably holds the doctrine of an Intuitive or Immediate Perception, never generalised, never articulately understood, the distinction of the two forms of the Representative Hypothesis. This was the cause of the most important errors on his part. In the first place, it prevented him from observing the obtrusive and vital distinction between Perception, to him a faculty immediately cognitive or presentative of external objects, and the faculties of Imagination and Memory, in which external objects can only be known to the mind mediately or in a representation. In the second place, this, as we shall see, causes him the greatest perplexity, and sometimes leads him into errors in his history of the opinions of previous philosophers, in regard to which he has, independently of this, been guilty of various mistakes.

Brown's general error in regard to Reid.

As to Brown, again,—he holds the simple doctrine of a representative perception,—a doctrine which Reid does not seem to have understood; and this opinion he not only holds himself, but attributes, with one or two exceptions, to all modern philosophers, nay even to Reid himself, whose philosophy he thus maintains to be one great blunder, both in regard to the new truths it professes to establish, and to the old errors it professes to refute. It turns out, however, that Brown in relation to Reid is curiously wrong from first to last,—not one of Reid's numerous mistakes, historical and

[a] See the Author's *Discussions*, p. 39 *et seq.*, and his Supplementary Dissertations to Reid, Notes B and C. —Ed.

LECT. XXI.

philosophical, does he touch, far less redargue; whereas in every point on which he assails Reid, he himself is historically or philosophically in error.

I meant to have first shown you Reid's misrepresentations of the opinions of other philosophers, and then to have shown you Brown's misrepresentations of Reid. I find it better to effect both purposes together, which, having now prepared you by a statement of Brown's general error, it will not, I hope, be difficult to do.

Reid's historical view of the theories of Perception. The Platonic.

This being premised, I now proceed to follow Reid through his historical view and scientific criticism of the various theories of Perception; and I accordingly commence with the Platonic. In this, however, he is unfortunate, for the simile of the cave which is applied by Plato in the seventh book of the Republic, was not intended by him as an illustration of the mode of our sensible perception at all. "Plato," says Reid,[a] "illustrates our manner of perceiving the objects of sense, in this manner. He supposes a dark subterraneous cave, in which men lie bound in such a manner that they can direct their eyes only to one part of the cave: far behind, there is a light, some rays of which come over a wall to that part of the cave which is before the eyes of our prisoners. A number of persons, variously employed, pass between them and the light, whose shadows are seen by the prisoners, but not the persons themselves.

"In this manner, that philosopher conceived that, by our senses, we perceive the shadows of things only, and not things themselves. He seems to have borrowed his notions on this subject from the Pythagoreans, and they very probably from Pythagoras himself.

a *Works*, p. 262.—ED.

If we make allowance for Plato's allegorical genius, his sentiments on this subject correspond very well with those of his scholar Aristotle, and of the Peripatetics. The shadows of Plato may very well represent the species and phantasms of the Peripatetic school, and the ideas and impressions of modern philosophers."

Reid wrong in regard to the Platonic theory of perception, and misapprehends Plato's simile of the cave.

Reid's account of the Platonic theory of perception is utterly wrong.[a] Plato's simile of the cave he completely misapprehends. By his cave, images, and shadows, this philosopher intended only to illustrate the great principle of his philosophy, that the sensible or ectypal world,—the world phænomenal, transitory, ever becoming but never being, (ἀεὶ γιγνόμενον, μηδέποτε ὄν), stands to the noetic or archetypal world,—the world substantial, permanent (ὄντως ὄν), in the same relation of comparative unreality, in which the shadows of the images of sensible existences themselves, stand to the objects of which they are the dim and distant adumbrations. The Platonic theory of these two worlds and their relations, is accurately stated in some splendid verses of Fracastorius,—a poet hardly inferior to Virgil, and a philosopher far superior to his age.

Fracastorius quoted.

"An nescis, quæcunque heic sunt, quæ hac nocte teguntur,
Omnia res prorsus veras non esse, sed umbras,
Aut specula, unde ad nos aliena elucet imago?
Terra quidem, et maria alta, atque his circumfluus aer,
Et quæ consistunt ex iis, hæc omnia tenues
Sunt umbræ, humanos quæ tanquam somnia quædam
Pertingunt animos, fallaci et imagine ludunt,
Nunquam eadem, fluxu semper variata perenni.
Sol autem, Lunæque globus, fulgentiaque astra
Cætera, sint quamvis meliori prædita vita,

a See the Author's note, *Reid's Works*, p. 262.—Ed.

LECT. XXI.

Et donata ævo immortali, hæc ipsa tamen sunt
Æterni specula, in quæ animus, qui est inde profectus,
Inspiciens, patriæ quodam quasi tactus amore,
Ardescit. Sed enim quoniam heic non perstat et ultra
Nescio quid sequitur secum, tacitusque requirit,
Nosse licet circum hæc ipsum consistere verum
Non finem: verum esse aliud quid, cujus imago
Splendet in iis, quod per se ipsum est, et principium esse
Omnibus æternum, ante omnem numerumque diemque;
In quo alium Solem atque aliam splendescere Lunam
Adspicias, aliosque orbes, alia astra manere,
Terramque, fluviosque alios, atque aera, et ignem,
Et nemora, atque aliis errare animalia silvis."[α]

Now, as well might it be said of these verses, that they are intended to illustrate a theory of perception, as of Plato's cave. But not only is Reid wrong in regard to the meaning of the cave, he is curiously wrong in regard to Plato's doctrine, at least of vision. For so far was Plato from holding that we only perceive in consequence of the representations of objects being thrown upon the percipient mind,—he, on the contrary, maintained in the *Timæus*,[β] that, in vision, a percipient power of the sensible soul sallies out towards the object, the images of which it carries back into the eye,—an opinion, by the way, held likewise by Empedocles,[γ] Alexander of Aphrodisias,[δ]

α These lines are given in the Author's note, *Reid's Works*, p. 262, and occur in the *Carmen ad M. Antonium Flaminium et Galeatium Florimontium*.—*Opera*, Venet., 1584, f. 206.—Ed.

β P. 45.—Ed.

γ "Visionem fieri per *extramissionem*" (as opposed to the *intromissionem* of Democritus, Leucippus, and Epicurus), "ait Empedocles, cui et Hipparchus astipulatus est, ita ut radii exeuntes quasi manu comprehendant imagines rerum quæ visionis sint effectrices." Gabriel Buratellus, *An Visio Fiat Extramittendo*, lib. v. Cf. *Empedoclis Fragmenta*, ed. Sturz, p. 416. Stallbaum, *In Plat. Timæum*, p. 45. Buratellus thus states Plato's doctrine of vision: "Visionem Plato fieri sentit ut oculi ex se naturam quandam lucidam habeant, ex qua visivi radii effluentes in extremum aeris lucem objectæ rei imaginem adducant, et in animo repræsentent, ex qua repræsentatione fit visus."—*Ibid.* Cf. Leo Hebræus, *De Amore*, Dial. iii.; Chalcidius, *In Timæum Platonis*, p. 388. See Bernardus, *Seminarium Philosophiæ Platonicæ*, p. 922.—Ed.

δ *In Arist. De Sensu*, ff. 95, 96, edit.

Seneca,[a] Chalcidius,[β] Euclid,[γ] Ptolemy,[δ] Alchindus,[ε] Galen,[ζ] Lactantius,[η] and Lord Monboddo.[θ]

Reid's account of the Aristotelic doctrine.

The account which Reid gives of the Aristotelic doctrine is, likewise, very erroneous. "Aristotle seems to have thought that the soul consists of two parts, or rather that we have two souls,—the animal and the rational; or, as he calls them, the soul and the intellect. To the *first* belong the senses, memory and imagination; to the *last*, judgment, opinion, belief, and reasoning. The first we have in common with brute animals; the last is peculiar to man. The animal soul he held to be a certain form of the body, which is inseparable from it, and perishes at death. To this soul the senses belong; and he defines a sense to be that which is capable of receiving the sensible forms or species of objects, without any of the matter of them; as wax receives the form of the seal without any of the matter of it. The forms of sound, of colour, of taste, and of other sensible qualities, are, in a manner, received by the senses. It seems to be a necessary consequence of Aristotle's doctrine, that bodies are constantly sending forth, in all directions, as many different kinds of forms without matter as they have different sensible qualities; for the forms of colour must enter by the eye, the forms of sound by the ear,

Ald. The Conimbricenses refer to the (probably spurious) *Problemata*, (lib. i. § 57, Lat. tr. 59, ed. Ald.)—ED.

a *Naturalium Quaestionum*, lib. i. c. 5-7.—ED.

β *In Timaeum Platonis*, p. 339. Cf. p. 329 *et seq.*, (edit. Leyden, 1617).—ED.

γ See Conimbricenses, *In De Anima*, lib. ii. c. vii. qu. 5, art. i. p. 231, (edit. 1629).—ED.

δ See Conimbricenses, *ibid.*—ED.

ε See Conimbricenses, *ibid.*—ED.

ζ *De Plac. Hippocratis et Platonis*, lib. vii. c. 5 (vol. v. p. 215, edit. Chartier).—ED.

η *De Opificio Dei*, c. viii. *Opera*, ii. (edit. 1754), where Lactantius, moreover, denies the necessity of visual species. See Conimbricenses, as above; and compare Stallbaum's note on the *Timaeus*, p. 45, B.—ED.

θ *Antient Metaphysics*, vol. i. book ii. chap. ii. p. 151. Cf. *Origin and Progress of Language*, vol. i. p. 26, (2d edit.)—ED.

LECT. XXI.

—and so of the other senses. This, accordingly, was maintained by the followers of Aristotle, though not, as far as I know, expressly mentioned by himself. They disputed concerning the nature of those forms of species, whether they were real beings or nonentities; and some held them to be of an intermediate nature between the two. The whole doctrine of the Peripatetics and schoolmen concerning forms, substantial and accidental, and concerning the transmission of sensible species from objects of sense to the mind, if it be at all intelligible, is so far above my comprehension that I should perhaps do it injustice, by entering into it more minutely." [a]

Only partially correct.

In regard to the statement of the Peripatetic doctrine of species, I must observe that it is correct only as applied to the doctrine taught as the Aristotelic in the schools of the middle ages; and even in these schools there was a large party who not only themselves disavowed the whole doctrine of species, but maintained that it received no countenance from the authority of Aristotle.[β] This opinion is correct; and I could easily prove to you, had we time, that there is nothing in the

[a] *Coll. Works*, p. 207.—Ed.

[β] [See Durandus, *In Sent.*, lib. II. dist. III. qu. 6, § 9: "Species originaliter introductæ videntur esse propter sensum visus, et sensibilia illius sensus. . . . . Sed quia quidam credunt quod species coloris *in oculo* representat visui colorem, cujus est species, ideo ponunt in intellectu quasdam species ad representandum res ut cognoscantur. § 10: Hoc autem non reputo verum nec in *sensu* nec in *intellectu*. Et quod non sit ponere speciem in *sensu*, patet sic:—Omne illud per quod tanquam per representativum potentia cognitiva fertur in alterum est primo cognitum; sed species coloris in oculo non est primo cognita seu visa ab eo, immo *nullo modo* est visa ab eo; ergo, per ipsam tanquam per representativum visus, non fertur in aliquid aliud. § 11: Quamvis enim color imprimat in medio et in oculo suam speciem propter similem dispositionem diaphaneitatis quæ est in eis, illa tamen nihil facit ad visionem, neque visui representat colorem ut videatur. § 21: Sensibilia secundum se præsentia sensui cognoscuntur per sensum, puta omnia colorata, et omnia lucentia quæ secundum se præsen-

metaphorical expressions of εἶδος and τύπος, which on one or two occasions he cursorily uses,[a] to warrant the attribution to him of the doctrine of his disciples. This is even expressly maintained by several of his Greek commentators,—as the Aphrodisian,[β] Michael

LECT. XXI.

tialiter objiciuntur visui, statim videntur, quia unum est *visivum* et aliud *visibile*, propter quod, vis approximatis, statim sequitur visio, a quocunque sit (fit ?) effective. Et similiter est de aliis sensibus." Durandus thus reduces *species* to the physical impression of the external object, which is unknown to the mind, and not like the object.] [See Conimbricenses, *In De Anima*, lib. ii. c. vi. qu. 2, p. 168. The Conimbricenses refer besides to Occam, Gregory (Ariminensis), and Biel, among the schoolmen, as concurring with Durandus on this point. The doctrine of species was also rejected by the Nominalists. See Toletus, *In De Anima*, lib. ii. c. xii. f. 109, (edit. 1594.) Cf. Plotinus, *Ennead*, iv. lib. iii. c. 26, p. 391, (edit. Basle, 1516): Τί οὖν; εἰ αὐτὴ μὲν μνημονεύει, τῷ δὲ ἐν σώματι εἶναι, τῷ μὴ καθαρᾷ (Codd. fere omnes καθαρά) εἶναι· ἀλλ' ὥσπερ ποιωθεῖσαν ἀναμάττεσθαι δύναται τοὺς τῶν αἰσθητῶν τύπους, καὶ τὸ οἷον ἕδραν ἐν τῷ σώματι πρὸς τὸ παραδέχεσθαι, καὶ μὴ ὥσπερ παραρρεῖν. Ἀλλὰ πρῶτον μὲν οἱ τύποι, οὐ μεγέθη· οὐδ' ὥσπερ αἱ ἐνσφραγίσεις, οὐδ' ἀντερείσεις, ἢ τυπώσεις, ὅτι μηδ' ὠθισμός, μηδ' ὥσπερ ἐν κηρῷ, ἀλλ' ὁ τρόπος οἷον νόησις, καὶ ἐπὶ τῶν αἰσθητῶν. See also Galen, *De Placitis Hippocratis et Platonis*, lib. vii. c. ix. It should be observed, however, that the great majority of the schoolmen attributed species both to the external and internal senses, and held that this was the doctrine of Aristotle. To this class belong Anselm, John of Damascus, Augustin, Aquinas, Alensis, Albertus Magnus, Bonaventura, Scotus, Argentinas, Richardus, Capreolus, Marsilius, Hervæus, and Ægidius. See Conimbricenses, *In De Anima*, p. 192, and Toletus, *In De Anima*, f. 109.—Ed.]

[a] See *De Anima*, lib. ii. c. xii. § 1, (edit. Trend.): Καθόλου δὲ περὶ πάσης αἰσθήσεως δεῖ λαβεῖν ὅτι ἡ μὲν αἴσθησίς ἐστι τὸ δεκτικὸν τῶν αἰσθητῶν εἰδῶν ἄνευ τῆς ὕλης, οἷον ὁ κηρὸς τοῦ δακτυλίου ἄνευ τοῦ σιδήρου καὶ τοῦ χρυσοῦ δέχεται τὸ σημεῖον, λαμβάνει δὲ τὸ χρυσοῦν ἢ τὸ χαλκοῦν σημεῖον, ἀλλ' οὐχ ᾗ χρυσὸς ἢ χαλκός. κ. τ. λ. *Ibid.*, lib. iii. c. ii. §§ 3, 4: Τὸ γὰρ αἰσθητήριον δεκτικὸν τοῦ αἰσθητοῦ ἄνευ τῆς ὕλης ἕκαστον· διὸ καὶ ἀπελθόντων τῶν αἰσθητῶν ἔνεισιν αἱ αἰσθήσεις καὶ φαντασίαι ἐν τοῖς αἰσθητηρίοις. Ἡ δὲ τοῦ αἰσθητοῦ ἐνέργεια καὶ τῆς αἰσθήσεως ἡ αὐτὴ μέν ἐστι καὶ μία, τὸ δ' εἶναι οὐ ταὐτὸν αὐταῖς. Cf. *De Memoria et Reminiscentia*, c. i., and *De An.*, lib. ii. c. iv.; lib. iii. c. viii.—Ed. [On Aristotle's doctrine in these passages; see Gassendi, *Physica*, Sect. iii., Mem. Post., lib. vi. c. ii., *Opera*, t. ii. p. 339, (edit. 1658). Cf. *Ibid.*, p. 337, and t. i. p. 443; t. iii. p. 467; Piccolomini, *In Phys.*, p. 1308; Zabarella, *De Rebus Naturalibus*, p. 989, *Liber, De Speciebus Intelligibilibus*; Devillemandy, *Scepticismus Debellatus*, c. xxiv. p. 165.] [Cf. Reid's *Works*, p. 827, note.—Ed.]

[β] [*In De Anima*, lib. i. f. 136a, (edit. Ald. 1534): Χρὴ δὲ τοῦ τύπου καταχρώμενον ἐπὶ τῆς φαντασίας ἀκούειν. Κυρίως μὲν γὰρ τύπος, τὸ κατ' εἰσοχήν τε καὶ ἐξοχήν, ἢ τὸ τοῦ τυποῦντος ἐν τῷ τυπουμένῳ σχῆμα γινόμενον, ὡς ὁρῶμεν τὰ ἐπὶ τῶν σφραγίδων ἔχοντα· οὐχ οὕτω δὲ τὰ ἀπὸ τῶν αἰσθητῶν

LECT. XXI.

Ephesius,[α] and Philoponus.[β] In fact, Aristotle appears to have held the same doctrine in regard to perception as Reid himself. He was a natural realist.[γ]

Theory of Democritus and Epicurus, omitted by Reid.

Reid gives no account of the famous doctrine of perception held by Epicurus, and which that philosopher had borrowed from Democritus,—namely, that the εἴδωλα, ἀπόῤῥοιαι, *imagines, simulacra rerum*, etc., are like pellicles continually flying off from objects; and that these material likenesses, diffusing themselves

ἐγκαταλείμματα γίνεται ἐν ἡμῖν. Οὐδὲ γὰρ τὴν ἀρχὴν κατὰ σχῆμά τι ἡ τῶν αἰσθητῶν ἀντίληψις. Ποῖον γὰρ σχῆμα τὸ λευκὸν, ἢ ὅλως τὸ χρῶμα; ἢ ποῖον σχῆμα ἡ ὀσμή; Ἀλλὰ δι᾽ ἀκυρίαν κυρίου τινὸς ὀνόματος, τὸ ἴχνος καὶ ἐγκατάλειμμα τὸ ὑπομένον ἀπὸ τῶν αἰσθητῶν ἐν ἡμῖν τύπον καλοῦμεν, μεταφέροντες τοὔνομα.] [Cf. *Ibid.*, lib. i. f. 135b: Δεῖ νοεῖν γίνεσθαι ἐν ἡμῖν ἀπὸ τῶν ἐνεργειῶν τῶν περὶ τὰ αἰσθητὰ, οἷον τύπον τινὰ καὶ ἀναζωγράφημα ἐν τῷ πρώτῳ αἰσθητηρίῳ . . . . Μήποτε δὲ οὐχ ὁ τύπος αὐτὸς ἡ φαντασία, ἀλλὰ ἡ περὶ τὸν τύπον τοῦτον τῆς φανταστικῆς δυνάμεως ἐνέργεια. The Aphrodisian is literally followed by Themistius, *In De Memoria et Reminiscentia*, c. i. f. 96b; cf. also the same, *In De Anima*, lib. ii. c. vi. ff. 78a, 83a, 93a, 96b, (edit. Ald. 1534); and by Simon Simonius, *In De Memoria et Reminiscentia*, c. i. §§ 12, 14, p. 290-91, (edit. 1566).—Ed.]

α [*In De Memoria et Reminiscentia*, Procem.] (fol. 127b, (edit. 1527).—Ed.]

β *In De Anima*, lib. ii. c. v. text 62: Δύναμις δέ ἐστι τὸ αἰσθητικὸν οἷον τὸ αἰσθητὸν κατὰ τὴν δευτέραν δύναμιν. Οὐ γὰρ πάσχοντα [παθόν τι?], οὐδὲ ὑπ᾽ ἐναντίας ἕξεως μεταβάλλον ὁμοιοῦται αὐτῷ, ἀλλὰ τὸ εἶδος αὐτοῦ δεξάμενον οὐχ ὡς ὕλη αὐτοῦ γινόμενον· οὐδὲ γὰρ λευκὴ γίνεται ἡ αἴσθησις δεξαμένη τὸ εἶδος τοῦ αἰσθητοῦ· (διὸ οὐδὲ πάσχειν οὐδὲ ἀλλοιοῦσθαι κυρίως λέγεται)· ἀλλὰ τὸν λόγον τοῦ εἴδους γνωστικῶς ἐν ἑαυτῇ δεχομένη. Ὥσπερ γὰρ τὸν κηρὸν φαμὲν δυνάμει εἶναι ὅπερ τὸν δακτύλιον, διότι παθὼν ὑπ᾽ αὐτοῦ γίνεται ὅπερ ἐστὶν ἐκεῖνος ἐνεργείᾳ· οὐ τὴν ὕλην αὐτοῦ δεξάμενος, ἀλλὰ μόνον τὸ εἶδος· οὕτω καὶ ἡ αἴσθησις παθοῦσα ὑπὸ τῶν αἰσθητῶν τὰ εἴδη αὐτῶν ἀπομάττει ἀναμάττεται. Διαφέρει δὲ, ὅτι ὁ μὲν κηρὸς αὐτὸς ὕλη γίνεται τοῦ εἴδους τοῦ ἐν τῷ δακτυλίῳ· ἡ δ᾽ αἴσθησις, οὐχ ὕλη γίνεται τοῦ αἰσθητοῦ, ἀλλὰ γνωστικῶς τὴν ἰδέαν αὐτοῦ ἐκμάττεται. Ἔχει δέ τι πλέον ἡ αἴσθησις παρὰ τὸν κηρόν· ὁ μὲν κηρὸς γὰρ, εἰ καὶ ὕλη γίνεται τοῦ εἴδους τοῦ ἐν τῷ δακτυλίῳ, ἀλλὰ οὐ δι᾽ ὅλου αὐτοῦ δέχεται τὸ εἶδος, ἀλλ᾽ ἐπιπολῆς· ἡ μέντοι αἰσθητικὴ δύναμις ὅλη δι᾽ ὅλης ζωτικῆς τὰς τῶν αἰσθητῶν ἀναμάττεται ἰδέας. Cf. *Ibid.*, c. xii. text 121. In this passage Philoponus closely approximates to the doctrine of the Platonists, as expounded by Priscianus Lydus, according to which, perception takes place on condition of an assimilation between the living organ and the object, by means of forms and immaterial reasons (κατὰ τὰ εἴδη καὶ τοὺς λόγους ἄνευ τῆς ὕλης). See Μετάφρασις τοῦ Θεοφράστου Περὶ Αἰσθήσεως, c. i. (version of Ficinus, s. 1 et seq.), and *Reid's Works*, p. 262, note. —Ed.

γ See above, vol. i. p. 296, note. —Ed.

everywhere in the air, are propagated to the perceptive organs. In the words of Lucretius,—

> "Quæ, quasi Membranæ, summo de cortice rerum
> Derptæ volitant ultro citroque per auras."[a]

Reid's statement of the Cartesian doctrine of Perception.

Reid's statement of the Cartesian doctrine of perception is not exempt from serious error. After giving a long, and not very accurate, account of the philosophy of Descartes in general, he proceeds :—"To return to Des Cartes's notions of the manner of our perceiving external objects, from which a concern to do justice to the merit of that great reformer in philosophy has led me to digress, he took it for granted, as the old philosophers had done, that what we immediately perceive must be either in the mind itself, or in the brain, to which the mind is immediately present. The impressions made upon our organs, nerves, and brain, could be nothing, according to his philosophy, but various modifications of extension, figure, and motion. There could be nothing in the brain like sound or colour, taste or smell, heat or cold ; these are sensations in the mind, which, by the laws of the union of soul and body, are raised on occasion of certain traces in the brain ; and although he gives the name of ideas to those traces in the brain, he does not think it necessary that they should be perfectly like to the things which they represent, any more than that words or signs should resemble the things they signify. But, says he, that we may follow the received opinion as far as is possible, we may allow a slight resemblance. Thus we know that a print in a book may represent houses, temples, and groves ; and so far is it from being necessary that the print should be perfectly like

a Lib. iv. 35. So quoted in the Author's *Discussions*, p. 71, but the usual reading is *corpore*, not *cortice*.—Ed.

LECT. XXI. the thing it represents, that its perfection often requires the contrary; for a circle must often be represented by an ellipse, a square by a rhombus, and so of other things. . . . .

"The writings of Des Cartes have, in general, a remarkable degree of perspicuity; and he undoubtedly intended that, in this particular, his philosophy should be a perfect contrast to that of Aristotle; yet, in what he has said, in different parts of his writings, of our perceptions of external objects, there seems to be some obscurity, and even inconsistency; whether owing to his having had different opinions on the subject at different times, or to the difficulty he found in it, I will not pretend to say.

"There are two points, in particular, wherein I cannot reconcile him to himself: the *first*, regarding the place of the ideas or images of external objects, which are the immediate objects of perception; the *second*, with regard to the veracity of our external senses.

"As to the *first*, he sometimes places the ideas of material objects in the brain, not only when they are perceived, but when they are remembered or imagined; and this has always been held to be the Cartesian doctrine; yet he sometimes says, that we are not to conceive the images or traces in the brain to be perceived, as if there were eyes in the brain; these traces are only occasions on which, by the laws of the union of soul and body, ideas are excited in the mind; and, therefore, it is not necessary that there should be an exact resemblance between the traces and the things represented by them, any more than that words or signs should be exactly like the things signified by them.

"These two opinions, I think, cannot be reconciled. For, if the images or traces in the brain are perceived, they must be the objects of perception, and not the occasions of it only. On the other hand, if they are only the occasions of our perceiving, they are not perceived at all. Des Cartes seems to have hesitated between the two opinions, or to have passed from the one to the other."[α]

I have quoted to you this passage in order that I may clearly exhibit to you, in the first place, Reid's misrepresentations of Descartes; and, in the second, Brown's misrepresentation of Reid.

Cardinal principle of the Cartesian philosophy.

In regard to the former, Reid's principal error consists in charging Descartes with vacillation and inconsistency, and in possibly attributing to him the opinion that the representative object of which the mind is conscious in perception, is something material,—something in the brain. This arose from his ignorance of the fundamental principle of the Cartesian doctrine.[β] By those not possessed of the key to the Cartesian theory, there are many passages in the writings of its author which, taken by themselves, might naturally be construed to import, that Descartes supposed the mind to be conscious of certain motions in the brain, to which, as well as to the modifications of the intellect itself, he applies the terms *image* and *idea*. Reid, who did not understand the Cartesian philosophy as a system, was puzzled by these superficial ambiguities. Not aware that the cardinal point of that system is, that mind and body, as essentially opposed, are naturally to each other as zero; and that their mutual

α *Intellectual Powers*, Essay II. chap. viii. *Coll. Works*, p. 272.

β The following remarks have been printed in the Author's article on Reid and Brown. See *Discussions*, p. 72.—Ed.

LECT. XXI.

intercourse can, therefore, only be supernaturally maintained by the concourse of the Deity, Reid was led into the error of attributing, by possibility, to Descartes, the opinion that the soul was immediately cognisant of material images in the brain. But in the Cartesian theory, mind is only conscious of itself; the affections of body may by the law of union be proximately the occasions, but can never constitute the immediate objects, of knowledge. Reid, however, supposing that nothing could obtain the name of *image*, which did not represent a prototype, or the name of *idea*, which was not an object of thought, wholly misinterpreted Descartes, who applies, abusively indeed, these terms to the occasion of perception, that is, the motion in the sensorium, unknown in itself, and representing nothing; as well as to the object of thought, that is, the representation of which we are conscious in the mind itself. In the Leibnitzio-Wolfian system, two elements, both also denominated *ideas*, are in like manner accurately to be contradistinguished in the process of perception. The idea in the brain, and the idea in the mind, are, to Descartes, precisely what the "*material idea*" and the "*sensual idea*" are to the Wolfians. In both philosophies, the two ideas are harmonic modifications, correlative and coexistent; but in neither is the organic affection or sensorial idea an object of consciousness. It is merely the unknown and arbitrary condition of the mental representation; and in the hypothesis both of Assistance and of Preestablished Harmony, the presence of the one idea implies the concomitance of the other, only by virtue of the hyperphysical determination.

Twofold use of the term idea by Descartes.

LECTURE XXII.

THE PRESENTATIVE FACULTY.—I. PERCEPTION.—REID'S HISTORICAL VIEW OF THE THEORIES OF PERCEPTION.

LECT. XXII.

Recapitulation.

In our last Lecture, after recapitulating, with varied illustrations, the Distribution of the Cognitive Faculties, which I had detailed to you in the Lecture before, I entered upon the particular consideration of the Special Faculties themselves, and commenced with that which stands first in order, and which I had denominated the Acquisitive, or Receptive, or Presentative. And as this faculty is again subdivided into two, according as it is conversant either about the phænomena of matter, or about the phænomena of mind, the non-ego, or the ego, I gave precedence to the former of these,—the faculty known under the name of External Perception. Perception, as matter of psychological consideration, is of the very highest importance in philosophy; as the doctrine in regard to the object and operation of this faculty, affords the immediate data for determining the great question touching the existence or non-existence of an external world; and there is hardly a problem of any moment in the whole compass of philosophy, of which it does not mediately affect the solution. The doctrine of perception may thus be viewed as a cardinal point of philosophy. It is also exclusively in relation to this faculty, that Reid must claim his great, his distinguishing glory, as a

The doctrine of Perception a cardinal point in Philosophy.

Its place in the philosophy of Reid.

LECT. XXII.

philosopher; and of this no one was more conscious than himself. "The merit," he says, in a letter to Dr James Gregory, "of what you are pleased to call my philosophy, lies, I think, chiefly in having called in question the common theory of ideas or images of things in the mind being the only objects of thought; a theory founded on natural prejudices, and so universally received as to be interwoven with the structure of language." "I think," he adds, "there is hardly anything that can be called science in the philosophy of the mind, which does not follow with ease from the detection of this prejudice."[a] The attempts, therefore, among others, of Priestley, Gleig, Beasley,[β] and, though last not least, of Brown, to show that Reid in his refutation of the previous theory of perception, was only fighting with a shadow,—was only combating philosophers who, on the point in question, really coincided with himself, would, if successful, prove not merely that the philosophical reputation of Reid is only based upon a blunder, but would, in fact, leave us no rational conclusion short, not of idealism only, but of absolute scepticism. For, as I have shown you, Brown's doctrine of perception, as founded on a refusal of the testimony of consciousness to our knowledge of an external world, virtually discredits consciousness as an evidence at all; and in place of his system being, as its author confidently boasts, the one "which allows the sceptic no place for his foot,—no fulcrum for the instrument he uses,"—it is, on the contrary, perhaps the system which, of all others, is the most contradictory and suicidal, and which, consequently, may most

a *Collected Works*, p. 88.—ED.

β See Priestley, *Examination of Reid, Beattie, and Oswald*, sect. iii. (p. 30, 2d edition); Bishop Gleig, art. *Metaphysics*, *Encycl. Britan.*, vol. xiv. p. 601, 7th edit.; Beasley, *Search of Truth in the Science of the Human Mind*, book II. c. iii. p. 123 *et seq.*; cf. cc. iv., v., vi. (Philadelphia, U.S., 1822.)—ED.

easily be developed into scepticism. The determination of this point is, therefore, a matter affecting the vital interests of philosophy; for if Reid, as Brown and his coadjutors maintain, accomplished nothing, then is all philosophical reputation empty, and philosophy itself a dream.

Reid, philosophically and historically, not free from errors.

In preparing you for the discussion that was to follow, I stated to you that it would not be in my power to maintain Reid's absolute immunity from error, either in his philosophical or in his historical views; on the contrary, I acknowledged that I found him frequently at fault in both. His mistakes, however, I hope to show you, are not of vital importance, and I am confident their exposure will only conduce to illustrate and confirm the truths which he has the merit, though amid cloud and confusion, to have established. But as to Brown's elaborate attack on Reid, —this, I have no hesitation in asserting, to be not only unsuccessful in its results, but that in all its details, without a single, even the most insignificant, exception, it has the fortune to be regularly and curiously wrong. Reid had errors enough to be exposed, but Brown has not been so lucky as to stumble even upon one. Brown, however, sung his pæan as if his victory were complete; and, what is singular, he found a general chorus to his song. Even Sir James Mackintosh talks of Brown's triumphant exposure of Reid's marvellous mistakes.

But Brown's criticism of Reid wholly wrong.

General source of Reid's errors,—which, however, are comparatively unimportant.

To enable you provisionally to understand Reid's errors, I showed you how, holding himself the doctrine of an intuitive or immediate perception of external things, he did not see that the counter doctrine of a mediate or representative perception admitted of a subdivision into two forms,—a simpler and a more complex. The simpler, that the immediate or repre-

LECT. XXII.

sentative object is a mere modification of the percipient mind,—the more complex, that this representative object is something different both from the reality and from the mind. His ignorance of these two forms has caused him great confusion, and introduced much subordinate error into his system, as he has often confounded the simpler form of the representative hypothesis with the doctrine of an intuitive perception; but if he be allowed to have held the essential doctrine of an immediate perception, his errors in regard to the various forms of the representative hypothesis must be viewed as accidental, and comparatively unimportant.

Brown's errors vital.

Brown's errors, on the contrary, are vital. In the first place, he is fundamentally wrong in holding, in the teeth of consciousness, that the mind is incapable of an immediate knowledge of aught but its own modes. He adopts the simpler form of a representative perception. In the second place, he is wrong in reversing Reid's whole doctrine, by attributing to him the same opinion on this point which he himself maintains. In the third place, he is wrong in thinking that Reid only attacked the more complex, and not the more dangerous, form of the representative hypothesis, and did not attack the hypothesis of representation altogether. In the fourth place, he is wrong in supposing that modern philosophers in general held the simpler form of the representative hypothesis, and that Reid was, therefore, mistaken in supposing them to maintain the more complex,—mistaken, in fact, in supposing them to maintain a doctrine different from his own.

Having thus prepared you for the subsequent discussion, I proceeded to consider Reid's historical account

General character of Reid's historical account of philosophical opinions on Perception.

of the opinions on Perception held by previous philosophers. This historical account is without order, and at once redundant and imperfect. The most important doctrines are altogether omitted; of others the statement is repeated over and over in different places, and yet never completely done at last; no chronological succession, no scientific arrangement, is followed, and with all this the survey is replete with serious mistakes. Without, therefore, following Reid's confusion, I took up the opinions on which he touched, in the order of time. Of these the first was the doctrine of Plato; in regard to which I showed you, that Reid was singularly erroneous in mistaking what Plato meant by the simile of the cave. Then followed the doctrine of Aristotle and his school, in relation to whom he was hardly more correct. Did our time allow me to attempt a history of the doctrines on perception, I could show you, that Aristotle must be presumed to have held the true opinion in regard to this faculty;[α] but in respect to a considerable number of the Aristotelic schoolmen, I could distinctly prove, not only that the whole hypothesis of species was by them rejected, but that their hitherto neglected theory of perception is, even at this hour, the most philosophical that exists.[β] I have no hesitation in saying that, on this point, they are incomparably superior to Reid: for while he excuses Brown's misinterpretation, and, indeed, all but annihilates his own doctrine of perception, by placing that power in a line with imagination and memory, as all faculties immediately cognisant of the reality; they, on the contrary, distinguish Perception as a faculty intuitive, Imagina-

α See vol. i. p. 296, and vol. ii. p. 36 *et seq.*—ED.

β See above, vol. ii. p. 36 *et seq.*, and below, p. 71.—ED.

LECT. XXII.

tion and Memory as faculties representative of their objects.

Following Reid in his descent to modern philosophers, I showed you how, in consequence of his own want of a systematic knowledge of the Cartesian philosophy, he had erroneously charged Descartes with vacillation and contradiction, in sometimes placing the idea of a representative image in the mind, and sometimes placing it in the brain.

Reid right in supposing that Descartes held the more complex hypothesis of Representative Perception.

Such is the error of Reid in relation to Descartes, which I find it necessary to acknowledge. But, on the other hand, I must defend him on another point from Brown's charge of having not only ignorantly misunderstood, but of having exactly reversed, the notorious doctrine of Descartes; in supposing that this philosopher held the more complex hypothesis of a representative perception, which views in the representative image something different from the mind, instead of holding, with Reid himself and Brown, the simpler hypothesis, which views in this image only a mode of the percipient mind itself.

Now here you must observe that it would not be enough to convict Reid and to justify Brown, if it were made out that the former was wrong, the latter right, in his statement of Descartes' opinion; and I might even hold with Brown that Descartes had adopted the simpler theory of representation, and still vindicate Reid against his reproach of ignorant misrepresentation,—of reading the acknowledged doctrine of a philosopher, whose perspicuity he himself admits, in a sense "exactly the reverse" of truth. To determine with certainty what Descartes' theory of perception actually is, may be difficult, perhaps impossible. It here suffices to show that his opinion on the point in question is doubtful,—is even one mooted among his disciples; and that Brown, wholly unacquainted with the doubts and difficulties of the problem, dogmatises on the basis of a single passage of Descartes,—nay, of a passage wholly irrelevant to the matter in dispute. The opinion attributed by Reid to Descartes is the one which was almost universally held in the Cartesian school as the doctrine of its founder; and Arnauld is the only Cartesian who adopted an opinion upon perception identical with Brown's, and who also assigned that opinion to Descartes. The doctrine of Arnauld was long regarded throughout Europe as a paradox, original and peculiar to himself.

Malebranche cited in regard to opinion of Descartes.

Malebranche,[a] the most illustrious name in the school, after its founder, and who, not certainly with less ability, may be supposed to have studied the writings of his master with far greater attention than either Reid or Brown, ridicules, as "contrary to common sense and justice," the supposition that Descartes had rejected ideas in "the ordinary acceptation," and adopted the hypothesis of their being representations, not really distinct from their perception. And while he "was certain as he possibly can be in such matters," that Descartes had not dissented from the general doctrine, he taunts Arnauld with resting his paradoxical interpretation of that philosopher's doctrine, "not on any passages of his Metaphysics contrary to the 'common opinion,' but on his own arbitrary limitation of 'the ambiguous term perception.'"[β] That ideas are "found in the mind, not formed by it," and, consequently, that in the act of knowledge, the representation is really distinct from the cognition proper, is

α Given in *Discussions*, p. 74.—Ed.

β *Réponse au Livre des Idées*, passim.—Arnauld, *Œuvres*, xxxviii. pp. 388, 389.

LECT. XXII.

strenuously asserted as the doctrine of his master by the Cartesian Röell,* in the controversy he maintained with the anti-Cartesian De Vries. But it is idle to multiply proofs. Brown's charge of ignorance falls back upon himself; and Reid may lightly bear the reproach of "exactly reversing" the notorious doctrine of Descartes, when thus borne along with him by the profoundest of that philosopher's disciples.

Reid's account of the opinion of Malebranche.

Malebranche and Arnauld are the next philosophers, in chronological order, of whom Reid speaks. Concerning the former, his statements, though not complete, cannot be considered as erroneous; and Dr Brown, admitting that Malebranche is one of the two, and only two modern philosophers, (Berkeley is the other), who held the more complex doctrine of representation, of course does not attempt to accuse Reid of misrepresentation in reference to him. One error, however, though only an historical one, Reid does commit, in regard to this philosopher. He explains the polemic which Arnauld waged with Malebranche, on the ground of the antipathy between Jansenist and Jesuit. Now Malebranche was not a Jesuit, but a priest of the Oratory.

Reid confused in his account of the view of Arnauld.

In treating of Arnauld's opinion, we see the confusion arising from Reid's not distinctly apprehending the two forms of the representative hypothesis. Arnauld held, and was the first of the philosophers noticed by Reid or Brown who clearly held, the simpler of these forms. Now in his statement of Arnauld's doctrine, Reid was perplexed,—was puzzled. As opposing the philosophers who maintained the more complex doctrine of representation, Arnauld seemed to Reid to coincide in opinion with himself; but yet, though he never rightly

* Cf. Röell, *Dissertationes Philosophicæ*, i. § 13; iii. § 64.—Ed.

understood the simpler doctrine of representation, he still feels that Arnauld did not hold with him an intuitive perception. Dr Brown is, therefore, wrong in asserting that Reid admits Arnauld's opinion on perception and his own to be identical.[a] "To these authors," says Dr Brown, "whose opinions on the subject of perception Dr Reid has misconceived, I may add one whom even he himself allows to have shaken off the ideal system, and to have considered the idea and the perception as not distinct, but the same,—a modification of the mind, and nothing more. I allude to the celebrated Jansenist writer, Arnauld, who maintains this doctrine as expressly as Dr Reid himself, and makes it the foundation of his argument in his controversy with Malebranche."[β] If this statement be true, then is Dr Brown's interpretation of Reid himself correct. A representative perception under its third and simplest modification, is held by Arnauld as by Brown; and his exposition is so clear and articulate that all essential misconception of these doctrines is precluded. In these circumstances, if Reid avow the identity of Arnauld's opinion and his own, this avowal is tantamount to a declaration that his peculiar doctrine of perception is a scheme of representation; whereas, on the contrary, if he signalise the contrast of their two opinions, he clearly evinces the radical antithesis, and his sense of the radical antithesis, of his doctrine of intuition, to every, even the simplest, form of the hypothesis of representation. And this last he does.

Reid not satisfied with Arnauld's opinion.

It cannot be maintained, that Reid admits a philosopher to hold an opinion convertible with his own, whom he states to "profess the doctrine, universally received, that we perceive not material things imme-

a See *Discussions*, p. 76.—Ed. β Lect. xxvii. p 173 (edit. 1830).

LECT. XXII. diately,—that it is their ideas that are the immediate objects of our thoughts,—and that it is in the idea of everything that we perceive its properties."[a] This fundamental contrast being established, we may safely allow that the original misconception, which caused Reid to overlook the difference of our intuitive and representative faculties, caused him, likewise, to believe that Arnauld had attempted to unite two contradictory theories of perception. Not aware, that it was possible to maintain a doctrine of perception in which the idea was not really distinguished from its cognition, and yet to hold that the mind had no immediate knowledge of external things, Reid supposes, in the first place, that Arnauld, in rejecting the hypothesis of ideas, as representative existences, really distinct from the contemplative act of perception, coincided with him in viewing the material reality, as the immediate object of that act; and, in the second, that Arnauld again deserted this opinion, when, with the philosophers, he maintained that the idea, or act of the mind representing the external reality, and not the external reality itself, was the immediate object of perception. But Arnauld's theory is one and indivisible; and, as such, no part of it is identical with Reid's. Reid's confusion, here as elsewhere, is explained by the circumstance, that he had never speculatively conceived the possibility of the simplest modification of the representative hypothesis. He saw no medium between rejecting ideas as something different from thought, and his own doctrine of an immediate knowledge of the material object. Neither does Arnauld, as Reid[β] supposes, ever assert against Malebranche, "that

a *Intellectual Powers*, Essay ii. ch. xiii. *Coll. Works*, p. 295.

β *Ibid.*, p. 296.

we perceive external things immediately," that is, in themselves: maintaining that all our perceptions are modifications essentially representative, he everywhere avows, that he denies ideas, only as existences distinct from the act itself of perception.[a]

LECT. XXII.

Reid was, therefore, wrong, and did Arnauld less than justice, in viewing his theory "as a weak attempt to reconcile two inconsistent doctrines:" he was wrong, and did Arnauld more than justice, in supposing that one of these doctrines was not incompatible with his own. The detection, however, of this error only tends to manifest more clearly how just, even when under its influence, was Reid's appreciation of the contrast, subsisting between his own and Arnauld's opinion, considered as a whole; and exposes more glaringly Brown's general misconception of Reid's philosophy, and his present gross misrepresentation, in affirming that the doctrines of the two philosophers were identical, and by Reid admitted to be the same.

Reid on Locke.

Locke is the philosopher next in order, and it is principally against Reid's statement of the Lockian doctrine of ideas, that the most vociferous clamour has been raised, by those who deny that the cruder form of the representative hypothesis was the one prevalent among philosophers, after the decline of the scholastic theory of species; and who do not see, that, though Reid's refutation, from the cause I have already noticed, was ostensibly directed only against that cruder form, it was virtually and in effect levelled against the doctrine of a representative perception altogether. Even supposing that Reid was wrong in attributing this particular modification of the representative hypothesis to Locke, and the philosophers in

a *Œuvres*, tom. xxxviii. 187, 198, 190, 389. [See *Discussions*, p. 77.—Ed.]

LECT. XXII.

general,—this would be a trivial error, provided it can be shown that he was opposed to every doctrine of perception, except that founded on the fact of the duality of consciousness. But let us consider whether Reid be really in error when he attributes to Locke the opinion in question. And let us first hear the charge of his opponents. Of these, I shall only particularly refer to the first and last,—to Priestley and to Brown,—though the same argument is confidently maintained by several other philosophers, in the interval between the publications of Priestley and of Brown.

Priestley quoted on Reid's view of Locke's opinion.

Priestley asserts, that Reid's whole polemic is directed against a phantom of his own creation, and that the doctrine of ideas which he combats was never seriously maintained by any philosopher, ancient or modern. "Before," says Priestley, "Dr Reid had rested so much upon this argument, it behoved him, I think, to have examined the strength of it a little more carefully than he seems to have done; for he appears to me to have suffered himself to be misled in the very foundation of it, merely by philosophers happening to call ideas *images* of external things; *as if this was not known to be a figurative expression* denoting, *not* that the actual shapes of things were delineated in the brain, or upon the mind, but only that impressions of some kind or other were conveyed to the mind by means of the organs of sense and their corresponding nerves, and that between these impressions and the sensations existing in the mind, there is a real and necessary, though at present an unknown, connection."[a]

[a] *Examination of Reid, Beattie, and Oswald*, sect. III., (p. 30, 2d edition). On Priestley, see Stewart, *Phil. Ess.*, Note H, *Works*, vol. v. p. 422.—Ed.

Brown does not go the length of Priestley; he admits that, in more ancient times, the obnoxious opinion was prevalent, and allows even two among modern philosophers, Malebranche and Berkeley, to have been guilty of its adoption. Both Priestley and Brown strenuously contend against Reid's interpretation of the doctrine of Locke, who states it as that philosopher's opinion, "that images of external objects are conveyed to the brain; but whether he thought with Descartes [*lege omnino* Dr Clarke] and Newton, that the images in the brain are perceived by the mind there present, or that they are imprinted on the mind itself, is not so evident."[a]

Brown coincides with Priestley in censuring Reid's view of Locke's opinion.

[β]This Brown, Priestley, and others, pronounce a flagrant misrepresentation. Not only does Brown maintain, that Locke never conceived the idea to be substantially different from the mind, as a material image in the brain; but, that he never supposed it to have an existence apart from the mental energy of which it is the object. Locke, he asserts, like Arnauld, considered the idea perceived and the percipient act, to constitute the same indivisible modification of the conscious mind. This we shall consider.

General character of Locke's philosophical style.

In his language, Locke is of all philosophers the most figurative, ambiguous, vacillating, various, and even contradictory; as has been noticed by Reid and Stewart, and Brown himself,—indeed, we believe, by every philosopher who has had occasion to animadvert on Locke. The opinions of such a writer are not, therefore, to be assumed from isolated and casual expressions, which themselves require to be interpreted

a *Intellectual Powers*, Essay II. ch. iv. *Coll. Works*, p. 256.

β See *Discussions*, p. 73.—ED.

LECT. XXII.

on the general analogy of the system; and yet this is the only ground on which Dr Brown attempts to establish his conclusions. Thus, on the matter under discussion, though really distinguishing, Locke verbally confounds, the objects of sense and of pure intellect, the operation and its object, the objects immediate and mediate, the object and its relations, the images of fancy and the notions of the understanding. Consciousness is converted with Perception; Perception with Idea; Idea with the object of Perception, and with Notion, Conception, Phantasm, Representation, Sense, Meaning, &c. Now, his language identifying ideas and perceptions, appears conformable to a disciple of Arnauld; and now it proclaims him a follower of Democritus and Digby, explaining ideas by mechanical impulse and the propagation of material particles from the external reality to the brain. In one passage, the idea would seem an organic affection,—the mere occasion of a spiritual representation; in another, a representative image, in the brain itself. In employing thus indifferently the language of every hypothesis, may we not suspect that he was anxious to be made responsible for none? One, however, he has formally rejected, and that is the very opinion attributed to him by Dr Brown,—that the idea, or object of consciousness in perception, is only a modification of the mind itself.

The interpretation adopted by Brown of Locke's opinion, explicitly contradicted by Locke himself.

I do not deny that Locke occasionally employs expressions, which, in a writer of more considerate language, would imply the identity of ideas with the act of knowledge; and, under the circumstances, I should have considered suspense more rational than a dogmatic confidence in any conclusion, did not the following passage, which has never, I believe, been noticed, afford a positive and explicit contradiction of Dr Brown's interpretation. It is from Locke's *Examination of Malebranche's Opinion*, which, as subsequent to the publication of the *Essay*, must be held decisive in relation to the doctrines of that work. At the same time, the statement is articulate and precise, and possesses all the authority of one cautiously emitted in the course of a polemical discussion. Malebranche coincided with Arnauld, Reid, and recent philosophers in general, and consequently with Locke, as interpreted by Brown, to the extent of supposing that *sensation proper* is nothing but a state or modification of the mind itself; and Locke had thus the opportunity of expressing, in regard to this opinion, his agreement or dissent. An acquiescence in the doctrine, that the secondary qualities, of which we are conscious in sensation, are merely mental states, by no means involves an admission that the primary qualities, of which we are conscious in perception, are nothing more. Malebranche, for example, affirms the one and denies the other. But if Locke be found to ridicule, as he does, even the opinion which merely reduces the secondary qualities to mental states, *a fortiori*, and this on the principle of his own philosophy, he must be held to reject the doctrine, which would reduce not only the non-resembling sensations of the secondary, but even the resembling, and consequently extended, ideas of the primary qualities of matter, to modifications of the immaterial unextended mind. In these circumstances, the following passage is superfluously conclusive against Brown; and equally so, whether we coincide or not in all the principles it involves. "But to examine their doctrine of *modification* a little farther.—Different sentiments

Locke quoted.

LECT. XXII. (sensations) are different modifications of the mind. The mind, or soul, that perceives, is one immaterial indivisible substance. Now I see the white and black on this paper; I hear one singing in the next room; I feel the warmth of the fire I sit by; and I taste an apple I am eating, and all this at the same time. Now, I ask, take modification for what you please, can the same unextended indivisible substance have different, nay, inconsistent and opposite (as these of white and black must be) modifications at the same time? Or must we suppose distinct parts in an indivisible substance, one for black, another for white, and another for red ideas, and so of the rest of those infinite sensations, which we have in sorts and degrees; all which we can distinctly perceive, and so are distinct ideas, some whereof are opposite, as heat and cold, which yet a man may feel at the same time? I was ignorant before, how sensation was performed in us: this they call an explanation of it! Must I say now I understand it better? If this be to cure one's ignorance, it is a very slight disease, and the charm of two or three insignificant words will at any time remove it; *probatum est*."[a] This passage is correspondent to the doctrine held, on this point, by Locke's personal friend and philosophical follower, Le Clerc.

But if it be thus evident that Locke held neither the third form of representation, that lent to him by Brown, nor even the second; it follows, that Reid did him anything but injustice, in supposing him to maintain that ideas are objects, either in the brain, or in the mind itself. Even the more material of these alternatives has been the one generally attributed to him by

a Section 39.

his critics,[a] and the one adopted from him by his disciples.[β] Nor is this to be deemed an opinion too monstrous to be entertained by so enlightened a philosopher. It was the common opinion of the age; the opinion, in particular, held by the most illustrious philosophers, his countrymen and contemporaries,—by Newton, Clarke, Willis, Hook, &c.[7]

Brown passes over Reid's interpretation of the opinions of certain philosophers.

Descartes, Arnauld, and Locke, are the only philosophers in regard to whom Brown attempts articulately to show, that Reid's account of their opinions touching the point at issue is erroneous. But there are others, such as Newton, Clarke, Hook, Norris, whom Reid charged with holding the obnoxious hypothesis, and whom Brown passes over without an attempt to vindicate, although Malebranche and Berkeley be the only two philosophers in regard to whom he explicitly avows that Reid is correct. But as an instance of Reid's error, Brown alleges Hobbes; and as an evidence of its universality, the authority of Le Clerc and Crousaz.

But adduces Hobbes as an instance of Reid's error.

[8] To adduce Hobbes as an instance of Reid's misrepresentation of the "common doctrine of ideas," betrays, on the part of Brown, a total misapprehension of the conditions of the question; or he forgets that Hobbes was a materialist. The doctrine of representation, under all its modifications, is properly subordinate to the doctrine of a spiritual principle of thought; and on the supposition, all but universally admitted among philosophers, that the relation of knowledge implied the analogy of existence, it was

a E.g. Sergeant and Cousin. See *Discussions*, p. 80, note *; and Stewart, *Phil. Essays*, Note H, *Works*, v. 422.—Ed.

β Tucker's *Light of Nature*, i. pp. 15, 18, (2d edit.) See *Discussions*, p. 80, note †.—Ed.

7 See *Discussions*, p. 80.—Ed.

8 See *Ibid.*, p. 75.—Ed.

mainly devised to explain the possibility of a knowledge by an immaterial subject, of an existence so disproportioned to its nature, as the qualities of a material object. Contending, that an immediate cognition of the accidents of matter, infers an essential identity of matter and mind, Brown himself admits, that the hypothesis of representation belongs exclusively to the doctrine of dualism;[a] whilst Reid, assailing the hypothesis of ideas only as subverting the reality of matter, could hardly regard it as parcel of that scheme, which acknowledges the reality of nothing else. But though Hobbes cannot be adduced as a competent witness against Reid, he is, however, valid evidence against Brown. Hobbes, though a materialist, admitted no knowledge of an external world. Like his friend Sorbiere, he was a kind of material idealist. According to him, we know nothing of the qualities or existence of any outward reality. All that we know is the "seeming," the "apparition," the "aspect," the "phænomenon," the "phantasm," within ourselves; and this subjective object, of which we are conscious, and which is consciousness itself, is nothing more than the "agitation" of our internal organism, determined by the unknown "motions," which are supposed, in like manner, to constitute the world without. Perception he reduces to Sensation. Memory and Imagination are faculties specifically identical with Sense, differing from it simply in the degree of their vivacity; and this difference of intensity, with Hobbes as with Hume, is the only discrimination between our dreaming and our waking thoughts.—A doctrine of perception identical with Reid's!

[β] Dr Brown at length proceeds to consummate his

a Lect. xxv. pp. 159, 160 (edit. 1830). β See *Discussions*, p. 81.—Ed.

victory, by "that most decisive evidence, found not in treatises, read only by a few, but in the popular elementary works of science of the time, the general text-books of schools and colleges." He quotes, however, only two,—the *Pneumatology* of Le Clerc, and the *Logic* of Crousaz.

Le Clerc and Crousaz, referred to by Brown.

Le Clerc.

"Le Clerc," says Dr Brown, "in his chapter on the nature of ideas, gives the history of the opinions of philosophers on this subject, and states among them the very doctrine which is most forcibly and accurately opposed to the ideal system of perception. '*Alii putant ideas et perceptiones idearum easdem esse, licet relationibus differant.* Idea, uti censent, proprie ad objectum refertur, quod mens considerat; perceptio vero ad mentem ipsam quæ percepit: sed duplex illa relatio ad unam modificationem mentis pertinet. Itaque, secundum hosce philosophos, nullæ sunt, proprie loquendo, ideæ a mente nostra distinctæ.' What is it, I may ask, which Dr Reid considers himself as having added to this very philosophical view of perception? and if he added nothing, it is surely too much to ascribe to him the merit of detecting errors, the counter-statement of which had long formed a part of the elementary works of the schools."[a]

In the first place, Dr Reid certainly "added" nothing "to this very philosophical view of perception," but he exploded it altogether. In the second, it is false either that this doctrine of perception "had long formed part of the elementary works of the schools," or that Le Clerc affords any countenance to this assertion. On the contrary, it is virtually stated by him to be the novel paradox of a single philosopher; nay, it is already, as such a singular opinion, discussed and

a Lect. xxvii. p. 171 (edit. 1830).—Ed.

LECT. XXII.

referred to its author by Reid himself. Had Dr Brown proceeded from the tenth paragraph, which he quotes, to the fourteenth, which he could not have read, he would have found that the passage extracted, so far from containing the statement of an old and familiar dogma in the schools, was neither more nor less than a statement of the contemporary hypothesis of Antony Arnauld, and of Antony Arnauld alone. In the third place, from the mode in which he cites Le Clerc, his silence to the contrary, and the general tenor of his statement, Dr Brown would lead us to believe that Le Clerc himself coincides in "this very philosophical view of perception." So far, however, from coinciding with Arnauld, he pronounces his opinion to be false; controverts it upon very solid grounds; and in delivering his own doctrine touching ideas, though sufficiently cautious in telling us what they are, he has no hesitation in assuring us, among other things which they cannot be, that they are not modifications or essential states of mind. "*Non est* (idea sc.) *modificatio aut essentia mentis:* nam præterquam quod sentimus ingens esse discrimen inter idem *perceptionem* et *sensationem*; quid habet mens nostra simile monti, aut innumeris ejusmodi ideis?"[α] Such is the judgment of that authority to which Dr Brown appealed as "the most decisive."

Crousaz.

In Crousaz, Dr Brown has actually succeeded in finding one example, (he might have found twenty), of a philosopher, before Reid, holding the same theory of ideas with Arnauld and himself.[β]

α *Pneumatologia*, sect. i. c. 5, § 2. —Ed.

β See this subject further pursued in *Discussions*, p. 62 *et seq.* —Ed.

LECTURE XXIII.

THE PRESENTATIVE FACULTY.—I. PERCEPTION,— WAS REID A NATURAL REALIST?

LECT. XXIII.

Ends proposed in the review of Reid's account of opinions on Perception.

In our last Lecture, I concluded the review of Reid's Historical Account of the previous Opinions on Perception. In entering upon this review, I proposed the following ends. In the first place, to afford you, not certainly a complete, but a competent, insight into the various theories on this subject; and this was sufficiently accomplished by limiting myself to the opinions touched upon by Reid. My aim, in the second place, was to correct some errors of Reid arising from, and illustrative of, those fundamental misconceptions which have infected his whole doctrine of the cognitive faculties with confusion and error; and, in the third place, I had in view to vindicate Reid from the attack made on him by Brown. I, accordingly, showed you, that though not without mistakes, owing partly to his limited acquaintance with the works of previous philosophers, and partly to not having generalised to himself the various possible modifications of the hypothesis of representative perception,—I showed you, I say, that Reid, though certainly anything but exempt from error, was, however, absolutely guiltless of all and every one of that marvellous tissue of mistakes, with which he is so recklessly accused by Brown,—whereas Brown's own

LECT. XXIII.

attack is, from first to last, itself that very series of misconceptions which he imputes to Reid. Nothing, indeed, can be more applicable to himself than the concluding observations which he makes in reference to Reid; and as these observations, addressed to his pupils, embody in reality an edifying and well-expressed advice, they will lose nothing of their relevancy or effect, if the one philosopher must be substituted for the other.[a] "That a mind so vigorous as that of Dr Reid should have been capable of the series of misconceptions which we have traced, may seem wonderful, and truly is so; and equally, or rather still more wonderful, is the general admission of his merit in this respect. I trust it will impress you with one important lesson—to consult the opinions of authors in their own works, and not in the works of those who profess to give a faithful account of them. From my own experience I can most truly assure you, that there is scarcely an instance in which I have found the view which I had received of them to be faithful. There is usually something more, or something less, which modifies the general result; and by the various additions and subtractions thus made, so much of the spirit of the original doctrine is lost, that it may, in some cases, be considered as having made a fortunate escape, if it be not at last represented as directly opposite to what it is."[β]

Reid right in attributing to philosophers in general the cruder doctrine of Representative Perception.

The mistakes of Dr Brown in relation to Reid, on which I have hitherto animadverted, are comparatively unimportant. Their refutation only evinces that Reid did not erroneously attribute to philosophers in general the cruder form of the representative hypothesis of

a See *Discussions*, p. 52.—Ed.

β *Philosophy of the Human Mind*, Lecture xxvii. p. 175 (edit. 1830).

perception; and that he was fully warranted in this attribution, is not only demonstrated by the disproval of all the instances which Brown has alleged against Reid, but might be shown by a whole crowd of examples, were it necessary to prove so undeniable a fact. In addition to what I have already articulately proved, it will be enough now simply to mention that the most learned and intelligent of the philosophers of last century might be quoted to the fact, that the opinion attributed by Reid to psychologists in general, was in reality the prevalent; and that the doctrine of Arnauld, which Brown supposes to have been the one universally received, was only adopted by the few. To this point Malebranche, Leibnitz, and Brucker, the younger Thomasius, 'S Gravesande, Genovesi, and Voltaire,[a] are conclusive evidence.

Was Reid himself a Natural Realist?

But a more important historical question remains, and one which even more affects the reputations of Reid and Brown. It is this,—Did Reid, as Brown supposes, hold, not the doctrine of Natural Realism, but the finer hypothesis of a Representative Perception?

If Reid did hold this doctrine, I admit at once that Brown is right.[β] Reid accomplished nothing; his philosophy is a blunder, and his whole polemic against the philosophers, too insignificant for refutation or comment. The one form of representation may be somewhat simpler and more philosophical than the other; but the substitution of the former for the latter is hardly deserving of notice; and of all conceivable hallucinations the very greatest would be that of Reid, in arrogating to himself the merit of thus subverting the foundation of Idealism and Scepticism,

a These testimonies are given in full, *Discussions*, p. 83.—Ed.

β See *Discussions*, p. 91.—Ed.

LECT. XXIII.

and of philosophers at large in acknowledging the pretension. The idealist and sceptic can establish their conclusions indifferently on either form of a representative perception; nay, the simpler form affords a securer, as the more philosophical, foundation. The idealism of Fichte is accordingly a system far more firmly founded than the idealism of Berkeley; and as the simpler involves a contradiction of consciousness more extensive and direct, so it furnishes to the sceptic a longer and more powerful lever.

The distinction of Intuitive and Representative Knowledge, to be first considered.

Before, however, discussing this question, it may be proper here to consider more particularly a matter of which we have hitherto treated only by the way,—I mean the distinction of Immediate or Intuitive, in contrast to Mediate or Representative Knowledge. This is a distinction of the most important kind, and it is one which has, however, been almost wholly overlooked by philosophers. This oversight is less to be wondered at in those who allowed no immediate knowledge to the mind, except of its proper modes; in their systems the distinction, though it still subsisted, had little relevancy or effect, as it did not discriminate the faculty by which we are aware of the presence of external objects, from that by which, when absent, these are imaged to the mind. In neither case, on this doctrine, are we conscious or immediately cognisant of the external reality, but only of the mental mode through which it is represented. But it is more astonishing that those who maintain, that the mind is immediately percipient of external things, should not have signalised this distinction; as on it is established the essential difference of Perception as a faculty of intuitive, Imagination as a faculty of representative, knowledge. But the marvel is still more enhanced when we find that Reid and Stewart, (if to them this opinion really belongs), so far from distinguishing Perception as an immediate and intuitive, from Imagination (and under Imagination, be it observed, I include both the Conception and the Memory of these Philosophers), as a mediate or representative, faculty,—in language make them both equally immediate. You will recollect the refutation I formerly gave you of Reid's self-contradictory assertion, that in Memory we are immediately cognisant of that which, as past, is not now existent, and cannot, therefore, be known in itself; and that, in Imagination, we are immediately cognisant of that which is distant, or of that which is not, and probably never was, in being.[a] Here the term *immediate* is either absurd, as contradictory; or it is applied only, in a certain special meaning, to designate the simpler form of representation, in which nothing is supposed to intervene between the mental cognition and the external reality; in contrast to the more complex, in which the representative or vicarious image is supposed to be something different from both. Thus, in consequence of this distinction not only not having been traced by Reid, as the discriminative principle of his doctrine, but having been even overlaid, obscured, and perplexed, his whole philosophy has been involved in haze and confusion; insomuch that a philosopher of Brown's acuteness could, (as we have seen and shall see), actually so far misconceive, as even to reverse, its import. The distinction is, therefore, one which, on every account, merits your most sedulous attention; but though of primary importance, it is fortunately not of any considerable difficulty.

Reid's view of this distinction obscure.

His whole philosophy hence involved in confusion.

a See Lect. xii. vol. i. p. 218 et seq.—Ed.

LECT. XXIII.

This distinction in general stated and illustrated.

As every cognitive act which, in one relation, is a mediate or representative, is, in another, an immediate or intuitive, knowledge, let us take a particular instance of such an act; as hereby we shall at once obtain an example of the one kind of knowledge, and of the other, and these also in proximate contrast to each other. I call up an image of the *High Church.* Now, in this act, what do I know immediately or intuitively? what mediately or by representation? It is manifest that I am conscious or immediately cognisant of all that is known as an act or modification of my mind, and, consequently, of the modification or act which constitutes the mental image of the Cathedral. But as, in this operation, it is evident, that I am conscious or immediately cognisant of the Cathedral, as imaged in my mind; so it is equally manifest, that I am not conscious or immediately cognisant of the Cathedral as existing. But still I am said to know it; it is even called the object of my thought. I can, however, only know it mediately,—only through the mental image which represents it to consciousness; and it can only be styled the object of thought, inasmuch as a reference to it is necessarily involved in the act of representation. From this example is manifest, what in general is meant by immediate or intuitive,—what by mediate or representative, knowledge. All philosophers are at one in regard to the immediate knowledge of our present mental modifications; and all are equally agreed, if we remove some verbal ambiguities, that we are only mediately cognisant of all past thoughts, objects, and events, and of every external reality not at the moment within the sphere of sense. There is but one point on which they are now at variance,—viz. whether the thinking subject is competent to an intuitive knowledge of aught but the modifications of the mental self, in other words, whether we can have any immediate perception of external things. Waiving, however, this question for the moment, let us articulately state what are the different conditions involved in the two kinds of knowledge.

The contrasts between Intuitive and Representative Cognition.

In the first place, considered as acts.—An act of immediate knowledge is simple; there is nothing beyond the mere consciousness, by that which knows, of that which is known. Here consciousness is simply contemplative. On the contrary, an act of mediate knowledge is complex; for the mind is conscious not only of the act as its own modification, but of this modification as an object representative of, or relative to, an object beyond the sphere of consciousness. In this act, consciousness is both representative and contemplative of the representation.

1. Considered as acts.

In the second place, in relation to their objects.—In an immediate cognition, the object is single, and the term unequivocal. Here the object in consciousness, and the object in existence, are the same; in the language of the schools, the *esse intentionale* or *repræsentativum*, coincides with the *esse entitativum*. In a mediate cognition, on the other hand, the object is twofold, and the term equivocal; the object known and representing being different from the object unknown, except as represented. The immediate object, or object known in this act, should be called the *subjective object*, or *subject-object*, in contradistinction to the mediate or unknown object, which might be discriminated as the *object-object*. A slight acquaintance with philosophical writings will show you how necessary such a distinction is; the want of it has

2. In relation to their objects.

LECT. XXIII. caused Reid to puzzle himself, and Kant to perplex his readers.

3. As judgments. In the third place, considered as judgments, (for you will recollect that every act of Consciousness involves an affirmation).—In an intuitive act, the object known is known as actually existing; the cognition, therefore, is assertory, inasmuch as the reality of that, its object, is given unconditionally as a fact. In a representative act, on the contrary, the represented object is unknown as actually existing; the cognition, therefore, is problematical, the reality of the object represented being only given as a possibility, on the hypothesis of the object representing.

4. In relation to their sphere. In the fourth place, in relation to their sphere.—Representative knowledge is exclusively subjective, for its immediate object is a mere mental modification, and its mediate object is unknown, except in so far as that modification represents it. Intuitive knowledge, on the other hand, if consciousness is to be credited, is either subjective or objective, for its single object may be a phænomenon either of the ego or of the non-ego,—either mental or material.

5. In reference to their perfection. In the fifth place, considered in reference to their perfection.—An intuitive cognition, as an act, is complete and absolute, as irrespective of aught beyond the dominion of consciousness; whereas a representative cognition, as an act, is incomplete, being relative to, and vicarious of, an existence beyond the sphere of actual knowledge. The object likewise of the former is complete, being at once known and real; whereas, in the latter, the object known is ideal, the real object unknown. In their relations to each other, immediate knowledge is complete, as self-sufficient;

mediate knowledge, on the contrary, is incomplete, as dependent on the other for its realisation.[a]

Such are the two kinds of knowledge which it is necessary to distinguish, and such are the principal contrasts they present. I said a little ago that this distinction, so far from being signalised, had been almost abolished by philosophers. I ought, however, to have excepted certain of the schoolmen,[β] by whom this discrimination was not only taken, but admirably applied; and, though I did not originally borrow it from them, I was happy to find that what I had thought out for myself, was confirmed by the authority of these subtle spirits. The names given in the schools to the immediate and mediate cognitions were *intuitive* and *abstractive*, (*cognitio intuitiva*, *cognitio abstractiva*), meaning by the latter term not merely what we, with them, call abstract knowledge, but also the representations of concrete objects in the imagination or memory.

This distinction taken by certain of the schoolmen.

Now, possessed of this distinction, of which Reid knew nothing, and asserting far more clearly and explicitly than he has ever done the doctrine of an

a For a fuller statement of the points of distinction between Immediate and Mediate Knowledge, see *Reid's Works*, Suppl. Dissert., Note B, p. 804-815.—Ed.

β [See Durandus, *In Sent.*, Prologus, qu. 3, § 6: "Cognitio *intuitiva*, illa quæ immediate tendit ad rem sibi præsentem objective, secundum ejus actualem existentiam: sicut cum video colorem existentem in pariete, vel rosam quam in manu teneo. *Abstractiva* dicitur omnis cognitio quæ habetur de re, non sic realiter præsente in ratione objecti immediate cogniti. § 9: Actus sensuum exteriorum sunt intuitivi, propter immediatum ordinem ad objecta sua." Cf. John Major, *In Sent.*, lib. i. dist. iii. qu. 2. f. 33, and Tellez, *Summa Philosophiæ*, tom. ii. p. 952.] [Besides Durandus, the Conimbricenses refer to Scotus, Ferrariensis, Anselm, Hugo a Sancto Victore, the Master of Sentences, Aquinas, Gregory Ariminensis, Paludanus, Cajetan, as distinguishing between knowledge *intuitive* and *abstractive*. See *In De Anima*, lib. ii. c. vi. qu. 3, p. 198, and *Reid's Works*, Suppl. Diss., Note B, p. 812.—See above, vol. ii. lect. xxi. p. 36, and lect. xxii. p. 47.—Ed.]

LECT. XXIII.

intuitive perception, I think the affirmation I made in my last Lecture is not unwarranted,—that a considerable section of the schoolmen were incomparably superior to Reid, or any modern philosopher, in their exposition of the true theory of that faculty. It is only wonderful that this, their doctrine, has not hitherto attracted attention, and obtained the celebrity it merits.

Order of the discussion.

Having now prepared you for the question concerning Reid, I shall proceed to its consideration; and shall, in the first place, state the arguments that may be adduced in favour of the opinion, that Reid did not assert a doctrine of Natural Realism,—did not accept the fact of the duality of consciousness in its genuine integrity, but only deluded himself with the belief that he was originating a new or an important opinion, by the adoption of the simpler form of Representation; and, in the second place, state the arguments that may be alleged in support of the opposite conclusion, that his doctrine is in truth the simple doctrine of Natural Realism.

1. Grounds on which Reid may be supposed not a Natural Realist. Brown's single argument in support of the view, that Reid was a Cosmothetic Idealist,—refuted.

But before proceeding to state the grounds on which alone I conceive any presumption can be founded, that Reid is not a Natural Realist, but, like Brown, a Cosmothetic Idealist, I shall state and refute the only attempt made by Brown to support this, his interpretation of Reid's fundamental doctrine. Brown's interpretation of Reid seems, in fact, not grounded on anything which he found in Reid, but simply on his own assumption of what Reid's opinion must be. For, marvellous as it may sound, Brown hardly seems to have contemplated the possibility of an immediate knowledge of anything beyond the sphere of self; and I should say, without qualification, that he had never at all imagined this possibility, were it not for the single attempt he makes at a proof of the impossibility of Reid holding such an opinion, when on one occasion Reid's language seems for a moment to have actually suggested to him the question,—Might that philosopher not perhaps regard the external object as identical with the immediate object in perception? In the following passage, you will observe, by anticipation, that by Sensation, which ought to be called Sensation Proper, is meant the subjective feeling,—the pleasure or pain involved in an act of sensible perception; and by Perception, which ought to be called Perception Proper, is meant the objective knowledge which we have, or think we have, of the external object in that act. "'Sensation,' says Dr Reid, 'can be nothing else than it is felt to be. Its very essence consists in being felt; and when it is not felt, it is not. There is no difference between the sensation and the feeling of it; they are one and the same thing.' But this is surely equally true of what he terms perception, which, as a state of the mind, it must be remembered, is, according to his own account of it, as different from the object perceived as the sensation is. We may say of the mental state of perception too, in his own language, as indeed we must say of all our states of mind, whatever they may be, that it can be nothing else than it is felt to be. Its very essence consists in being felt; and when it is not felt, it is not. There is no difference between the perception and the feeling of it; they are one and the same thing. The sensation, indeed, which is mental, is different from the object exciting it, which we term material; but so also is the state of mind which constitutes perception; for Dr Reid was surely too zealous an opponent of the systems which

Brown quoted.

LECT. XXIII.

ascribe everything to mind alone, or to matter alone, to consider the perception as itself the object perceived. That in sensation, as contradistinguished from perception, there is no reference made to an external object, is true; because, when the reference is made, we then use the new term of perception; but that in sensation there is no object distinct from that act of the mind by which it is felt,—no object independent of the mental feeling, is surely a very strange opinion of this philosopher; since what he terms perception is nothing but the reference of this very sensation to its external object. The sensation itself he certainly supposes to depend on the presence of an external object, which is all that can be understood in the case of perception, when we speak of its objects, or, in other words, of those external causes to which we refer our sensations; for the material object itself he surely could not consider as forming a part of the perception, which is a state of the mind alone. To be the object of perception, is nothing more than to be the foreign cause or occasion, on which this state of the mind directly or indirectly arises; and an object, in this only intelligible sense, as an occasion or cause of a certain subsequent effect, must, on his own principles, be equally allowed to sensation. Though he does not inform us what he means by the term *object*, as peculiarly applied to perception,—(and, indeed, if he had explained it, I cannot but think that a great part of his system, which is founded on the confusion of this single word, as something different from a mere external cause of an internal feeling, must have fallen to the ground),—he yet tells us very explicitly, that to be the object of perception, is something more than to be the external occasion on which that state of the mind arises which he terms perception; for, in arguing against the opinion of a philosopher who contends for the existence of certain images or traces in the brain, and yet says, 'that we are not to conceive the images or traces in the brain to be perceived, as if there were eyes in the brain; these traces are only occasions, on which, by the laws of the union of soul and body, ideas are excited in the mind; and therefore it is not necessary that there should be an exact resemblance between the traces and the things represented by them, any more than that words or signs should be exactly like the things signified by them,' he adds: 'These two opinions, I think, cannot be reconciled. For if the images or traces in the brain are perceived, they must be the objects of perception, and not the occasions of it only. On the other hand, if they are only the occasions of our perceiving, they are not perceived at all.' Did Dr Reid, then, suppose that the feeling, whatever it may be, which constitutes perception as a state of the mind, or, in short, all of which we are conscious in perception, is not strictly and exclusively mental, as much as all of which we are conscious in remembrance, or in love, or hate; or did he wish us to believe that matter itself, in any of its forms, is, or can be, a part of the phænomena or states of the mind, —a part, therefore, of that mental state or feeling which we term a perception? Our sensations, like our remembrances or emotions, we refer to some cause or antecedent. The difference is, that in the one case we consider the feeling as having for its cause some previous feeling or state of the mind itself; in the other case we consider it as having for its cause

LECT. XXIII.

something which is external to ourselves, and independent of our transient feelings,—something which, in consequence of former feelings suggested at the moment, it is impossible for us not to regard as extended and resisting. But still what we thus regard as extended and resisting, is known to us only by the feelings which it occasions in our mind. What matter, in its relation to percipient mind, can be, but the cause or occasion, direct or indirect, of that class of feelings which I term sensations or perceptions, it is absolutely impossible for me to conceive.

"The percipient mind, in no one of its affections, can be said to be the mass of matter which it perceives, unless the separate existence, either of matter or of mind, be abandoned by us, the existence of either of which, Dr Reid would have been the last of philosophers to yield. He acknowledges that our perceptions are consequent on the presence of external bodies, not from any necessary connection subsisting between them, but merely from the arrangement which the Deity, in his wisdom, has chosen to make of their mutual phænomena; which is surely to say, that the Deity has rendered the presence of the external object the occasion of that affection of the mind which is termed perception; or, if it be not to say this, it is to say nothing. Whatever state of mind perception may be; whether a primary result of a peculiar power, or a mere secondary reference of association that follows the particular sensation, of which the reference is made, it is itself, in either view of it, but a state of the mind; and to be the external occasion or antecedent of this state of mind, since it is to produce, directly or indirectly, all which constitutes perception, is surely, therefore, to be perceived, or there must be something in the mere word perceived, different from the physical reality which it expresses."[a]

LECT. XXIII.

[β] Now the sum and substance of this reasoning is, as far as I can comprehend it, to the following effect:— To assert an immediate perception of material qualities, is to assert an identity of matter and mind; for that which is immediately known must be the same in nature as that which immediately knows.

Brown's reasoning stated and refuted.

But Reid was not a materialist, was a sturdy spiritualist; therefore, he could not really maintain an immediate perception of the qualities of matter.

The whole validity of this argument consists in the truth of the major proposition, (for the minor proposition that Reid was not a materialist is certain),—To assert an immediate perception of material qualities, is to assert an identity of matter and mind; for that which is immediately known must be the same in essence as that which immediately knows.

Now in support of the proposition which constitutes the foundation of his argument, Brown offers no proof. He assumes it as an axiom. But so far from his being entitled to do so, by its being too evident to fear denial, it is, on the contrary, not only not obtrusively true, but, when examined, precisely the reverse of truth.

His fundamental proposition assumed.

In the first place, if we appeal to the only possible arbiter in the case,—the authority of consciousness,—we find that consciousness gives as an ultimate fact, in the unity of knowledge, the duality of existence; that is, it assures us that, in the act of perception, the percipient subject is at once conscious of something

In the first place, disproved by consciousness.

a *Lectures on the Philosophy of the Human Mind*, Lect. xxv. pp. 159, 160.

β See *Discussions*, p. 60.—Ed.

which it distinguishes as a modification of self, and of something which it distinguishes as a modification of not-self. Reid, therefore, as a dualist, and a dualist founding not on the hypotheses of philosophers, but on the data of consciousness, might safely maintain the fact of our immediate perception of external objects, without fear of involving himself in an assertion of the identity of mind and matter.

In the second place, would prove the converse of what Brown employs it to establish.

But, in the second place, if Reid did not maintain this immediacy of perception, and assert the veracity of consciousness, he would at once be forced to admit one or other of the unitarian conclusions of materialism or idealism. Our knowledge of mind and matter, as substances, is merely relative; they are known to us only in their qualities; and we can justify the postulation of two different substances, exclusively on the supposition of the incompatibility of the double series of phænomena to coinhere in one. Is this supposition disproved?— The presumption against dualism is again decisive. Entities are not to be multiplied without necessity; a plurality of principles is not to be assumed, where the phænomena can be explained by one. In Brown's theory of perception, he abolishes the incompatibility of the two series; and yet his argument, as a dualist, for an immaterial principle of thought, proceeds on the ground that this incompatibility subsists.[a] This philosopher denies us an immediate knowledge of aught beyond the accidents of mind. The accidents which we refer to body, as known to us, are only states or modifications of the percipient subject itself; in other words, the qualities we call *material*, are known by us to exist, only as they are known by us to inhere in the same substance as the qualities we denominate

a *Philosophy of the Human Mind*, Lect. xcvi. pp. 646, 647.

mental. There is an apparent antithesis, but a real identity. On this doctrine, the hypothesis of a double principle losing its necessity, becomes philosophically absurd; on the law of parcimony, a psychological unitarianism is established. To the argument, that the qualities of the object, are so repugnant to the qualities of the subject, of perception, that they cannot be supposed the accidents of the same substance, the unitarian,—whether materialist, idealist, or absolutist, has only to reply:—that so far from the attributes of the object being exclusive of the attributes of the subject, in this act, the hypothetical dualist himself establishes, as the fundamental axiom of his philosophy of mind, that the object known is universally identical with the subject knowing. The materialist may now derive the subject from the object, the idealist derive the object from the subject, the absolutist sublimate both into indifference, nay, the nihilist subvert the substantial reality of either;—the hypothetical realist, so far from being able to resist the conclusion of any, in fact accords their assumptive premises to all.

So far, therefore, is Brown's argument from inferring the conclusion, that Reid could not have maintained our immediate perception of external objects, that not only is its inference expressly denied by Reid, but if properly applied, it would prove the very converse of what Brown employs it to establish.

Reid's equalising Perception and Imagination,—a ground on which he may be supposed not a Natural Realist.

But there is a ground considerably stronger than that on which Brown has attempted to evince the identity of Reid's opinion on perception with his own. This ground is his equalising Perception and Imagination. (Under Imagination you will again observe, that I include Reid's Conception and Memory.) Other philosophers brought perception into unison with ima-

LECT. XXIII.

gination, by making perception a faculty of mediate knowledge. Reid, on the contrary, has brought imagination into unison with perception, by calling imagination a faculty of immediate knowledge. Now, as it is manifest that, in an act of imagination, the object-object is and can possibly be known only mediately, through a representation, it follows that we must perforce adopt one of two alternatives,—we may either suppose that Reid means by immediate knowledge only that simpler form of representation from which the idea or *tertium quid*, intermediate between the external reality and the conscious mind, is thrown out, or that, in his extreme horror of the hypothesis of ideas, he has altogether overlooked the fundamental distinction of mediate and immediate cognition, by which the faculties of perception and imagination are discriminated; and that thus his very anxiety to separate more widely his own doctrine of intuition from the representative hypothesis of the philosophers, has, in fact, caused him almost inextricably to confound the two opinions.

But may be explained consistently with his doctrine of Natural Realism.

Positive evidence that Reid held Natural Realism.

That this latter alternative is greatly the more probable, I shall now proceed to show you; and in doing this, I beg you to keep in mind the necessary contrasts by which an immediate or intuitive is opposed to a mediate or representative cognition. The question to be solved is,—Does Reid hold that in perception we immediately know the external reality, in its own qualities, as existing; or only mediately know them, through a representative modification of the mind itself? In the following proof, I select only a few out of a great number of passages which might be adduced from the writings of Reid, in support of the same conclusions. I am, however, confident that they are sufficient; and quotations longer or more numerous would tend rather to obscure than to illustrate.[α]

Application of the conditions of Immediate Knowledge to Reid's statements.

In the first place, knowledge and existence are then only convertible when the reality is known in itself; for then only can we say, that it is known because it exists, and exists since it is known. And this constitutes an immediate or intuitive cognition, rigorously so called. Nor did Reid contemplate any other. "It seems admitted," he says, "as a first principle, by the learned and the unlearned, that what is really perceived must exist, and that to perceive what does not exist is impossible. So far the unlearned man and the philosopher agree."[β]

In the second place, philosophers agree, that the idea or representative object, in their theory, is, in the strictest sense, immediately perceived. And so Reid understands them. "I perceive not, says the Cartesian, the external object itself; (so far he agrees with the Peripatetic, and differs from the unlearned man); but I perceive an image, or form, or idea, in my own mind, or in my brain. I am certain of the existence of the idea, because I immediately perceive it."[γ]

In the third place, philosophers concur in acknowledging that mankind at large believe, that the external reality itself constitutes the immediate and only object of perception. So also Reid: "On the same principle, the unlearned man says, I perceive the external object, and I perceive it to exist."—"The vulgar undoubtedly believe, that it is the external object which we immediately perceive, and not

α See this question discussed in *Reid's Works*, Suppl. Dissert., Note C, § ii. p. 819 *et seq.* Compare *Discussions*, p. 58 *et seq.*—Ed.

β *Works*, p. 271.—Ed.

γ *Ibid.*—Ed.

LECT. XXIII. a representative image of it only. It is for this reason, that they look upon it as perfect lunacy to call in question the existence of external objects."[α]—"The vulgar are firmly persuaded that the very identical objects which they perceive continue to exist when they do not perceive them: and are no less firmly persuaded, that when ten men look at the sun or the moon they all see the same individual object."[β] Speaking of Berkeley,—"The vulgar opinion he reduces to this, that the very things which we perceive by our senses do really exist. This he grants"[γ]—"It is, therefore, acknowledged by this philosopher to be a natural instinct or prepossession, an universal and primary opinion of all men, that the objects which we immediately perceive by our senses are not images in our minds, but external objects, and that their existence is independent of us and our perception."[δ]

In the fourth place, all philosophers agree that consciousness has an immediate knowledge, and affords an absolute certainty of the reality, of its object. Reid, as we have seen, limits the name of consciousness to self-consciousness, that is, to the immediate knowledge we possess of the modifications of self; whereas, he makes perception the faculty by which we are immediately cognisant of the qualities of the not-self.

In these circumstances, if Reid either, 1°, Maintain that his immediate perception of external things is convertible with their reality; or, 2°, Assert, that, in his doctrine of perception, the external reality stands to the percipient mind face to face, in the same im-

α *Works*, p. 274.—ED.
β *Ibid.*, p. 284.—ED.
γ *Works*, p. 284.—ED.
δ *Ibid.*, p. 299.—ED.

mediacy of relation which the idea holds in the representative theory of the philosophers; or, 3°, Declare the identity of his own opinion with the vulgar belief, as thus expounded by himself and the philosophers; or, 4°, Declare, that his Perception affords us equal evidence of the existence of external phænomena, as his Consciousness affords us of the existence of internal;—in all and each of these suppositions, he would unambiguously declare himself a natural realist, and evince that his doctrine of perception is one not of a mediate or representative, but of an immediate or intuitive knowledge. And he does all four.

The first and second.—"We have before examined the reasons given by philosophers to prove that ideas, and not external objects, are the immediate objects of perception. We shall only here observe, that if external objects be perceived immediately," [and he had just before asserted for the hundredth time that they were so perceived,] "we have the same reason to believe their existence, as philosophers have to believe the existence of ideas, while they hold them to be the immediate objects of perception."[a]

The third.—Speaking of the perception of the external world,—"We have here a remarkable conflict between two contradictory opinions, wherein all mankind are engaged. On the one side, stand all the vulgar, who are unpractised in philosophical researches, and guided by the uncorrupted primary instincts of nature. On the other side, stand all the philosophers, ancient and modern; every man, without exception, who reflects. In this division, to my great humiliation, I find myself classed with the vulgar."[β]

[a] *Works*, p. 446. Cf. pp. 263, 272.—ED. [β] *Works*, p. 302.—ED.

LECT. XXIII.

The fourth.—"Philosophers sometimes say that we perceive ideas,—sometimes that we are conscious of them. I can have no doubt of the existence of anything which I either perceive, or of which I am conscious; but I cannot find that I either perceive ideas or am conscious of them."[a]

Various other proofs of the same conclusion could be adduced; these, for brevity, we omit.

General conclusion, and caution.

On these grounds, therefore, I am confident that Reid's doctrine of Perception must be pronounced a doctrine of Intuition, and not of Representation; and though, as I have shown you, there are certainly some plausible arguments which might be alleged in support of the opposite conclusion, still these are greatly overbalanced by stronger positive proofs, and by the general analogy of his philosophy. And here I would impress upon you an important lesson. That Reid, a distinguished philosopher, and even the founder of an illustrious school, could be so greatly misconceived, as that an eminent disciple of that school itself should actually reverse the fundamental principle of his doctrine,—this may excite your wonder, but it ought not to move you to disparage either the talent of the philosopher misconceived, or of the philosopher misconceiving. It ought, however, to prove to you the permanent importance, not only in speculation, but in practice, of precise thinking. You ought never to rest content, so long as there is aught vague or indefinite in your reasonings,—so long as you have not analysed every notion into its elements, and excluded the possibility of all lurking ambiguity in your expressions. One great,—perhaps the one greatest advantage, re-

a *Works*, p. 373.—Ed.

sulting from the cultivation of Philosophy, is the habit it induces of vigorous thought, that is, of allowing nothing to pass without a searching examination, either in your own speculations, or in those of others. We may never, perhaps, arrive at truth, but we can always avoid self-contradiction.

LECTURE XXIV.

THE PRESENTATIVE FACULTY.—I. PERCEPTION.—THE DISTINCTION OF PERCEPTION PROPER FROM SENSATION PROPER.

LECT. XXIV.

Recapitulation.

In my last Lecture, having concluded the review of Reid's Historical Account of Opinions on Perception, and of Brown's attack upon that account, I proceeded to the question,—Is Reid's own doctrine of perception a scheme of Natural Realism, that is, did he accept in its integrity the datum of consciousness,—that we are immediately cognitive both of the phænomena of matter and of the phænomena of mind; or did he, like Brown, and the greater number of more recent philosophers, as Brown assumes, hold only the finer form of the representative hypothesis, which supposes that, in perception, the external reality is not the immediate object of consciousness, but that the ego is only determined in some unknown manner to represent the non-ego, which representation, though only a modification of mind or self, we are compelled, by an illusion of our nature, to mistake for a modification of matter, or not-self? I stated to you how, on the determination of this question, depended nearly the whole of Reid's philosophical reputation; his philosophy professes to subvert the foundations of idealism and scepticism, and it is as having accomplished what he thus attempted, that any principal or peculiar glory can be awarded to him. But if all he did was merely to explode the cruder hypothesis of representation, and to adopt in its place the finer,—why, in the first place, so far from depriving idealism and scepticism of all basis, he only placed them on one firmer and more secure; and, in the second, so far from originating a new opinion, he could only have added one to a class of philosophers, who, after the time of Arnauld, were continually on the increase, and who, among the contemporaries of Reid himself, certainly constituted the majority. His philosophy would thus be at once only a silly blunder; its pretence to originality only a proclamation of ignorance; and so far from being an honour to the nation from which it arose, and by whom it was respected, it would, in fact, be a scandal and a reproach to the philosophy of any country in which it met with any milder treatment than derision.

Previously, however, to the determination of this question, it was necessary to place before you, more distinctly than had hitherto been done, the distinction of Mediate or Representative from Immediate or Intuitive knowledge,—a distinction which, though overlooked, or even abolished, in the modern systems of philosophy, is, both in itself and in its consequences, of the highest importance in psychology. Throwing out of view, as a now exploded hypothesis, the cruder doctrine of representation,—that, namely, which supposes the immediate, or representative object to be something different from a mere modification of mind,—from the mere energy of cognitions,—I articulately displayed to you these two kinds of knowledge in their contrasts and correlations. They are thus defined. Intuitive or immediate knowledge is that in which

LECT. XXIV. there is only one object, and in which that object is known in itself, or as existing. Representative or mediate knowledge, on the contrary, is that in which there are two objects,—an immediate and a mediate object;—the immediate object or that known in itself, being a mere subjective or mental mode relative to and representing a reality beyond the sphere of consciousness; the mediate object being that reality, thus supposed and represented. As an act of representative knowledge involves an intuitive cognition, I took a special example of such an act. I supposed that we called up to our minds the image of the *High Church*. Now here the immediate object,—the object of consciousness, is the mental image of that edifice. This we know, and know not as an absolute object, but as a mental object relative to a material object which it represents; which material object, in itself, is, at present, beyond the reach of our faculties of immediate knowledge, and is, therefore, only mediately known in its representation. You must observe that the mental image,—the immediate object, is not really different from the cognitive act of imagination itself. In an act of mediate or representative knowledge, the cognition and the immediate object are really an identical modification,—the cognition and the object,—the imagination and the image, being nothing more than the mental representation,—the mental reference itself. The indivisible modification is distinguished by two names, because it involves a relation between two terms, (the two terms being the mind knowing and the thing represented), and may, consequently, be viewed in more proximate reference to the one or to the other of these. Looking to the mind knowing, it is called a cognition, an act of knowledge, an imagination, etc.;—looking to the thing represented, it is called a representation, an object, an image, an idea, etc.

All philosophers admit that the knowledge of our present mental states is immediate; if we discount some verbal ambiguities, all would admit that our actual knowledge of all that is not now existent, or not now existent within the sphere of consciousness, must be mediate or representative. The only point on which any serious difference of opinion can obtain, is,—Whether the ego or mind can be more than mediately cognisant of the phænomena of the non-ego or matter.

Summary of the reasons for holding Reid a Natural Realist.

I then detailed to you the grounds on which it ought to be held that Reid's doctrine of Perception is one of Natural Realism, and not a form of Cosmothetic Idealism, as supposed by Brown. An immediate or intuitive knowledge is the knowledge of a thing as existing,—consequently, in this case, knowledge and existence infer each other. On the one hand, we know the object, because it exists,—and, on the other, the object exists, since it is known. This is expressly maintained by Reid, and universally admitted by philosophers. In the first place, on this principle, the philosophers hold that ideas, (whether on the one hypothesis of representation, or on the other), necessarily exist, because immediately known. Now, if Reid, fully aware of this, assert that, on his doctrine, the external reality holds, in the act of perception, the same immediate relation to the mind, in which the idea or representative image stands in the doctrine of philosophers; and that, consequently, on the one opinion, we have the same assurance of the existence of the material world, as, on the other, of the reality of the ideal

LECT. XXIV.

world;—if, I say, he does this, he unambiguously proclaims himself a natural realist. And that this he actually does, I showed you by various quotations from his writings.

In the second place, upon the same principle, mankind at large believe in the existence of the external universe, because they believe that the external universe is by them immediately perceived. This fact, I showed you, is acknowledged both by the philosophers, who regard the common belief itself as an illusion, and by Reid. In these circumstances, if Reid declares that he coincides with the vulgar, in opposition to the learned, belief, he must again be held unambiguously to pronounce his doctrine of perception a scheme of natural realism. And that he emphatically makes this declaration, I also proved to you by sundry passages.

In the third place, Reid and all philosophers are at one in maintaining, that self-consciousness, as immediately cognisant of our mental modifications, affords us an absolute assurance of their existence. If then Reid hold that perception is as immediately cognisant of the external modification, as self-consciousness is of the internal, and that the one cognition thus affords us an equal certainty of the reality of its object as does the other,—on this supposition, it is manifest that Reid, a third time, unambiguously declares his doctrine of perception a doctrine of natural realism. And that he does so, I proved by various quotations.

I might have noticed, in the fourth place, that Reid's assertion, that our belief in the existence of external things is immediate, and not the result of inference or reasoning, is wholly incompatible with the doctrine of a representative perception. I do not, however, lay much stress on this argument, because we may possibly suspect that he makes the same mistake in regard to the term *immediate*, as applied to this belief, which he does in its application to our representative cognitions. But, independently of this, the three former arguments are amply sufficient to establish our conclusion.

These are the grounds on which I would maintain that Brown has not only mistaken, but absolutely reversed the fundamental principle of Reid's philosophy; although it must be confessed, that the error and perplexity of Reid's exposition, arising from his non-distinction of the two possible forms of representation, and his confusion of representative and of intuitive knowledge, afford a not incompetent apology for those who might misapprehend his meaning. In this discussion, it may be matter of surprise, that I have not called in the evidence of Mr Stewart. The truth is,—his writings afford no applicable testimony to the point at issue. His own statements of the doctrine of perception are brief and general, and he is content to refer the reader to Reid for the details.

Reid the first champion of Natural Realism in these latter times.

Of the doctrine of an intuitive perception of external objects,—which, as a fact of consciousness, ought to be unconditionally admitted,—Reid has the merit, in these latter times, of being the first champion. I have already noticed that, among the scholastic philosophers, there were some who maintained the same doctrine, and with far greater clearness and comprehension than Reid.[a] These opinions are, however, even at this moment, I may say, wholly unknown; and it would be ridiculous to suppose that their speculations had exerted any influence, direct or indirect, upon a thinker so imperfectly acquainted with what had been

[a] See above, vol. ii. pp. 36, 47, 71, notes.—Ed.

LECT. XXIV.

Two modern philosophers, previously to Reid, held Intuitive Perception.

done by previous philosophers, as Reid. Since the revival of letters, I have met with only two anterior to Reid, whose doctrine on the present question coincided with his. One of these may, indeed, be discounted; for he has stated his opinions in so paradoxical a manner, that his authority is hardly worthy of notice.[α] The other,[β] who flourished about a century before Reid, has,

α The philosopher here meant is probably John Sergeant, who inculcated a doctrine of Realism against modern philosophers generally, and Locke in particular,—in his *Method to Science* (1696) and *Solid Philosophy asserted against the Fancies of the Ideists* (1697). See, of the latter work, Preface, especially §§ 7, 18, 19; pp. 23, 42-44, 58 *et seq.*, 142, 338 *et seq.* See below, vol. ii. p. 123-124.—Ed.

β The latter of the two philosophers here referred to, is doubtless Peter Poiret. He is mentioned in the Author's Common-Place Book, as holding a more correct opinion than Reid on the point raised in the text. Poiret was born in 1646, and died in 1719. He states his doctrine as follows: "In nobis duplicis generis (saltem quantum ad cognitionem, voce hac late sumpta) facultates inesse; reales alteras, quæ res ipsas; alteras umbratiles, quæ rerum picturas, umbrasve sive *ideas* exhibent: et utrasque quidem facultates illas iterum duplices existere; nempe, vel reales spiritales, pro rebus spiritalibus; vel reales corporeas, pro rebus materialibus. *Spiritales reales* sunt passivus intellectus sensusque spiritales et intimi, qui ab objectis ipsis realibus ac spiritalibus, eorumve effluviis veris afficiuntur. . . . *Corporeæ reales* facultates sunt (hoc in negotio) visus sensusque cæteri corporei qui ab objectis ipsis corporeis affecti, eorum exhibent nobis *cognitionem sensualem*. *Umbratiles* autem facultates (quæ sunt ipsa hominis *Ratio*, sive intellectus activus) comparent maxime, quando objectis sive rebus quæ facultates reales affecerunt, eorumque affectione et effluviis absentibus, mens activitate sua eorumdem imagines sive ideas in se excitat et considerat. Et hoc quidem modo *idealiter* sive per *ideam* possunt quoque cognosci, *Deus, Mentes, Corpora*." *Cogitationes Rationales*, lib. ii. c. iv. p. 176, (edit. 1715)—first published apparently in 1675. Again he says: "Intellectus triplex. . . . . Intellectus, sive facultas percipiendi, cujus objectum ipsemet Deus est ejusque divinæ operationes ac emanationes, dicitur a me *intellectus divinus*, ac mere *passivus* sive receptivus; qui etiam *intelligentia* dici potest. Intellectus, sive facultas percipiendi, cujus objectum sunt res hujus mundi naturales earumque realia effluvia, dicitur a me intellectus animalis sive *sensualis*, qui quoque mere *passivus* est. Intellectus vero cujus objecta sunt picturæ et imagines ac ideæ rerum, quas ipsemet format et varie regit, sive imagines illæ ideæve sint de rebus spiritalibus sive de corporeis, dicitur a me *Ratio humana* vel intellectus *activus* et *picturarius* . . . intellectus *idealis*." *Defensio Methodi Inveniendi Verum*, sect. ii. § 4; cf. sect. iii. § 5; *Opera Posthuma*, pp. 113, 127, (edit. 1721). Cf. his *De Vera Methodo Inveniendi Verum*, pars i. §§ 20, 21, pp. 23, 24, (1st edit. 1692),—prefixed to his *De Eruditione*. See vol. i. p. 293, note β.—Ed.

on the contrary, stated the doctrine of an intuitive, and refuted the counter hypothesis of a representative perception, with a brevity, perspicuity, and precision, far superior to the Scottish philosopher. Both of these authors, I may say, are at present wholly unknown.

Having concluded the argument by which I endeavoured to satisfy you that Reid's doctrine is Natural Realism, I should now proceed to show that Natural Realism is a more philosophical doctrine than Hypothetical Realism. Before, however, taking up the subject, I think it better to dispose of certain subordinate matters, with which it is proper to have some preparatory acquaintance.

The distinction of Perception Proper from Sensation Proper.

Of these the first is the distinction of Perception Proper from Sensation Proper.

Use of the term Perception previously to Reid.

I have had occasion to mention, that the word *Perception* is, in the language of philosophers previous to Reid, used in a very extensive signification. By Descartes, Malebranche, Locke, Leibnitz, and others, it is employed in a sense almost as unexclusive as consciousness in its widest signification. By Reid, this word was limited to our faculty acquisitive of knowledge, and to that branch of this faculty whereby, through the senses, we obtain a knowledge of the external world. But his limitation did not stop here. In the act of external perception, he distinguished two elements, to which he gave the names of Perception and Sensation. He ought, perhaps, to have called these *perception proper* and *sensation proper*, when employed in his special meaning; for, in the language of other philosophers, *sensation* was a term which included his Perception, and *perception* a term comprehensive of what he called Sensation.

LECT. XXIV.

Reid's account of Perception.

There is a great want of precision in Reid's account of Perception and Sensation. Of Perception he says:—"If, therefore, we attend to that act of our mind, which we call the perception of an external object of sense, we shall find in it these three things. *First*, Some conception or notion of the object perceived. *Secondly*, A strong and irresistible conviction and belief of its present existence. And, *Thirdly*, That this conviction and belief are immediate, and not the effect of reasoning.

"*First*, it is impossible to perceive an object without having some notion or conception of what we perceive. We may indeed conceive an object which we do not perceive; but when we perceive the object, we must have some conception of it at the same time; and we have commonly a more clear and steady notion of the object while we perceive it, than we have from memory or imagination, when it is not perceived. Yet, even in perception, the notion which our senses give of the object may be more or less clear, more or less distinct, in all possible degrees."*

Wanting in precision.

Now here you will observe that the "having a notion or conception," by which he explains the act of perception, might at first lead us to conclude that he held, as Brown supposes, the doctrine of a representative perception; for notion and conception are generally used by philosophers for a representation or mediate knowledge of a thing. But, though Reid cannot escape censure for ambiguity and vagueness, it appears, from the analogy of his writings, that by *notion* or *conception* he meant nothing more than knowledge or cognition.

Sensation.

Sensation he thus describes:—"Almost all our per-

* *Intellectual Powers*, Essay II. ch. v. *Works*, p. 258.

ceptions have corresponding sensations, which constantly accompany them, and, on that account, are very apt to be confounded with them. Neither ought we to expect that the sensation, and its corresponding perception, should be distinguished in common language, because the purposes of common life do not require it. Language is made to serve the purposes of ordinary conversation; and we have no reason to expect that it should make distinctions that are not of common use. Hence it happens that a quality perceived, and the sensation corresponding to that perception, often go under the same name.

"This makes the names of most of our sensations ambiguous, and this ambiguity hath very much perplexed the philosophers. It will be necessary to give some instances, to illustrate the distinction between our sensations and the objects of perception.

"When I smell a rose, there is in this operation both sensation and perception. The agreeable odour I feel, considered by itself, without relation to any external object, is merely a sensation. It affects the mind in a certain way; and this affection of the mind may be conceived, without a thought of the rose or any other object. This sensation can be nothing else than it is felt to be. Its very essence consists in being felt; and when it is not felt, it is not. There is no difference between the sensation and the feeling of it; they are one and the same thing. It is for this reason, that we before observed, that in sensation, there is no object distinct from that act of the mind by which it is felt; and this holds true with regard to all sensations.

"Let us next attend to the perception which we have in smelling a rose. Perception has always an

LECT. XXIV.

external object; and the object of my perception, in this case, is that quality in the rose which I discern by the sense of smell. Observing that the agreeable sensation is raised when the rose is near, and ceases when it is removed, I am led, by my nature, to conclude some quality to be in the rose which is the cause of this sensation. This quality in the rose is the object perceived; and that act of the mind, by which I have the conviction and belief of this quality, is what in this case I call perception."[a]

Reid anticipated in his distinction of Perception from Sensation.

By *perception*, Reid, therefore, means the objective knowledge we have of an external reality, through the senses; by *sensation*, the subjective feeling of pleasure or pain, with which the organic operation of sense is accompanied. This distinction of the objective from the subjective element in the act, is important. Reid is not, however, the author of this distinction. He himself notices of Malebranche that "he distinguished more accurately than any philosopher had done before, the objects which we perceive from the sensations in our own minds, which, by the laws of nature, always accompany the perception of the object. As in many things, so particularly in this, he has great merit; for this, I apprehend, is a key that opens the way to a right understanding both of our external senses, and of other powers of the mind."[β] I may notice that

Malebranche.

Malebranche's distinction is into *Idée*, corresponding to Reid's Perception, and *Sentiment*, corresponding to his Sensation; and this distinction is as precisely marked in Malebranche[γ] as in Reid. Subsequently to

[a] *Intellectual Powers*, Essay ii. ch. xvi. *Coll. Works*, p. 310.

[β] *Intellectual Powers*, Essay ii. ch. vii. *Coll. Works*, p. 265.

[γ] *Recherche de la Vérité*, liv. iii. part ii. ch. 6, and 7, with Eclaircissement on text. See *Reid's Works*, pp. 834, 887.—Ed.

Malebranche, the distinction became even common; and there is no reason for Mr Stewart[a] being struck when he found it in Crousaz and Hutcheson. It is to be found in Le Clerc,[β] in Sinsart,[γ] in Buffier,[δ] in Genovesi,[ε] and in many other philosophers. It is curious that Malebranche's distinction was apprehended neither by Locke nor by Leibnitz, in their counter examinations of the theory of that philosopher. Both totally mistake its import. Malebranche, however, was not the original author of the distinction. He himself professedly evolves it out of Descartes.[ζ] But long previously to Descartes, it had been clearly established. It formed a part of that admirable doctrine of perception maintained by the party of the Schoolmen to whom I have already alluded.[η] I find it, however, long prior to them. It is, in particular, stated with great precision by Plotinus,[θ] and even some inferences drawn from it, which are supposed to be the discoveries of modern philosophy.

Crousaz, Hutcheson, Le Clerc, Sinsart, Buffier, Genovesi.

Descartes.

Plotinus.

Before proceeding to state to you the great law which regulates the mutual relation of these phænomena,—a law which has been wholly overlooked by our psychologists,—it is proper to say a few words, illustrative of the nature of the phænomena themselves; for what you will find in Reid, is by no means either complete or definite.

The nature of the phænomena,—Perception and Sensation, illustrated.

The opposition of Perception and Sensation is true,

a *Philosophical Essays*, Notes F and G. The passages from Hutcheson and Crousaz are given in Sir W. Hamilton's edition of the *Collected Works*, vol. v. p. 420.—Ed.

β *Pneumatologia*, § I. ch. v. *Opera Philosophica*, tom. ii. p. 31, (edit. 1726).—Ed.

γ [*Recueil des Pensées sur l'Immortalité de l'Ame*, p. 119.]

δ *First Truths*, part. I. ch. xiv. §§ 109-111. Cf. *Remarks on Crousaz*, art. viii. p. 427 (Eng. Trans.)—Ed.

ε [*Elementa Metaphysicæ*, pars ii. p. 12.]

ζ See *Reid's Works*, p. 831.—Ed.

η See above, vol. ii. lect. xxiii. p. 71, and *Reid's Works*, p. 887.—Ed.

θ *Enn.* iii. lib. vi. c. 2. See *Reid's Works*, p. 887.—Ed.

LECT. XXIV.

The contrast of Perception, and Sensation, the special manifestation of a contrast which universally divides Knowledge and Feeling.

but it is not a statement adequate to the generality of the contrast. Perception is only a special kind of knowledge, and sensation only a special kind of feeling; and *Knowledge* and *Feeling*, you will recollect, are two out of the three great classes, into which we primarily divided the phænomena of mind. *Conation* was the third. Now, as perception is only a special mode of knowledge, and sensation only a special mode of feeling, so the contrast of perception and sensation is only the special manifestation of a contrast, which universally divides the generic phænomena themselves. It ought, therefore, in the first place, to have been noticed, that the generic phænomena of knowledge and feeling are always found coexistent, and yet always distinct; and the opposition of perception and sensation should have been stated as an obtrusive, but still only a particular, example of the general law.

Perception Proper and Sensation Proper, precisely distinguished.

But not only is the distinction of perception and sensation not generalised,—not referred to its category, by our psychologists; it is not concisely and precisely stated. A cognition is objective, that is, our consciousness is then relative to something different from the present state of the mind itself; a feeling, on the contrary, is subjective, that is, our consciousness is exclusively limited to the pleasure or pain experienced by the thinking subject. Cognition and feeling are always coexistent. The purest act of knowledge is always coloured by some feeling of pleasure or pain; for no energy is absolutely indifferent, and the grossest feeling exists only as it is known in consciousness. This being the case of cognition and feeling in general, the same is true of perception and sensation in particular. Perception proper is the consciousness, through the senses, of the qualities of an object known as different from self; Sensation proper is the consciousness of the subjective affection of pleasure or pain, which accompanies that act of knowledge. Perception is thus the objective element in the complex state, —the element of cognition; sensation is the subjective element,—the element of feeling.

The grand law by which the phænomena of Knowledge and Feeling,—Perception and Sensation, are governed in their reciprocal relation.

The most remarkable defect, however, in the present doctrine upon this point, is the ignorance of our psychologists in regard to the law by which the phænomena of cognition and feeling,—of perception and sensation, are governed, in their reciprocal relation. This law is simple and universal; and, once enounced, its proof is found in every mental manifestation. It is this:—Knowledge and Feeling,—Perception and Sensation, though always coexistent, are always in the inverse ratio of each other.[a] That these two elements are always found in coexistence, as it is an old and a notorious truth, it is not requisite for me to prove. But that these elements are always found to coexist in an inverse proportion,—in support of this universal fact, it will be requisite to adduce proof and illustration.

Established and illustrated.

In doing this I shall, however, confine myself to the relation of Perception and Sensation. These afford the best examples of the generic relation of knowledge and feeling; and we must not now turn aside from the special faculty with which we are engaged.

1. From a comparison of the several senses.

The first proof I shall take from a comparison of the several senses; and it will be found that, precisely as a sense has more of the one element, it has less of the other. Laying Touch aside for the moment, as

a This law is thus enunciated by Kant:—"Je stärker die Sinne, bei eben demselben Grade des auf sie geschehenen Einflusses, sich afficirt fühlen, desto weniger *lehren* sie. Umgekehrt; wenn sie viel lehren sollen, müssen sie mässig afficiren." *Anthropologie*, § 20, (*Werke*, edit. Rosenkranz and Schubert, vii. part 2, p. 51) § 20 of this edition corresponds to § 19, edit. 1800.—Ed.

this requires a special explanation, the other four Senses divide themselves into two classes, according as perception, the objective element, or sensation, the subjective element, predominates. The two in which the former element prevails, are Sight and Hearing; the two in which the latter, are Taste and Smell.[a]

Sight.

Now, here, it will be at once admitted, that Sight, at the same instant, presents to us a greater number and a greater variety of objects and qualities, than any other of the senses. In this sense, therefore, perception,—the objective element, is at its maximum. But sensation,—the subjective element, is here at its minimum; for, in the eye, we experience less organic pleasure or pain from the impressions of its appropriate objects (colours), than we do in any other sense.

Hearing.

Next to Sight, Hearing affords us, in the shortest interval, the greatest variety and multitude of cognitions; and as sight divides space almost to infinity, through colour, so hearing does the same to time, through sound. Hearing is, however, much less extensive in its sphere of knowledge or perception than sight; but in the same proportion is its capacity of feeling or sensation more intensive. We have greater pleasure and greater pain from single sounds than from single colours; and, in like manner, concords and discords, in the one sense, affect us more agreeably or disagreeably, than any modifications of light in the other.[β]

Taste and Smell.

In Taste and Smell, the degree of sensation, that is, of pleasure or pain, is great in proportion as the percep-

[a] Compare Kant, *Anthropologie*, § 15.—Ed.

[β] [In regard to the subjective and objective nature of the sensations of the several senses, or rather the perceptions we have through them, it may be observed, that what is more objective is more easily remembered; whereas, what is more subjective affords a much less distinct remembrance. Thus, what we perceive by the eye, is better remembered than what we hear.]—*Oral Interpolation.*

tion, that is, the information they afford, is small. In all these senses, therefore,—Sight, Hearing, Taste, Smell, it will be admitted that the principle holds good.

Touch.

The sense of Touch, or Feeling strictly so called, I have reserved, as this requires a word of comment. Some philosophers include under this name all our sensitive perceptions, not obtained through some one of the four special organs of sense, that is, sight, hearing, taste, smell; others, again, divide the sense into several. To us at present this difference is of no interest: for it is sufficient for us to know, that in those parts of the body where sensation predominates, perception is feeble; and in those where perception is lively, sensation is obtuse. In the finger points, tactile perception is at its height; but there is hardly another part of the body in which sensation is not more acute. Touch, or Feeling strictly so called, if viewed as a single sense, belongs, therefore, to both classes,—the objective and subjective. But it is more correct, as we shall see, to regard it as a plurality of senses, in which case Touch, properly so called, having a principal organ in the finger points, will belong to the first class,—the class of objective senses,—the perceptions,—that class in which perception proper predominates.

Touch properly a plurality of Senses.

2. From the several impressions of the same sense.

The analogy, then, which we have thus seen to hold good in the several senses in relation to each other, prevails likewise among the several impressions of the same sense. Impressions in the same sense, differ both in degree and in quality or kind. By *impression* you will observe that I mean no explanation of the mode in which the external reality acts upon the sense, (the metaphor you must disregard), but simply the fact of the agency itself. Taking, then, their difference in degree, and supposing that the degree of the im-

Difference in degree.

LECT. XXIV.

pression determines the degree of the sensation, it cannot certainly be said, that the minimum of sensation infers the maximum of perception; for perception always supposes a certain quantum of sensation: but this is undeniable, that, above a certain limit, perception declines, in proportion as sensation rises. Thus, in the sense of sight, if the impression be strong we are dazzled, blinded, and consciousness is limited to the pain or pleasure of the sensation, in the intensity of which, perception has been lost.

Difference in kind. Sight; Colour, and Figure, as sources of pleasure.

Take now the difference, in kind, of impressions in the same sense. Of the senses, take again that of Sight. Sight, as will hereafter be shown, is cognisant of colour, and, through colour, of figure. But though figure is known only through colour, a very imperfect cognisance of colour is necessary, as is shown in the case, (and it is not a rare one), of those individuals who have not the faculty of discriminating colours. These persons, who probably perceive only a certain difference of light and shade, have as clear and distinct a cognisance of figure, as others who enjoy the sense of sight in absolute perfection. This being understood, you will observe, that, in the vision of colour, there is more of sensation; in that of figure, more of perception. Colour affords our faculties of knowledge a far smaller number of differences and relations than figure; but, at the same time, yields our capacity of feeling a far more sensual enjoyment. But if the pleasure we derive from colour be more gross and vivid, that from figure is more refined and permanent. It is a law of our nature, that the more intense a pleasure, the shorter is its duration. The pleasures of sense are grosser and more intense than those of intellect; but, while the former alternate speedily with disgust, with the latter we are never satiated. The same analogy holds among the senses themselves. Those in which sensation predominates, in which pleasure is most intense, soon pall upon us; whereas those in which perception predominates, and which hold more immediately of intelligence, afford us a less exclusive but a more enduring gratification. How soon are we cloyed with the pleasures of the palate, compared with those of the eye; and, among the objects of the former, the meats that please the most are soonest objects of disgust. This is too notorious in regard to taste to stand in need of proof. But it is no less certain in the case of vision. In Painting, there is a pleasure derived from a vivid and harmonious colouring, and a pleasure from the drawing and grouping of the figures. The two pleasures are distinct, and even, to a certain extent, incompatible. For if we attempt to combine them, the grosser and more obtrusive gratification, which we find in the colouring, distracts us from the more refined and intellectual enjoyment we derive from the relation of figure; while, at the same time, the disgust we soon experience from the one tends to render us insensible to the other. This is finely expressed by a modern Latin poet of high genius;—

LECT. XXIV.

Joannes Secundus quoted.

"Mensura rebus est sua dulcibus;
Ut quodque mentes suavius afficit,
Fastidium sic triste secum
Limite proximiore ducit."[α]

"Est modus et dulci: nimis immoderata voluptas
Tædia finitimo limite semper habet.
Cerne novas tabulas; vident florente colore,
Picta velut primo vere coruscat humus.
Cerne diu tamen has, hebetataque lumina flecies,
Et tibi conspectus nausea mollis erit;
Subque tuos oculos aliquid revocare libebit,
Prisca quod inculta secla tulere manu."[β]

α Joannes Secundus, *Basia*, ix. [*Opera*, p. 65, (edit. 1631).—Ed.]

β Joannes Secundus, *Epigrammata*, liii. [*Opera*, p. 115.—Ed.]

LECT. XXIV.

Paraphrases Cicero.

His learned commentator, Bosscha, has not, however, noticed that these are only paraphrases of a remarkable passage of Cicero.[α] Cicero and Secundus have not, however, expressed the principle more explicitly than Shakespeare;—

Shakespeare.

> "These violent delights have violent ends,
> And in their triumph die. The sweetest honey
> Is loathsome in its own deliciousness,
> And in the taste confounds the appetite.
> Therefore, love moderately; long love doth so.
> Too swift arrives as tardy as too slow."[β]

Result in sum of foregoing discussion.

The result of what I have now stated, therefore, is, in the first place, that, as philosophers have observed, there is a distinction between Knowledge and Feeling,—Perception and Sensation, as between the objective and the subjective element; and, in the second, that this distinction is, moreover, governed by the law,—that the two elements, though each necessarily supposes the other, are still always in a certain inverse proportion to each other.[γ]

The distinction of Perception from Sensation, of importance only in the doctrine of Intuitive Perception.

Before leaving this subject, I may notice that the distinction of perception proper and sensation proper, though recognised as phænomenal by philosophers who hold the doctrine of a representative perception, rises into reality and importance only in the doctrine of an intuitive perception. In the former doctrine, perception is supposed to be only apparently objective; being, in reality, no less subjective than sensation proper,—the subjective element itself. Both are

α *De Oratore*, iii. 25: "Difficile enim dictu est, quaenam causa sit, cur ea, quae maxime sensus nostros impellunt voluptate, et specie prima acerrime commovent, ab iis celerrime fastidio quodam et satietate abalienemur." &c.—ED.

β *Romeo and Juliet*, Act II. scene 6.

γ For historical notices of approximations to this Law, see *Reid's Works*, Note D*, p. 887.—ED.

LECT. XXIV.

nothing more than mere modes of the ego. The philosophers who hold the hypothesis of a representative perception, make the difference of the two to consist only in this;—that in perception proper, there is reference to an unknown object, different from me; in sensation, there is no reference to aught beyond myself. Brown, on the supposition that Reid held that doctrine in common with himself and philosophers at large, states sensation, as understood by Reid, to be "the simple feeling that immediately follows the action of an external body on any of our organs of sense, considered merely as a feeling of the mind; the corresponding perception being the reference of this feeling to the external body as its cause."[a] The distinction he allows to be a convenient one, if the nature of the complex process which it expresses be rightly understood. "The only question," he says, "that seems, philosophically, of importance, with respect to it, is whether the perception in this sense,—the reference of the sensation to its external corporeal cause,—implies, as Dr Reid contends, a peculiar mental power, coextensive with sensation, to be distinguished by a peculiar name in the catalogue of our faculties; or be not merely one of the results of a more general power, which is afterwards to be considered by us,—the power of association,—by which one feeling suggests, or induces, other feelings that have formerly coexisted with it."[β]

That Reid laid stress on this distinction, serves to determine the nature of his doctrine of Perception.

If Brown be correct in his interpretation of Reid's general doctrine of perception, his criticism is not only true but trite. In the hands of a cosmothetic idealist, the distinction is only superficial, and manifestly of no import; and the very fact, that Reid laid

[a] Lecture xxvi. p. 162 (edit. 1830).—Ed.

[β] *Ibid.*—Ed.

LECT. XXIV.

so great a stress on it, would tend to prove, independently of what we have already alleged, that Brown's interpretation of his doctrine is erroneous. You will remark, likewise, that Brown, (and Brown only speaks the language of all the philosophers who do not allow the mind a consciousness of aught beyond its own states), misstates the phænomenon, when he asserts that, in perception, there is a reference from the internal to the external, from the known to the unknown. That this is not the fact, an observation of the phænomenon will at once convince you. In an act of perception, I am conscious of something as self, and of something as not self:—this is the simple fact. The philosophers, on the contrary, who will not accept this fact, misstate it. They say that we are there conscious of nothing but a certain modification of mind; but this modification involves a reference to,—in other words, a representation of,—something external, as its object. Now this is untrue. We are conscious of no reference,—of no representation; we believe that the object of which we are conscious is the object which exists. Nor could there possibly be such reference or representation; for reference or representation supposes a knowledge already possessed of the object referred to or represented; but perception is the faculty by which our first knowledge is acquired, and, therefore, cannot suppose a previous knowledge as its condition. But this I notice only by the way; this matter will be regularly considered in the sequel.

No reference from the internal to the external in Perception, as Brown states.

I may here notice the false analysis, which has endeavoured to take perception out of the list of our faculties, as being only a compound and derivative power. Perception, say Brown and others, supposes memory and comparison and judgment; therefore, it is not a primary faculty of mind. Nothing can be more erroneous than this reasoning. In the first place, I have formerly shown you that consciousness supposes memory, and discrimination, and judgment;[a] and, as perception does not pretend to be simpler than consciousness, but in fact only a modification of consciousness, that, therefore, the objection does not apply. But, in the second place, the objection is founded on a misapprehension of what a faculty properly is. It may be very true that an act of perception cannot be realised simply and alone. I have often told you that the mental phænomena are never simple, and that as tissues are woven out of many threads, so a mental phænomenon is made up of many acts and affections, which we can only consider separately by abstraction, but can never even conceive as separately existing. In mathematics, we consider a triangle or a square, the sides and the angles apart from each other, though we are unable to conceive them existing independently of each other. But because the angles and sides exist only through each other, would it be correct to deny their reality as distinct mathematical elements? As in geometry, so is it in psychology. We admit that no faculty can exist itself alone; and that it is only by viewing the actual manifestations of mind in their different relations, that we are able by abstraction to analyse them into elements, which we refer to different faculties. Thus, for example, every judgment, every comparison, supposes two terms to be compared, and, therefore, supposes an act of representative, or an act of acquisitive, cognition. But go back to one or other of these acts, and you will find

Perception taken out of the list of primary faculties, through a false analysis.

a See above, vol. I. pp. 202-205.—Ed.

LECT. XXIV. that each of them supposes a judgment and a memory. If I represent in imagination the terms of comparison, there is involved a judgment; for the fact of their representation supposes the affirmation or judgment that they are called up, that they now ideally exist; and this judgment is only possible, as the result of a comparison of the present consciousness of their existence with a past consciousness of their non-existence, which comparison, again, is only possible through an act of memory.

The Primary and Secondary Qualities of matter. Connected with the preceding distinction of Perception and Sensation, is the distinction of the Primary and Secondary Qualities of matter. This distinction cannot be omitted; but I shall not attempt to follow out the various difficult and doubtful problems which it presents.*

Historical notices of this distinction. It would only confuse you were I to attempt to determine, how far this distinction was known to the Atomic Physiologists, prior to Aristotle, and how far Aristotle himself was aware of the principle on which it proceeds. It is enough to notice, as the most remarkable opinion of antiquity, that of Democritus, (Democritus.) who, except the common qualities of body which are known by Touch, denied that the senses afforded us any information concerning the real properties of matter. (Descartes.) Among modern philosopers, Descartes was the first who recalled attention to the distinction. According to him, the primary qualities differ from the secondary in this,—that our knowledge of the former is more clear and distinct than of the latter. "Longe alio modo cognoscimus quid sit in corpore magnitudo vel figura quam quid sit, in eodem corpore, color, vel

* For a fuller and more accurate account of the history of this distinction, see *Reid's Works*, Note D.—Ed.

odor, vel sapor.—Longe evidentius cognoscimus quid sit in corpore esse figuratum quam quid sit esse coloratum."[α]

"The qualities of external objects," says Locke,[β] "are of two sorts; first, Original or Primary; such are solidity, extension, motion or rest, number and figure. These are inseparable from body, and such as it constantly keeps in all its changes and alterations. Thus take a grain of wheat, divide it into two parts; each part has still solidity, extension, figure, mobility; divide it again, and it still retains the same qualities; and will do still, though you divide it on till the parts become insensible.

Locke.

"Secondly, Secondary qualities, such as colours, smells, tastes, sounds, &c., which, whatever reality we by mistake may attribute to them, are in truth nothing in the objects themselves, but powers to produce various sensations in us; and depend on the qualities before mentioned.

"The ideas of primary qualities of bodies are resemblances of them; and their patterns really exist in bodies themselves: but the ideas produced in us by secondary qualities, have no resemblance of them at all: and what is sweet, blue, or warm in the idea, is but the certain bulk, figure, and motion of the insensible parts in the bodies themselves, which we call so."

Reid.

Reid adopted the distinction of Descartes: he holds that our knowledge of the primary qualities is clear and distinct, whereas our knowledge of the secondary qualities is obscure.[γ] "Every man," he says, "capable of reflection, may easily satisfy himself, that he has a

α *Principia*, i. § 69.—Ed.

β *Essay*, book ii. ch. viii. §§ 9-15. The text is an abridgment of Locke, not an exact quotation.—Ed.

γ *Intellectual Powers*, Essay ii. ch. xvii. *Works*, p. 311.—Ed.

LECT. XXIV.

perfectly clear and distinct notion of extension, divisibility, figure, and motion. The solidity of a body means no more, but that it excludes other bodies from occupying the same place at the same time. Hardness, softness, and fluidity, are different degrees of cohesion in the parts of a body. It is fluid, when it has no sensible cohesion; soft when the cohesion is weak; and hard when it is strong: of the cause of this cohesion we are ignorant, but the thing itself we understand perfectly, being immediately informed of it by the sense of touch. It is evident, therefore, that of the primary qualities we have a clear and distinct notion; we know what they are, though we may be ignorant of the causes." But he did more, he endeavoured to show that this difference arises from the circumstance,—that the perception, in the case of the primary qualities, is direct; in the case of the secondary, only relative. This he explains: "I observe further that the notion we have of primary qualities is direct and not relative only. A relative notion of a thing is, strictly speaking, no notion of the thing at all, but only of some relation which it bears to something else.

"Thus gravity sometimes signifies the tendency of bodies towards the earth; sometimes it signifies the cause of that tendency. When it means the first, I have a direct and distinct notion of gravity; I see it, and feel it, and know perfectly what it is; but this tendency must have a cause. We give the same name to the cause; and that cause has been an object of thought and of speculation. Now what notion have we of this cause when we think and reason about it? It is evident we think of it as an unknown cause of a known effect. This is a relative notion; and it must be obscure, because it gives us no conception of what the thing is, but of what relation it bears to something else. Every relation which a thing unknown bears to something that is known, may give a relative notion of it; and there are many objects of thought and of discourse, of which our faculties can give no better than a relative notion.

"Having premised these things to explain what is meant by a relative notion, it is evident, that our notion of Primary Qualities is not of this kind; we know what they are, and not barely what relation they bear to something else.

"It is otherwise with Secondary Qualities. If you ask me, what is that quality or modification in a rose which I call its smell, I am at a loss what to answer directly. Upon reflection I find, that I have a distinct notion of the sensation which it produces in my mind. But there can be nothing like to this sensation in the rose, because it is insentient. The quality in the rose is something which occasions the sensation in me; but what that something is, I know not. My senses give me no information upon this point. The only notion, therefore, my senses give is this, that smell in the rose is an unknown quality or modification, which is the cause or occasion of a sensation which I know well. The relation which this unknown quality bears to the sensation with which nature hath connected it, is all I learn from the sense of smelling; but this is evidently a relative notion. The same reasoning will apply to every secondary quality.

"Thus I think it appears, that there is a real foundation for the distinction of primary from secondary qualities; and that they are distinguished by this, that of the primary we have by our senses a direct and distinct notion; but of the secondary only a

LECT. XXIV. relative notion, which must, because it is only relative, be obscure; they are conceived only as the unknown causes or occasions of certain sensations, with which we are well acquainted."

The list of primary qualities given by Locke, and that of Reid, do not coincide. You will observe that the lists of the primary qualities given by Locke and Reid do not coincide. According to Locke, these are Solidity, Extension, Motion, Hardness, Softness, Roughness, Smoothness, and Fluidity.

Stewart. Mr Stewart proposes another line of demarcation. "I distinguish," he says, "Extension and Figure by the title of the *Mathematical Affections* of matter; restricting the phrase *Primary Qualities*, to Hardness and Softness, Roughness and Smoothness, and other properties of the same description. The line which I would draw between *Primary* and *Secondary Qualities* is this, that the former necessarily involve the notion of *Extension*, and consequently of *externality* or *outness;* whereas the latter are only conceived as the unknown causes of known sensations; and *when first apprehended by the mind*, do not imply the existence of anything locally distinct from the subjects of its own consciousness." [a]

The Primary Qualities reducible to two,—Extension and Solidity. All these Primary Qualities, including Mr Stewart's Mathematical Affections of matter, may easily be reduced to two,—Extension and Solidity. Thus:—Figure is a mere limitation of extension; Hardness, Softness, Fluidity, are only Solidity variously modified,—only its different degrees; while Roughness and Smoothness denote only the sensations connected with certain perceptions of Solidity. On the other hand, in regard to Divisibility, (which is proper to Reid), and to Motion,—these can hardly be mere data of sense.

[a] *Phil. Essays, Works*, vol. v. pp. 116, 117.

Divisibility supposes division, and a body divided supposes memory, for if we did not remember that it had been one, we should not know that it is now two; we could not compare its present with its former state; and it is by this comparison alone that we learn the fact of division. As to Motion, this supposes the exercise of memory, and the notion of time, and, therefore, we do not owe it exclusively to sense. Finally as to Number, which is peculiar to Locke, it is evident that this, far from being a quality of matter, is only an abstract notion,—the fabrication of the intellect, and not a datum of sense.[a]

This reduction involves a difficulty.

Thus, then, we have reduced all primary qualities to Extension and Solidity, and we are, moreover, it would seem, beginning to see light, inasmuch as the primary qualities are those in which perception is dominant, the secondary those in which sensation prevails. But here we are again thrown back: for extension is only another name for space, and our notion of space is not one which we derive exclusively from sense,—not one which is generalised only from experience; for it is one of our necessary notions,—in fact, a fundamental condition of thought itself. The analysis of Kant, independently of all that has been done by other philosophers, has placed this truth beyond the possibility of doubt, to all those who understand the meaning and conditions of the problem. For us, however, this is not the time to discuss the subject. But, taking it for granted that the notion of space is native or *a priori*, and not adventitious

What, and how solved.

a In this reduction of the primary qualities to Extension and Solidity, the author follows Royer-Collard, whose remarks will be found quoted in *Reid's Works*, p. 844. From the notes appended to that quotation, it will be seen that Sir W. Hamilton's final opinion differs in some respects from that expressed in the present text.—ED.

Space known *a priori*; Extension *a posteriori*.

or *a posteriori*, are we not at once thrown back into idealism? For if extension itself be only a necessary mental mode, how can we make it a quality of external objects, known to us by sense; or how can we contrast the outer world, as the extended, with the inner, as the unextended world? To this difficulty, I see only one possible answer. It is this:—It cannot be denied that space, as a necessary notion, is native to the mind; but does it follow, that, because there is an *a priori* space, as a form of thought, we may not also have an empirical knowledge of extension, as an element of existence? The former, indeed, may be only the condition through which the latter is possible. It is true that, if we did not possess the general and necessary notion of space anterior to, or as the condition of, experience, from experience we should never obtain more than a generalised and contingent notion of space. But there seems to me no reason to deny, that because we have the one, we may not also have the other. If this be admitted, the whole difficulty is solved; and we may designate by the name of *extension* our empirical knowledge of space, and reserve the term *space* for space considered as a form or fundamental law of thought.[a] This matter will, however, come appropriately to be considered, in treating of the Regulative Faculty.

General result.—In the Primary Qualities, Perception predominates; in the Secondary, Sensation.

The following is the result of what I think an accurate analysis would afford, though there are no doubt many difficulties to be explained.—That our knowledge of all the qualities of matter is merely relative. But though the qualities of matter are all known only in relation to our faculties, and the total or absolute cog-

[a] Here, on blank leaf of MS., are jotted the words, "So Causality." [Causality depends, first, on the *a priori* necessity in the mind to think some cause; and, second, on experience, as revealing to us the particular cause of any effect.]—*Oral Interpolation*, but not at this passage.—ED.

nition in perception is only matter in a certain relation to mind, and mind in a certain relation to matter; still, in different perceptions, one term of the relation may predominate, or the other. Where the objective element predominates,—where matter is known as principal in its relation to mind, and mind only known as subordinate in its correlation to matter,—we have Perception Proper, rising superior to sensation; this is seen in the Primary Qualities. Where, on the contrary, the subjective element predominates,—where mind is known as principal in its relation to matter, and matter is only known as subordinate in its relation to mind,—we have Sensation Proper rising superior to perception; and this is seen in the Secondary Qualities. The adequate illustration of this would, however, require both a longer, and a more abstruse, discussion than we can afford.[a]

[a] Cf. *Reid's Works*, Notes D and D*.—Ed.

LECTURE XXV.

THE PRESENTATIVE FACULTY.—I. PERCEPTION.—OBJECTIONS TO THE DOCTRINE OF NATURAL REALISM.

Objections to the doctrine of Natural Realism.

FROM our previous discussions, you are now, in some measure, prepared for a consideration of the grounds on which philosophers have so generally asserted the scientific necessity of repressing the testimony of consciousness to the fact of our immediate perception of external objects, and of allowing us only a mediate knowledge of the material world: a procedure by which they either admit or cannot rationally deny, that Consciousness is a mendacious witness; that Philosophy and the Common Sense of mankind are placed in contradiction; nay, that the only legitimate philosophy is an absolute and universal scepticism. That consciousness, in perception, affords us, as I have stated, an assurance of an intuitive cognition of the non-ego, is not only notorious to every one who will interrogate consciousness as to the fact, but is, as I have already shown you, acknowledged not only by cosmothetic idealists, but even by absolute idealists and sceptics. "It seems evident," says Hume, who in this concession must be allowed to express the common acknowledgment of philosophers, "that when men follow this blind and powerful instinct of nature, they always suppose the very images, presented by the senses, to be the external objects, and never entertain any suspicion, that the one are nothing but

The testimony of Consciousness in perception, notorious, and acknowledged by philosophers of all classes. Hume quoted.

representations of the other. This very table, which we see white, and which we feel hard, is believed to exist, independent of our perception, and to be something external to our mind, which perceives it. Our presence bestows not being on it: our absence does not annihilate it. It preserves its existence, uniform and entire, independent of the situation of intelligent beings, who perceive or contemplate it. But this universal and primary opinion of all men is soon destroyed by the slightest philosophy, which teaches us that nothing can ever be present to the mind but an image or perception, and that the senses are only the inlets, through which these images are received, without being ever able to produce any immediate intercourse between the mind and the object."[a]

The discussion divided into two parts.

In considering this subject, it is manifest that, before rejecting the testimony of consciousness to our immediate knowledge of the non-ego, the philosophers were bound, in the first place, to evince the absolute necessity of their rejection; and, in the second place, in substituting an hypothesis in the room of the rejected fact, they were bound to substitute a legitimate hypothesis,—that is, one which does not violate the laws under which an hypothesis can be rationally proposed. I shall, therefore, divide the discussion into two sections. In the former, I shall state the reasons, as far as I have been able to discover them, on which philosophers have attempted to manifest the impossibility of acquiescing in the testimony of consciousness, and the general belief of mankind; and, at the same time, endeavour to refute these reasons, by showing that they do not establish the necessity required. In the

[a] *Enquiry concerning Human Understanding*, § xii., *Essays*, &c. [*Of the Academical or Sceptical Philosophy*, *Essays*, p. 367, edit. 1758. *Philosophical Works*, vol. iv. p. 177.—Ed.

LECT. XXV.

latter, I shall attempt to prove that the hypothesis proposed by the philosophers, in place of the fact of consciousness, does not fulfil the conditions of a legitimate hypothesis,—in fact, violates them almost all.

I. Reasons for rejecting the testimony of Consciousness in perception, detailed and criticised.

In the first place, then, in regard to the reasons assigned by philosophers for their refusal of the fact of our immediate perception of external things,—of these I have been able to collect in all five. As they cannot be very briefly stated, I shall not first enumerate them together, and then consider each in detail; but shall consider them one after the other, without any general and preliminary statement.

The first ground of rejection.

The first, and highest, ground on which it may be held, that the object immediately known in perception is a modification of the mind itself, is the following: Perception is a cognition or act of knowledge; a cognition is an immanent act of mind; but to suppose the cognition of anything external to the mind would be to suppose an act of the mind going out of itself, in other words, a transeunt act; but action supposes existence, and nothing can act where it is not; therefore, to act out of self is to exist out of self, which is absurd.[a]

Refuted. 1. Our inability to conceive how the fact of consciousness is possible, no ground for denying its possibility.

This argument, though I have never met with it explicitly announced, is still implicitly supposed in the arguments of those philosophers who hold, that the mind cannot be conscious of aught beyond its own modifications. It will not stand examination. It is very true that we can neither prove, nor even conceive, how the ego can be conscious or immediately cognitive of the non-ego; but this, our ignorance, is no sufficient reason on which to deny the possibility of

a See Biunde, *Versuch einer systematischen Behandlung der empirischen Psychologie*, vol. i. § 31, p. 139. [Biunde refers to Fichte as holding the principle of this argument.—Ed.] Cf. Schulze, *Anthropologie*, § 53, p. 107, (edit. 1820.) [Cicero, *Acad. Quaest.*, iv. 24.—Ed.]

the fact. As a fact, and a primary fact, of consciousness, we must be ignorant of the why and how of its reality, for we have no higher notion through which to comprehend it, and, if it involve no contradiction, we are, philosophically, bound to accept it. But if we examine the argument a little closer, we shall find that it proves too much; for, on the same principle, we should establish the impossibility of any overt act of volition,—nay, even the impossibility of all agency and mutual causation. For if, on the ground that nothing can act out of itself, because nothing exists out of itself, we deny to mind the immediate knowledge of things external; on the same principle, we must deny to mind the power of determining any muscular movement of the body. And if the action of every existence were limited to the sphere of that existence itself, then, no one thing could act upon any other thing, and all action and reaction, in the universe, would be impossible. This is a general absurdity, which follows from the principle in question. But there is a peculiar and proximate absurdity into which this theory runs, in the attempt it makes to escape the inexplicable. It is this:—The cosmothetic idealists, who found their doctrine on the impossibility of mind acting out of itself, in relation to matter, are obliged to admit the still less conceivable possibility of matter acting out of itself, in relation to mind. They deny that mind is immediately conscious of matter; and, to save the phænomenon of perception, they assert that the non-ego, as given in that act, is only an illusive representation of the non-ego, in, and by, the ego. Well, admitting this, and allowing them to belie the testimony of consciousness to the reality of the non-ego as perceived, what do they gain by this? They surrender the simple datum of consciousness,—that the

2. The reason adduced involves a general absurdity.

3. Involves a special absurdity.

LECT. XXV.

external object is immediately known; and, in lieu of that real object, they substitute a representative object. But still they hold (at least those who do not fly to some hyperphysical hypothesis), that the mind is determined to this representation by the material reality, to which material reality they must, therefore, accord the very transeunt efficiency which they deny to the immaterial principle. This first and highest ground, therefore, on which it is attempted to establish the necessity of a representative perception, is not only insufficient, but self-contradictory.

The second ground of rejection.

The second ground on which it has been attempted to establish the necessity of this hypothesis, is one which has been more generally and more openly founded on than the preceding. Mind and matter, it is said, are substances, not only of different, but of the most opposite, natures; separated, as some philosophers express it, by the whole diameter of being: but what immediately knows must be of a nature correspondent, analogous, to that which is known; mind cannot, therefore, be conscious or immediately cognisant of what is so disproportioned to its essence as matter.

This principle has influenced the whole history of philosophy.

This principle is one whose influence is seen pervading the whole history of philosophy, and the tracing of this influence would form the subject of a curious treatise.* To it we principally owe the doctrine of a *representative perception*, in one or other of its forms; and in a higher or lower potence, according as the representative object was held to be, in relation to mind, of a nature either the same or similar. Derivative from the principle in its lower potence or degree, (that is, the immediate object being supposed to be only something similar to the mind,)

* Cf. *Reid's Works*, p. 300, note, and *Discussions*, p. 61.—ED.

we have, among other less celebrated and less definite theories, the *intentional species* of the schoolmen, (at least as generally held,) and the *ideas* of Malebranche and Berkeley. In its higher potence, (that is, where the representative object is supposed to be of a nature not merely similar to, but identical with, mind, though it may be numerically different from individual minds,) it affords us, among other modifications, the *gnostic reasons* (λόγοι γνωστικοί) of the Platonists, the *pre-existing species* of Avicenna and other Arabian Aristotelians, the *ideas* of Descartes, Arnauld, Leibnitz, Buffier, and Condillac, the *phænomena* of Kant, and the *external states* of Dr Brown. It is doubtful to which head we should refer Locke, and Newton, and Clarke,—nay, whether we should not refer them to the class of those who, like Democritus, Epicurus, and Digby, viewed the representative or immediate object, as a material efflux or propagation from the external reality to the brain.

This principle also indirectly determined many celebrated theories in philosophy, as the *hierarchical gradation of souls or substantial faculties*, held by many followers of Aristotle, the ὄχοι or *vehicular media* of the Platonists, the *plastic medium* of Cudworth and Le Clerc, the doctrine of the *community, oneness*, or *identity of the human intellect* in all men, maintained by the Aphrodisian, Themistius, Averroes, Cajetanus, and Zabarella, the *vision of all things in the Deity* of Malebranche, and the Cartesian and Leibnitzian doctrine of *assistance* and *pre-established harmony*. To the influence of the same principle, through the refusal of the testimony of consciousness to the duality of our knowledge, are also mediately to be traced the unitarian systems of *absolute identity, materialism*, and *idealism*.

LECT. XXV.

But, 1. Is perfectly arbitrary.

But, if no principle was ever more universal in its effects, none was ever more arbitrarily assumed. It not only can pretend to no necessity; it has absolutely no probability in its favour. Some philosophers, as Anaxagoras, Heraclitus, Alcmæon, have even held that the relation of knowledge supposes, not a similarity or sameness between subject and object, but, in fact, a contrariety or opposition; and Aristotle himself is sometimes in favour of this opinion, though, sometimes, it would appear, in favour of the other.* But, however this may be, each assertion is just as likely, and just as unphilosophical, as its converse. We know, and can know, nothing *a priori* of what is possible or impossible to mind, and it is only by observation and by generalisation *a posteriori*, that we can ever hope to attain any insight into the question. But the very first fact of our experience contradicts the assertion, that mind, as of an opposite nature, can have no immediate cognisance of matter; for the primary datum of consciousness is, that, in perception, we have an intuitive knowledge of the ego and of the non-ego, equally and at once. This second ground, therefore, affords us no stronger necessity than the first, for denying the possibility of the fact of which consciousness assures us.

2. Is unphilosophical.

3. Contradicted by the first fact of our experience.

The third ground of rejection.

The third ground on which the representative hypothesis of perception is founded, and that apparently alone contemplated by Reid and Stewart, is, that the mind can only know immediately that to which it is immediately present; but as external objects can neither themselves come into the mind, nor the mind go out to them, such presence is impossible; therefore, external objects can only be mediately known, through some representative object, whether that object be a

* See above, vol. I. lect. xvi. p. 296, note.—Ed.

modification of mind, or something in immediate relation to the mind. It was this difficulty of bringing the subject and object into proximate relation, that, in part, determined all the various schemes of a representative perception; but it seems to have been the one which solely determined the peculiar form of that doctrine in the philosophy of Democritus, Epicurus, Digby, and others, under which it is held, that the immediate or internal object is a representative emanation, propagated from the external reality to the sensorium.

Has been redargued in three different ways.

Now, this objection to the immediate cognition of external objects, has, as far as I know, been redargued in three different ways. In the first place, it has been denied, that the external reality cannot itself come into the mind. In the second, it has been asserted, that a faculty of the mind itself does actually go out to the external reality; and, in the third place, it has been maintained that, though the mind neither goes out, nor the reality comes in, and though subject and object are, therefore, not present to each other, still that the mind, through the agency of God, has an immediate perception of the external object.

The first by Sergeant.

The first mode of obviating the present objection to the possibility of an immediate perception, might be thought too absurd to have been ever attempted. But the observation of Varro,[a] that there is nothing so absurd which has not been asserted by some philosopher, is not destined to be negatived in the present instance. In opposition to Locke's thesis, "that the

[a] In a fragment of his satire *Eumenides*, preserved by Nonius Marcellus, *De Proprietate Sermonis*, c. I. n. 276, v. *Infans*:—

"Postremo nemo aegrotus quicquam somniat
Tam infandum quod non aliquis dicat philosophus."

But the words in the text occur more exactly in Cicero, *De Divinatione*, II. 58: "Sed, nescio quomodo, nihil tam absurde dici potest, quod non dicatur ab aliquo philosophorum."—Ed.

LECT. XXV.

mind knows not things immediately, but only by the intervention of the ideas it has of them," and in opposition to the whole doctrine of representation, it is maintained, in terms, by Sergeant, that "I know the very thing; therefore, the very thing is in my act of knowledge; but my act of knowledge is in my understanding; therefore, the thing which is in my knowledge, is also in my understanding."[a] We may suspect that this is only a paradoxical way of stating his opinion; but though this author, the earliest and one of the most eloquent of Locke's antagonists, be destitute neither of learning nor of acuteness, I must confess, that Locke and Molyneux cannot be blamed in pronouncing his doctrine unintelligible.

The second by Empedocles, the Platonists, &c.

The second mode of obviating the objection,—by allowing to the mind a power of sallying out to the external reality, has higher authority in its favour. That vision is effected by a perceptive emanation from the eye, was held by Empedocles, the Platonists, and Stoics, and was adopted also by Alexander the Aphrodisian, by Euclid, Ptolemy, Galen, and Alchindus.[β] This opinion, as held by these philosophers, was limited; and, though erroneous, is not to be viewed as irrational. But in the hands of Lord Monboddo, it is carried to an absurdity which leaves even Sergeant far behind. "The mind," says the learned author of *Antient Metaphysics*, "is not where the body is, when it perceives what is distant from the body, either in time or place, because nothing can act but when and where it is. Now the mind acts when it perceives. The mind, therefore, of every animal who has memory or imagination, acts, and, by consequence, exists, when

a *Solid Philosophy*, p. 29. [See above, vol. ii. lect. xxiv. p. 92.—Ed.]

β See above, vol. ii. lect. xxi. pp. 34, 35.—Ed.

and where the body is not; for it perceives objects distant from the body, both in time and place."[a] ✕✕

The third by Reid and Stewart.

The third mode is apparently that adopted by Reid and Stewart, who hold, that the mind has an immediate knowledge of the external reality, though the subject and object may not be present to each other; and, though this be not explicitly or obtrusively stated, that the mind obtains this immediate knowledge through the agency of God. Dr Reid's doctrine of perception is thus summed up by Mr Stewart: "To what then, it may be asked, does this statement amount? Merely to this; that the mind is so formed that certain impressions produced on our organs of sense by external objects, are followed by correspondent sensations, and that these sensations, (which have no more resemblance to the qualities of matter than the words of a language have to the things they denote), are followed by a perception of the existence and qualities of the bodies by which the impressions are made; that all the steps of this process are equally incomprehensible; and that, for anything we can prove to the contrary, the connection between the sensation and the perception, as well as that between the impression and the sensation, may be both arbitrary; that it is therefore by no means impossible, that our sensations may be merely the occasions on which the correspondent perceptions are excited; and that, at any rate, the consideration of these sensations, which are attributes of mind, can throw no light on the manner in which we acquire our knowledge of the existence and qualities of body. From this view of the subject it follows, that it is the external objects themselves, and not any

a See *Antient Metaphysics*, vol. ii. 85.—Ed.
p. 306, and above, vol. ii, lect. xxi. p.

LECT. XXV.

species or images of the objects, that the mind perceives; and that, although, by the constitution of our nature, certain sensations are rendered the constant antecedents of our perceptions, yet it is just as difficult to explain how our perceptions are obtained by their means, as it would be upon the supposition that the mind were all at once inspired with them, without any concomitant sensations whatever."[α]

Their opinion almost identical with the doctrine of Occasional Causes.

This statement, when illustrated by the doctrine of these philosophers in regard to the distinction of Efficient and Physical Causes, might be almost identified with the Cartesian doctrine of Occasional Causes. According to Reid and Stewart,[β] and the opinion has been more explicitly asserted by the latter, there is no really efficient cause in nature but one—viz. the Deity. What are called physical causes and effects being antecedents and consequents, but not in virtue of any mutual and necessary dependence;—the only efficient being God, who, on occasion of the antecedent, which is called the physical cause, produces the consequent, which is called the physical effect. So in the case of perception; the cognition of the external object is not, or may not be, a consequence of the immediate and natural relation of that object to the mind, but of the agency of God, who, as it were, reveals the outer existence to our perception. A similar doctrine is held by a great German philosopher, Frederick Henry Jacobi.[γ]

And exposed to many objections.

To this opinion many objections occur. In the first place, so far is it from being, as Mr Stewart

α *Elements*, vol. i. c. i. § 3; *Works*, vol. ii. pp. 111, 112.

β Reid, *Intellectual Powers*, Essay ii. c. vi.; *Active Powers*, Essay i. c. v. vi.; Essay iv. c. ii. iii. Stewart, *Elements*, vol. i. c. i. § 2; vol. ii. c. iv. § 1.—Ed.

γ *David Hume, über den Glauben*,—*Werke*, ii. p. 165; *Über die Lehre des Spinoza*,—*Werke*, iv. p. 210. Quoted by Sir W. Hamilton, *Reid's Works*, p. 793.—Ed.

LECT. XXV.

affirms, a plain statement of the fact, apart from all hypothesis, it is manifestly hypothetical. In the second place, the hypothesis assumes an occult principle,—it is mystical. In the third place, the hypothesis is hyperphysical,—calling in the proximate assistance of the Deity, while the necessity of such intervention is not established. In the fourth place, it goes even far to frustrate the whole doctrine of the two philosophers in regard to perception, as a doctrine of intuition. For if God has bestowed on me the faculty of immediately perceiving the external object, there is no need to suppose the necessity of an immediate intervention of the Deity to make that act effectual; and if, on the contrary, the perception I have of the reality is only excited by the agency of God, then I can hardly be held to know that reality, immediately and in itself, but only mediately, through the notion of it determined in my mind.

1. Hypothetical. 2. Mystical. 3. Hyperphysical. 4. Goes to frustrate a doctrine of Intuitive Perception.

Let us try, then, whether it be impossible, not to explain, (for that it would be ridiculous to dream of attempting), but to render intelligible the possibility of an immediate perception of external objects; without assuming any of the three preceding hypotheses, and without postulating aught that can fairly be refused.

The possibility of an immediate perception of external objects intelligible.

Now, in the first place, there is no good ground to suppose, that the mind is situate solely in the brain, or exclusively in any one part of the body. On the contrary, the supposition that it is really present wherever we are conscious that it acts,—in a word, the Peripatetic aphorism, the soul is all in the whole and all in every part,[a]—is more philosophical, and, consequently, more

1. No ground to suppose that the mind is situated solely in any one part of the body.

[a] Aristotle, *De Anima*, i. 5, 26 (ed. Trend.): ἅπαντ' ἐνυπάρχει τὰ μόρια τῆς ψυχῆς. Ἐν ἑκατέρῳ τῶν μορίων Augustin, *De Trinitate*, vi. 6: "Idem

LECT. XXV.

We materialise mind in attributing to it the relations of matter.

probable than any other opinion. It has not been always noticed, even by those who deem themselves the chosen champions of the immateriality of mind, that we materialise mind when we attribute to it the relations of matter. Thus, we cannot attribute a local seat to the soul, without clothing it with the properties of extension and place, and those who suppose this seat to be but a point, only aggravate the difficulty. Admitting the spirituality of mind, all that we know of the relation of soul and body is, that the former is connected with the latter in a way of which we are wholly ignorant; and that it holds relations, different both in degree and kind, with different parts of the organism. We have no right, however, to say that it is limited to any one part of the organism; for even if we admit that the nervous system is the part to which it is proximately united, still the nervous system is itself universally ramified throughout the body; and we have no more right to deny that the mind feels at the finger-points, as consciousness assures us, than to assert that it thinks exclusively in the brain. The

Sum of our knowledge of the connection of mind and body.

sum of our knowledge of the connection of mind and body is, therefore, this,—that the mental modifications are dependent on certain corporeal conditions; but of the nature of these conditions we know nothing. For example, we know, by experience, that the mind perceives only through certain organs of sense, and that, through these different organs, it perceives in a different manner. But whether the senses be instruments, whether they be media, or whether they be only partial outlets to the mind incarcerated in the body,—on

simplicior est corpore, quia non mole diffunditur per spatium loci, sed in unoquoque corpore et in toto tota est, et in qualibet ejus parte tota est." See above, vol. ii. lect. xx. p. 6, note *; and *Reid's Works*, p. 861, note.

LECT. XXV.

all this we can only theorise and conjecture. We have no reason whatever to believe, contrary to the testimony of consciousness, that there is an action or affection of the bodily sense previous to the mental perception; or that the mind only perceives in the head, in consequence of the impression on the organ. On the other hand, we have no reason whatever to doubt the report of consciousness, that we actually perceive at the external point of sensation, and that we perceive the material reality. But what is meant by perceiving the material reality?

What is meant by perceiving the material reality?

The total and real object of Perception, what.

In the first place, it does not mean that we perceive the material reality absolutely and in itself, that is, out of relation to our organs and faculties; on the contrary, the total and real object of perception, is the external object under relation to our sense and faculty of cognition. But though thus relative to us, the object is still no representation,—no modification of the ego. It is the non-ego,—the non-ego modified, and relative, it may be, but still the non-ego. I formerly illustrated this to you by a supposition. Suppose that the total object of consciousness in perception is = 12; and suppose that the external reality contributes 6, the material sense 3, and the mind 3;—this may enable you to form some rude conjecture of the nature of the object of perception.[a]

What is meant by the external object perceived?

But, in the second place, what is meant by the external object perceived? Nothing can be conceived more ridiculous than the opinion of philosophers in regard to this. For example, it has been curiously held, (and Reid is no exception), that in looking at the sun, moon, or any other object of sight, we are, on the one doctrine, actually conscious of these distant objects;

[a] See above, vol. I, lect. viii. p. 147.—ED.

LECT. XXV.

or, on the other, that these distant objects are those really represented in the mind. Nothing can be more absurd: we perceive, through no sense, aught external but what is in immediate relation and in immediate contact with its organ; and that is true which Democritus of old asserted, that all our senses are only modifications of touch.[a] Through the eye we perceive nothing but the rays of light in relation to, and in contact with, the retina; what we add to this perception must not be taken into account. The same is true of the other senses. Now, what is there monstrous or inconceivable in this doctrine of an immediate perception? The objects are neither carried into the mind, nor the mind made to sally out to them; nor do we require a miracle to justify its possibility. In fact, the consciousness of external objects, on this doctrine, is not more inconceivable than the consciousness of species or ideas on the doctrine of the schoolmen, Malebranche, or Berkeley. In either case, there is a consciousness of the non-ego, and, in either case, the ego and non-ego are in intimate relation. There is, in fact, on this hypothesis, no greater marvel, that the mind should be cognisant of the external reality, than that it should be connected with a body at all. The latter being the case, the former is not even improbable; all inexplicable as both equally remain. "We are unable," says Pascal, "to conceive what is mind; we are unable to conceive what is matter; still less are we able to conceive how these are united:—yet this is our proper nature."[β] So much in refutation of the third ground of difficulty to the doctrine of an immediate perception.

Nothing especially inconceivable in the doctrine of an immediate perception.

[a] See below, vol. II., lect. xxvii. p. 152.—Ed.

[β] *Pensées*, [partie I. art. vi. § 26; vol. II. p. 74, edit. Faugère.—Ed.]

LECT. XXV.

The fourth ground of rejection. Hume quoted.

The fourth ground of rejection is that of Hume. It is alleged by him in the sequel of the paragraph of which I have already quoted to you the commencement: "This universal and primary opinion of all men is soon destroyed by the slightest philosophy, which teaches us, that nothing can ever be present to the mind but an image or perception, and that the senses are only the inlets, through which these images are conveyed, without being ever able to produce any immediate intercourse between the mind and the object. The table, which we see, seems to diminish, as we remove farther from it; but the real table which exists independent of us suffers no alteration: it was, therefore, nothing but its image, which was present to the mind. These are the obvious dictates of reason; and no man, who reflects, ever doubted that the existences, which we consider, when we say *this house*, and *that tree*, are nothing but perceptions in the mind, and fleeting copies or representations of other existences, which remain uniform and independent."[a]

Proceeds on a mistake of what the object in perception is.

This objection to the veracity of consciousness will not occasion us much trouble. Its refutation is, in fact, contained in the very statement of the real external object of perception. The whole argument consists in a mistake of what that object is. That a thing, viewed close to the eye, should appear larger and differently figured, than when seen at a distance, and that, at too great a distance, it should even become for us invisible altogether;—this only shows that what changes the real object of sight,—the reflected rays in contact with the eye,—also changes, as it ought to change, our perception of such object. This ground

[a] *Enquiry concerning Human Understanding*, sect. xii. [*Of the Academical or Sceptical Philosophy*, pp. 367, 368, edit. 1758.—ED.]

LECT. XXV.

of difficulty could be refuted through the whole senses; but its weight is not sufficient to entitle it to any further consideration.[a]

The fifth ground of rejection.

The fifth ground on which the necessity of substituting a representative for an intuitive perception has been maintained, is that of Fichte.[β] It asserts that the nature of the ego, as an intelligence endowed with will, makes it absolutely necessary, that, of all external objects of perception, there should be representative modifications in the mind. For as the ego itself is that which wills; therefore, in so far as the will tends toward objects, these must lie within the ego. An external reality cannot lie within the ego; there must, therefore, be supposed, within the mind, a representation of this reality different from the reality itself.

Involves sundry vices.

This fifth argument involves sundry vices, and is not of greater value than the four preceding.

1. Asserts that the objects on which the will is directed must lie within the ego.

In the first place, it proceeds on the assertion, that the objects on which the will is directed, must lie within the willing ego itself. But how is this assertion proved? That the will can only tend toward those things of which the ego has in itself a knowledge, is undoubtedly true. But from this it does not follow, that the object to which the knowledge is relative, must at the same time be present with it in the ego; but if there be a perceptive cognition, that is, a consciousness of some object external to the ego, this perception is competent to excite, and to direct, the will, notwithstanding that its object lies without the ego. That, therefore, no immediate knowledge of external objects is pos-

a Vide Schulze, *Anthropologie*, II. 40.

β See especially his *Grundlage der gesammten Wissenschaftslehre*, §§ 4, 10. *Werke*, I. pp. 134, 313 *et seq.*; and his *Bestimmung des Menschen*. *Werke*, II. p. 217 *et seq.*—Ed.

sible, and that consciousness is exclusively limited to the ego, is not evinced, by this argument of Fichte, but simply assumed.

2. Takes no account of the difference between cognitions.

In the second place, this argument is faulty, in that it takes no account of the difference between those cognitions which lie at the root of the energies of will, and the other kinds of knowledge. Thus, our will never tends to what is present,—to what we possess, and immediately cognise; but is always directed on the future, and is concerned either with the continuance of those states of the ego, which are already in existence, or with the production of wholly novel states. But the future cannot be intuitively, immediately, perceived, but only represented, and mediately conceived. That a mediate cognition is necessary, as the condition of an act of will,—this does not prove, that every cognition must be mediate.[a]

These grounds of rejection are thus, one and all, incompetent.

We have thus found by an examination of the various grounds on which it has been attempted to establish the necessity of rejecting the testimony of consciousness to the intuitive perception of the external world, that these grounds are, one and all, incompetent. I shall proceed in my next Lecture to the second section of the discussion,—to consider the nature of the hypothesis of Representation or Cosmothetic Idealism, by which it is proposed to replace the fact of consciousness, and the doctrine of Natural Realism; and shall show you that this hypothesis, though, under various modifications, adopted in almost every system of philosophy, fulfils none of the conditions of a legitimate hypothesis.

[a] Vide Schulze, *Anthropologie*, ii. p. 62. [Cf. § 53 third edit.—Ed.]

LECTURE XXVI.

THE PRESENTATIVE FACULTY.—I. PERCEPTION.—THE REPRESENTATIVE HYPOTHESIS.

LECT. XXVI. Recapitulation.

No opinion has perhaps been so universally adopted in the various schools of philosophy, and more especially of modern philosophy, as the doctrine of a Representative Perception; and, in our last Lecture, I was engaged in considering the grounds on which this doctrine reposes. The order of the discussion was determined by the order of the subject. It is manifest, that, in rejecting the testimony of consciousness to our immediate knowledge of the non-ego, the philosophers were bound to evince the absolute necessity of their rejection; and, in the second place, in substituting an hypothesis in the room of the rejected fact, they were bound to substitute a legitimate hypothesis, that is, one which does not violate the laws under which an hypothesis can be rationally proposed. I stated, therefore, that I should divide the criticism of their doctrine into two sections:—that, in the former, I should state the reasons which have persuaded philosophers of the impossibility of acquiescing in the evidence of consciousness, endeavouring at the same time to show that these reasons afford no warrant to the conclusion which they are supposed even to necessitate; and, in the latter, attempt to prove, that the hypothesis proposed by philosophers in lieu of the fact of consciousness, does not fulfil the conditions of a legitimate hypothesis, and is, therefore, not only unnecessary but inadmissible. The first of these sections terminated the Lecture. I stated that there are in all five grounds, on which philosophers have deemed themselves compelled to reject the fact of our immediate consciousness of the non-ego in perception, and to place philosophy in contradiction of the common-sense of mankind. The grounds I considered in detail, and gave you some of the more manifest reasons which went to prove their insufficiency. This discussion I shall not attempt to recapitulate; and now proceed to the second section of the subject,—to consider the nature of the hypothesis of a Representative Perception, by which it is proposed to replace the fact of consciousness which testifies to our immediate perception of the external world. On the *hypothesis*, the doctrine of Cosmothetic Idealism is established:—on the *fact*, the doctrine of Natural Dualism.

LECT. XXVI.

II. The nature of the hypothesis of a Representative Perception. It violates all the conditions of a legitimate hypothesis.

[a] In the first place, from the grounds on which the cosmothetic idealist would vindicate the necessity of his rejection of the datum of consciousness, the hypothesis itself is unnecessary. The examination of these grounds proves, that the fact of consciousness is not shown to be impossible. So far, therefore, there is no necessity made out for its rejection. But it is said the fact of consciousness is inexplicable;—we cannot understand how the immediate perception of an external object is possible: whereas the hypothesis of representation enables us to comprehend and explain the phænomenon, and is, therefore, if not absolutely

Conditions of a legitimate hypothesis.—First,—That it be necessary. The hypothesis in question unnecessary.

a See *Discussions*, p. 63.

necessary, at least entitled to favour and preference. But even on this lower,—this precarious ground, the hypothesis is absolutely unnecessary. That, on the incomprehensibility of the fact of consciousness, it is allowable to displace the fact by an hypothesis, is of all absurdities the greatest. As a fact,—an ultimate fact of consciousness, it must be incomprehensible; and were it comprehensible, that is, did we know it in its causes,—did we know it as contained in some higher notion,—it would not be a primary fact of consciousness,—it would not be an ultimate datum of intelligence. Every *how* (διότι) rests ultimately on a *that* (ὅτι), every demonstration is deduced from something given and indemonstrable; all that is comprehensible hangs from some revealed[a] fact, which we must believe as actual, but cannot construe to the reflective intellect in its possibility. In consciousness, in the original spontaneity of intelligence (νοῦς, *locus principiorum*), are revealed the primordial facts of our intelligent nature.

But the cosmothetic idealist has no right to ask the natural realist for an explanation of the fact of consciousness; supposing even that his own hypothesis were in itself both clear and probable, —supposing that the consciousness of self were intelligible, and the consciousness of the not-self the reverse. For, on this supposition, the intelligible consciousness of self could not be an ultimate fact, but must be comprehended through a higher cognition,—a higher consciousness, which would again be itself either comprehensible or not. If comprehensible, this would of

a [This expression is not meant to imply anything hyperphysical. It is used to denote the ultimate and incomprehensible nature of the fact; of the fact which must be believed, though it cannot be understood, cannot be explained.] *Discussions*, p. 63, note.—ED.

course require a still higher cognition, and so on till we arrive at some datum of intelligence, which, as highest, we could not understand through a higher; so that, at best, the hypothesis of representation, proposed in place of the fact of consciousness, only removes the difficulty by one or two steps. The end to be gained is thus of no value; and, for this end, as we have seen and shall see, there would be sacrificed the possibility of philosophy as a rational knowledge altogether; and, in the possibility of philosophy, of course, the possibility of the very hypothesis itself.

The hypothesis not more intelligible than the fact which it displaces.

But is the hypothesis really in itself a whit more intelligible than the fact which it displaces? The reverse is true. What does the hypothesis suppose? It supposes that the mind can represent that of which it knows nothing,—that of which it is ignorant. Is this more comprehensible than the simple fact, that the mind immediately knows what is different from itself, and what is really an affection of the bodily organism? It seems, in truth, not only incomprehensible, but contradictory. The hypothesis of a representative perception thus violates the first condition of a legitimate hypothesis,—it is unnecessary;—nay, not only unnecessary, it cannot do what it professes, —it explains nothing, it renders nothing comprehensible.

Second,—That the hypothesis shall not subvert that which it is devised to explain.

The second condition of a legitimate hypothesis is, that it shall not subvert that which it is devised to explain,—that it shall not explode the system of which it forms a part. But this, the hypothesis in question does; it annihilates itself in the destruction of the whole edifice of knowledge. Belying the testimony of consciousness to our immediate perception of an outer world, it belies the veracity of consciousness altogether;

LECT. XXVI.

and the truth of consciousness is the condition of the possibility of all knowledge.

Third,—That the fact or facts in explanation of which it is devised, be not hypothetical.

The third condition of a legitimate hypothesis is, that the fact or facts, in explanation of which it is devised, be ascertained really to exist, and be not themselves hypothetical. But so far is the principal fact which the hypothesis of a representative perception is proposed to explain, from being certain, that its reality is even rendered problematical by the proposed explanation itself. The facts which this hypothesis supposes to be ascertained and established are two—first, the fact of an external world existing; second, the fact of an internal world knowing. These, the hypothesis takes for granted. For it is asked, How are these connected?—How can the internal world know the external world existing? And, in answer to this problem, the hypothesis of representation is advanced as explaining the mode of their correlation. This hypothesis denies the immediate connection of the two facts; it denies that the mind, the internal world, can be immediately cognisant of matter, the external; and between the two worlds it interpolates a representation, which is at once the object known by mind, and, as known, an image vicarious or representative of matter, *ex hypothesi*, in itself unknown.

Two facts supposed by the hypothesis in question, and their connection sought to be explained by it.

The procedure vicious.

But mark the vice of the procedure. We can only, 1°, Assert the existence of an external world, inasmuch as we know it to exist; and we can only, 2°, Assert that one thing is representative of another, inasmuch as the thing represented is known, independently of the representation. But how does the hypothesis of a representative perception proceed? It actually converts the fact into an hypothesis; actually converts the hypothesis into a fact. On this theory, we do not know the existence of an external world, except on the supposition that that which we do know truly represents it as existing. The hypothetical realist cannot, therefore, establish the fact of the external world, except upon the fact of its representation. This is manifest. We have, therefore, next to ask him, how he knows the fact, that the external world is actually represented. A representation supposes something represented, and the representation of the external world supposes the existence of that world. Now the hypothetical realist, when asked how he proves the reality of the outer world, which, *ex hypothesi*, he does not know, can only say that he infers its existence from the fact of its representation. But the fact of the representation of an external world supposes the existence of that world; therefore, he is again at the point from which he started. He has been arguing in a circle. There is thus a see-saw between the hypothesis and the fact; the fact is assumed as an hypothesis; the hypothesis explained as a fact; each is established, each is expounded, by the other. To account for the possibility of an unknown external world, the hypothesis of representation is devised; and to account for the possibility of representation, we imagine the hypothesis of an external world. LECT. XXVI.

The cosmothetic idealist thus begs the fact which he would explain. And, on the hypothesis of a representative perception, it is admitted by the philosophers themselves who hold it, that the descent to absolute idealism is a logical precipice from which they can alone attempt to save themselves by appealing to the natural beliefs,—to the common-sense, of mankind, that is, to the testimony of that very consciousness to which their own hypothesis gives the lie.

LECT. XXVI.

Fourth,—That it save the phænomena which it is invented to explain.

In the fourth place, a legitimate hypothesis must save the phænomena which it is invented to explain, that is, it must account for them adequately and without exclusion, distortion, or mutilation. But the hypothesis of a representative perception proposes to accomplish its end only by first destroying, and then attempting to recreate, the phænomena, for the fact of which it should, as a legitimate hypothesis, only afford a reason. The total, the entire phænomenon to be explained, is the phænomenon given in consciousness of the immediate knowledge by me, or mind, of an existence different from me, or mind. This phænomenon, however, the hypothesis in question does not preserve entire. On the contrary, it hews it into two;—into the immediate knowledge by me, and into the existence of something different from me,—or more briefly, into the intuition and the existence. It separates, in its explanation, what is given it to explain as united. This procedure is at best monstrous; but this is not the worst. The entire phænomenon being cut in two, you will observe how the fragments are treated. The existence of the non-ego,—the one fragment, it admits; its intuition, its immediate cognition by the ego,—the other fragment, it disallows. Now mark what is the character of this proceeding. The former fragment of the phænomenon,—the fragment admitted, to us exists only through the other fragment which is rejected. The existence of an external world is only given us through its intuition,—we only believe it to exist because we believe that we immediately know it to exist, or are conscious of it as existing. The intuition is the *ratio cognoscendi*, and, therefore, to us the *ratio essendi*, of a material universe. Prove to me that I am wrong

The hypothesis in question sunders and subverts the phænomenon to be explained.

LECT. XXVI.

in regard to my intuition of an outer world, and I will grant at once, that I have no ground for supposing I am right in regard to the existence of that world. To annihilate the intuition is to annihilate what is prior and constitutive in the phænomenon; and to annihilate what is prior and constitutive in the phænomenon, is to annihilate the phænomenon altogether. The existence of a material world is no longer, therefore, even a truncated, even a fractional, fact of consciousness; for the fact of the existence of a material world, given in consciousness, necessarily vanished with the fact of the intuition on which it rested. The absurdity is about the same as if we should attempt to explain the existence of colour, on an hypothesis which denied the existence of extension. A representative perception is thus an hypothetical explanation of a supposititious fact; it creates the nature it interprets.[a]

Fifth.—That the fact to be explained lie within the sphere of experience.

In the fifth place, the fact which a legitimate hypothesis explains, must be within the sphere of experience; but the fact of an external world, for which the cosmothetic idealist would account, transcends, *ex hypothesi*, all experience, being unknown in itself, and a mere hyperphysical assumption.

[a] [With the hypothetical realist or cosmothetic idealist, it has been a puzzling problem to resolve how, on their doctrine of a representative perception, the mind can attain the notion of externality, or outness, far more be impressed with the invincible belief of the reality, and known reality, of an external world. Their attempts at this solution, are as unsatisfactory as they are operose. On the doctrine of an intuitive perception, all this is given in the fact of an immediate knowledge of the non-ego. To us, therefore, the problem does not exist; and Mr Stewart appears to me to have misunderstood the conditions of his own doctrine, or rather not to have formed a very clear conception of an intuitive perception, when he endeavours to explain, by inference and hypothesis, a knowledge and belief in the outness of the objects of sense, and when he desires the reality of our sensations at the points where we are conscious that they are.] [See Stewart, *Phil. Essays*; *Works*, v. 101 *et seq.*—ED.]

LECT. XXVI.

Sixth,—The hypothesis must be single.

In the sixth place, an hypothesis is probable in proportion as it works simply and naturally; that is, in proportion as it is dependent on no subsidiary hypothesis,—as it involves nothing petitory, occult, supernatural, as part and parcel of its explanation. In this respect, the doctrine of a representative perception is not less vicious than in others; to explain at all, it must not only postulate subsidiary hypotheses, but subsidiary miracles. The doctrine in question attempts to explain the knowledge of an unknown world, by the ratio of a representative perception: but it is impossible by any conceivable relation, to apply the ratio to the facts. The mental modification, of which, on the doctrine of representation, we are exclusively conscious in perception, either represents a real external world, or it does not. The latter is a confession of absolute idealism; we have, therefore, only to consider the former.

The hypothesis of Representation dependent on subsidiary hypotheses.

The hypothesis of a representative perception supposes, that the mind does not know the external world, which it represents; for this hypothesis is expressly devised only on the supposed impossibility of an immediate knowledge of aught different from, and external to, the mind. The percipient mind must, therefore, be, somehow or other, determined to represent the reality of which it is ignorant. Now, here one of two alternatives is necessary;—either the mind blindly determines itself to this representation, or it is determined to it by some intelligent and knowing cause, different from itself. The former alternative would be preferable, inasmuch as it is the more simple, and assumes nothing hyperphysical, were it not irrational, as wholly incompetent to account for the phænomenon. On this alternative, we should suppose, that the mind represented, and truly represented, that of whose existence and qualities it knew nothing. A great effect is here assumed, absolutely without a cause; for we could as easily conceive the external world springing into existence without a creator, as mind representing that external world to itself, without a knowledge of that which it represented. The manifest absurdity of this first alternative has accordingly constrained the profoundest cosmothetic idealists to call in supernatural aid by embracing the second. To say nothing of less illustrious schemes, the systems of Divine Assistance, of a Pre-established Harmony, and of the Vision of all things in the Deity, are only so many subsidiary hypotheses;—so many attempts to bridge, by supernatural machinery, the chasm between the representation and the reality, which all human ingenuity had found, by natural means, to be insuperable. The hypothesis of a representative perception thus presupposes a miracle to let it work. Dr Brown and others, indeed, reject, as unphilosophical, these hyperphysical subsidiaries; but they only saw less clearly the necessity for their admission. The rejection, indeed, is another inconsequence added to their doctrine. It is undoubtedly true that, without necessity, it is unphilosophical to assume a miracle, but it is doubly unphilosophical first to originate this necessity, and then not to submit to it. It is a contemptible philosophy that eschews the *Deus ex machina*, and yet ties the knot which can only be loosed by his interposition. Nor will it here do for the cosmothetic idealist to pretend that the difficulty is of nature's, not of his, creation. In fact, it only arises, because he has closed his eyes upon the light of nature, and refused the guidance of consciousness: but having swamped himself in following the *ignis fatuus* of a theory, he has no right to refer

LECT. XXVI.

its private absurdities to the imbecility of human reason, or to excuse his self-contracted ignorance by the narrow limits of our present knowledge.[a]

So much for the merits of the hypothesis of a Representative Perception,—an hypothesis which begins by denying the veracity of consciousness, and ends, when carried to its legitimate issue, in absolute idealism, in utter scepticism. This hypothesis has been, and is, one more universally prevalent among philosophers than any other; and I have given to its consideration a larger share of attention than I should otherwise have done, in consequence of its being one great source of the dissensions in philosophy, and of the opprobrium thrown on consciousness as the instrument of philosophical observation, and the standard of philosophical certainty and truth.

Other questions connected with the faculty of External Perception.

With this terminates the most important of the discussions to which the Faculty of Perception gives rise: the other questions are not, however, without interest, though their determination does not affect the vital interests of philosophy. Of these the first that I shall touch upon, is the problem;—Whether, in Perception, do we first obtain a general knowledge of the complex wholes presented to us by sense, and then, by analysis and limited attention, obtain a special knowledge of their several parts; or do we not first obtain a particular knowledge of the smallest parts to which sense is competent, and then, by synthesis, collect them into greater and greater wholes?

1. Whether we first obtain a knowledge of the whole, or of the parts, of the object in Perception.

Second alternative adopted by Mr Stewart.

The second alternative in this question is adopted by Mr Stewart; it is, indeed, involved in his doctrine in regard to Attention,—in holding that we recollect

a See *Discussions*, pp. 67, 68.—Ed.

nothing without attention, that we can attend only to a single object at once, which one object is the very smallest that is discernible through sense. "It is commonly," he says, "understood, I believe, that, in a concert of music, a good ear can attend to the different parts of the music separately, or can attend to them all at once, and feel the full effect of the harmony. If the doctrine, however, which I have endeavoured to establish, be admitted, it will follow, that in the latter case the mind is constantly varying its attention from the one part of the music to the other, and that its operations are so rapid, as to give us no perception of an interval of time.

LECT. XXVI.

Stewart quoted.

"The same doctrine leads to some curious conclusions with respect to vision. Suppose the eye to be fixed in a particular position, and the picture of an object to be painted on the retina. Does the mind perceive the complete figure of the object at once, or is this perception the result of the various perceptions we have of the different points in the outline? With respect to this question, the principles already stated lead me to conclude, that the mind does at one and the same time perceive every point in the outline of the object, (provided the whole of it be painted on the retina at the same instant,) for perception, like consciousness, is an involuntary operation. As no two points, however, of the outline are in the same direction, every point by itself constitutes just as distinct an object of attention to the mind, as if it were separated by an interval of empty space from all the rest. If the doctrine, therefore, formerly stated, be just, it is impossible for the mind to attend to more than one of these points at once; and as the perception of the figure of the object implies a knowledge of the relative

LECT. XXVI.

situation of the different points with respect to each other, we must conclude, that the perception of figure by the eye, is the result of a number of different acts of attention. These acts of attention, however, are performed with such rapidity, that the effect, with respect to us, is the same as if the perception were instantaneous.

* * * * * * *

"It may perhaps be asked, what I mean by a *point* in the outline of a figure, and what it is that constitutes this point *one* object of attention. The answer, I apprehend, is, that this point is the *minimum visibile*. If the point be less, we cannot perceive it; if it be greater, it is not all seen in one direction.

"If these observations be admitted, it will follow, that, without the faculty of memory, we could have had no perception of visible figure."[a]

The same view maintained by James Mill.

The same conclusion is attained, through a somewhat different process, by Mr James Mill, in his ingenious *Analysis of the Phænomena of the Human Mind*. This author, following Hartley and Priestley, has pushed the principle of Association to an extreme which refutes its own exaggeration,—analysing not only our belief in the relation of effect and cause into that principle, but even the primary logical laws. According to Mr Mill, the necessity under which we lie of thinking that one contradictory excludes another,—that a thing cannot at once be and not be, is only the result of association and custom.[β] It is not, therefore, to be marvelled at, that he should account for our knowledge of complex wholes in perception, by the same universal principle; and this he accordingly does.[γ] "Where two or more ideas have been

a *Elements of the Philosophy of the Human Mind*, vol. i. c. iii. *Works*, vol. ii. p. 141-143.

β Chap. iii. p. 75.—Ed.

γ Chap. iii. p. 68.—Ed.

often repeated together, and the association has become very strong, they sometimes spring up in such close combination as not to be distinguishable. Some cases of sensation are analogous. For example; when a wheel, on the seven parts of which the seven prismatic colours are respectively painted, is made to revolve rapidly, it appears not of seven colours, but of one uniform colour, white. By the rapidity of the succession, the several sensations cease to be distinguishable; they run, as it were, together, and a new sensation, compounded of all the seven, but apparently a simple one, is the result. Ideas, also, which have been so often conjoined, that whenever one exists in the mind, the others immediately exist along with it, seem to run into one another, to coalesce, as it were, and out of many to form one idea; which idea, however in reality complex, appears to be no less simple than any one of those of which it is compounded."

LECT. XXVI. — Mill quoted.

. . . .

* "It is to this great law of association that we trace the formation of our ideas of what we call external objects; that is, the ideas of a certain number of sensations, received together so frequently that they coalesce as it were, and are spoken of under the idea of unity. Hence, what we call the idea of a tree, the idea of a stone, the idea of a horse, the idea of a man.

"In using the names, tree, horse, man, the names of what I call objects, I am referring, and can be referring, only to my own sensations; in fact, therefore, only naming a certain number of sensations, regarded as in a particular state of combination; that is, concomitance. Particular sensations of sight, of touch, of the muscles, are the sensations, to the ideas of

* Chap. iii. p. 70.—Ed.

LECT. XXVI. which, colour, extension, roughness, hardness, smoothness, taste, smell, so coalescing as to appear one idea, I give the name, idea of a tree.

* * * * * * *

"Some ideas are by frequency and strength of association so closely combined, that they cannot be separated. If one exists, the other exists along with it, in spite of whatever effort we make to disjoin them.

"For example; it is not in our power to think of colour, without thinking of extension; or of solidity, without figure. We have seen colour constantly in combination with extension,—spread, as it were, upon a surface. We have never seen it except in this connection. Colour and extension have been invariably conjoined. The idea of colour, therefore, uniformly comes into the mind, bringing that of extension along with it; and so close is the association, that it is not in our power to dissolve it. We cannot, if we will, think of colour, but in combination with extension. The one idea calls up the other, and retains it, so long as the other is retained.

"This great law of our nature is illustrated in a manner equally striking, by the connection between the ideas of solidity and figure. We never have the sensations from which the idea of solidity is derived, but in conjunction with the sensations whence the idea of figure is derived. If we handle anything solid, it is always either round, square, or of some other form. The ideas correspond with the sensations. If the idea of solidity rises, that of figure rises along with it. The idea of figure which rises, is, of course, more obscure than that of extension; because figures being innumerable, the general idea is exceedingly complex, and hence, of necessity, obscure. But, such as it is, the idea of figure is always present when that of solidity is present; nor can we, by any effort, think of the one without thinking of the other at the same time."

The counter alternative maintained against Stewart and Mill.

Now in opposition to this doctrine, nothing appears to me clearer than the first alternative,—and that, in place of ascending upward from the minimum of perception to its maxima, we descend from masses to details. If the opposite doctrine were correct, what would it involve? It would involve as a primary inference, that, as we know the whole through the parts, we should know the parts better than the whole. Thus, for example, it is supposed that we know the face of a friend, through the multitude of perceptions which we have of the different points of which it is made up; in other words, that we should know the whole countenance less vividly than we know the forehead and eyes, the nose and mouth, &c., and that we should know each of these more feebly than we know the various ultimate points, in fact, unconscious minima, of perceptions, which go to constitute them. According to the doctrine in question, we perceive only one of these ultimate points at the same instant, the others by memory incessantly renewed. Now let us take the face out of perception into memory altogether. Let us close our eyes, and let us represent in imagination the countenance of our friend. This we can do with the utmost vivacity; or, if we see a picture of it, we can determine, with a consciousness of the most perfect accuracy, that the portrait is like or unlike. It cannot, therefore, be denied that we have the fullest knowledge of the face as a whole,—that we are familiar with its expression, with the general result of its parts. On the hypothesis, then, of Stewart and Mill, how accurate should be our knowledge of these parts them-

The doctrine of these philosophers implies, that we know the parts better than the whole.

This supposition shown to be erroneous.

selves. But make the experiment. You will find that, unless you have analysed,—unless you have descended from a conspectus of the whole face to a detailed examination of its parts,—with the most vivid impression of the constituted whole, you are almost totally ignorant of the constituent parts. You may probably be unable to say what is the colour of the eyes, and if you attempt to delineate the mouth or nose, you will inevitably fail. Or look at the portrait. You may find it unlike, but unless, as I said, you have analysed the countenance, unless you have looked at it with the analytic scrutiny of a painter's eye, you will assuredly be unable to say in what respect the artist has failed, —you will be unable to specify what constituent he has altered, though you are fully conscious of the fact and effect of the alteration. What we have shown from this example may equally be done from any other, —a house, a tree, a landscape, a concert of music, &c. But it is needless to multiply illustrations. In fact, on the doctrine of these philosophers, if the mind, as they maintain, were unable to comprehend more than one perceptible minimum at a time, the greatest of all inconceivable marvels would be, how it has contrived to realise the knowledge of wholes and masses which it has. Another refutation of this opinion might be drawn from the doctrine of latent modifications,—the obscure perceptions of Leibnitz,—of which we have recently treated. But this argument I think unnecessary.[a]

[a] Show this also, 1°, By the millions of acts of attention requisite in each of our perceptions. [Cf. Dr T. Young's *Lectures on Natural Philosophy*, vol. ii. Ess. v., *The Mechanism of the Eye*, § iii. p. 574, edit. 1807. —Ed.] 2°, By imperfection of Touch, which is a synthetic sense, as Sight is analytic.—*Marginal Jotting*.

LECTURE XXVII.

THE PRESENTATIVE FACULTY.—I. PERCEPTION.—GENERAL QUESTIONS IN RELATION TO THE SENSES.

LECT. XXVII.

Recapitulation.

In my last Lecture, I was principally occupied in showing that the hypothesis of a Representative Perception, considered in itself, and apart from the grounds on which philosophers have deemed themselves authorised to reject the fact of consciousness, which testifies to our immediate perception of external things, violates, in many various ways, the laws of a legitimate hypothesis; and having, in the previous Lecture, shown you that the grounds on which the possibility of an intuitive cognition of external objects had been superseded, are hollow, I thus, if my reasoning be not erroneous, was warranted in establishing the conclusion that there is nothing against, but everything in favour of, the truth of consciousness, and the doctrine of an immediate perception. At the conclusion of the Lecture, I endeavoured to prove, in opposition to Mr Stewart and Mr Mill, that we are not percipient, at the same instant, only of certain *minima*, our cognitions of which are afterwards, by memory or association, accumulated into masses; but that we are at once and primarily percipient of masses, and only require analysis to obtain a minute and more accurate knowledge of their parts,—that, in short, we can,

LECT. XXVII.

within certain limits, make a single object out of many. For example, we can extend our attentive perception to a house, and to it as only one object; or we can contemplate its parts, and consider each of these as separate objects.[a]

Resuming consideration of the more important psychological questions that have been agitated concerning the Senses, I proceed to take up those connected with the sense of Touch.

Two problems under sense of Touch.

The problems which arise under this sense, may be reduced to two opposite questions. The first asks, May not all the Senses be analysed into Touch? The second asks, Is not Touch or Feeling, considered as one of the five Senses, itself only a bundle of various senses?

1. May all the Senses be analysed into Touch? Democritus.

In regard to the first of these questions,—it is an opinion as old at least as Democritus, and one held by many of the ancient physiologists, that the four senses of Sight, Hearing, Taste, and Smell, are only modifications of Touch. This opinion Aristotle records in the fourth chapter of his book *On Sense and the Object of Sense* (*De Sensu et Sensili*), and contents himself with refuting it by the assertion, that its impossibility is manifest. So far, however, from being manifestly impossible, and, therefore, manifestly absurd, it can now easily be shown to be correct, if by touch is understood the contact of the external object of perception with the organ of sense. The opinion of

Aristotle.

In what sense the affirmative correct.

[a] Sir W. Hamilton here occasionally introduced an account of the mechanism of the organs of Sense; observing the following order,—Sight, Hearing, Taste, Smell, and Touch. This, he remarks, is the reverse of the order of nature, and is adopted by him because under Touch certain questions arise, the discussion of which requires some preliminary knowledge of the nature of the senses. As the Lecture devoted to this subject mainly consists of a series of extracts from Young and Bostock, and is purely physiological, it is here omitted. See Young's *Lectures on Natural Philosophy*, vol. I. pp. 387, 447 *et seq.*; vol. ii. p. 574, (4to edit.); Bostock's *Physiology*, pp. 692 *et seq.*, 723, 729-733, (3d edit.)—ED.

LECT. XXVII.

Telesius.

The proper object of Perception.

Democritus was revived, in modern times, by Telesius,[a] an Italian philosopher of the sixteenth century, and who preceded Bacon and Descartes, as a reformer of philosophical methods. I say, the opinion of Democritus can easily be shown to be correct; for it is only a confusion of ideas, or of words, or of both together, to talk of the perception of a distant object, that is, of an object not in relation to our senses. An external object is only perceived inasmuch as it is in relation to our sense, and it is only in relation to our sense, inasmuch as it is present to it. To say, for example, that we perceive by sight the sun or moon, is a false, or an elliptical, expression. We perceive nothing but certain modifications of light in immediate relation to our organ of vision; and so far from Dr Reid being philosophically correct, when he says that "when ten men look at the sun or moon, they all see the same individual object," the truth is that each of these persons sees a different object, because each person sees a different complement of rays, in relation to his individual organ.[β] In fact, if we look alternately with each, we have a different object in our right, and a different object in our left, eye. It is not by perception, but by a process of reasoning, that we connect the objects of sense with existences beyond the sphere of immediate knowledge. It is enough that perception affords us the

a [*De Rerum Natura*, lib. vii. c. viii.] From this reduction Telesius excepts Hearing. With regard to the senses of Taste, Smell, and Sight, he says:—"Non recte iidem . . . . gustum, olfactumque et visum a tactu diversum ponere, qui non tactus modo sunt omnes, sed multo etiam quam qui tactus dicitur exquisitiores. Non scilicet ea modo, quæ universo in corpore percipiuntur, et quæ tactilia (ut dictum est) dicuntur, propterea percipiuntur, quod eorum actio et vis substantiaque spiritum contingit, sed magis quæ in lingua, et multo etiam magis quæ per nares, et quæ in oculis percipiuntur."—*Loc. cit.*—ED.

β On this point, see Adam Smith, *Essays on Philosophical Subjects—Ancient Logics and Metaphysics*, p. 153. Cf. *Of the External Senses*, p. 250, (edit. 1800.)—ED.

LECT. XXVII.

knowledge of the non-ego at the point of sense. To arrogate to it the power of immediately informing us of external things, which are only the causes of the object we immediately perceive, is either positively erroneous, or a confusion of language, arising from an inadequate discrimination of the phænomenon. Such assumptions tend only to throw discredit on the doctrine of an intuitive perception; and such assumptions you will find scattered over the works both of Reid and Stewart. I would, therefore, establish as a fundamental position of the doctrine of an immediate perception, the opinion of Democritus, that all our senses are only modifications of touch; in other words, that the external object of perception is always in contact with the organ of sense.

2. Does Touch comprehend a plurality of Senses?

This determination of the first problem does not interfere with the consideration of the second; for, in the second, it is only asked, Whether, considering Touch or Feeling as a special sense, there are not comprehended under it varieties of perception and sensation so different, that these varieties ought to be viewed as constituting so many special senses. This question, I think, ought to be answered in the affirmative; for though I hold that the other senses are not to be discriminated from Touch, in so far as Touch signifies merely the contact of the organ and the object of perception, yet, considering Touch as a special sense distinguished from the other four by other and peculiar characters, it may easily, I think, be shown, that, if Sight and Hearing, if Smell and Taste, are to be divided from each other and from Touch Proper, under Touch there must, on the same analogy, be distinguished a plurality of special senses. This problem, like the other, is of ancient date. It is mooted by Aristotle in

Affirmative maintained.

the eleventh chapter of the second book *De Anima*, but his opinion is left doubtful. His followers were consequently divided upon the point.[a] Among his Greek interpreters, Themistius[β] adopts the opinion, that there is a plurality of senses under touch. Alexander[γ] favours, but not decidedly, the opposite opinion, which was espoused by Simplicius[δ] and Philoponus.[ε] The doctrine of Themistius was, however, under various modifications, adopted by Averroes and Avicenna among the Arabian, and by Apollinaris, Albertus Magnus, Ægidius, Jandunus, Marcellus, and many others, among the Latin, schoolmen.[ζ] These, however, and succeeding philosophers, were not at one in regard to the number of the senses, which they would distinguish. Themistius[η] and Avicenna[θ] allowed as many senses as there were different qualities of tactile feeling; but the number of these they did not specify. Avicenna, however, appears to have distinguished as one sense the feeling of pain from the lesion of a wound, and as another, the feeling of titillation.[ι] Others, as Ægidius,[κ] gave two senses, one for the hot

LECT. XXVII.

Historical notices of this problem.

Aristotle.

Greek commentators.

Arabian and Latin Schoolmen.

Themistius and Avicenna.

Ægidius.

a See Conimbricenses, *In Arist. de Anima*, [lib. ii. c. xi. p. 326.—Ed.]

β *In De Anima*, lib. ii. c. xi. fol. 82[a], (edit. Ald., 1534.) Οὐκ ἔστι μία αἴσθησις ἡ ἁφὴ σημεῖον δ' ἄν τις νομίζοι, τὸ μὴ μιᾶς ἐναντιώσεως . . . . κριτικὴν ταύτην τὴν αἴσθησιν ὥσπερ τὴν ὄψιν λευκοῦ καὶ μέλανος μόνον, καὶ τῶν μεταξὺ καὶ τὴν ἀκοὴν, ὀξέος καὶ βαρέος, καὶ τῶν μεταξὺ καὶ τὴν γεῦσιν, πικροῦ καὶ γλυκέος ἐν δὲ τοῖς ἁπτοῖς, πολλαί εἰσιν ἐναντιώσεις καὶ πᾶσαι ἔμμεσοι, μεσότητος καθ' ἑκάστην οὔσης θεωρουμένης οἷον θερμὸν, ψυχρὸν ξηρὸν, ὑγρὸν σκληρὸν, μαλακὸν βαρὺ, κοῦφον λεῖον, τραχύ. Cf. Aristotle, texts 106, 107.—Ed.

γ *Problemata*, ii. 62, (probably spurious).—Ed.

δ *In De Anima*, lib. ii. c. xi. text 106, fol. 44[ab], (edit. Ald. 1527).—Ed.

ε *In De Anima*, lib. ii. c. xi. texts 106, 107.—Ed.

ζ See Conimbricenses, *In De Anima*, lib. ii. c. xi. p. 326.—Ed.

η See above, note β, and Conimbricenses, as above, p. 327.—Ed.

θ See Conimbricenses, as above, p. 327.—Ed.

ι See *ibid.*—Ed.

κ See *ibid.*—Ed. [Cf. De Raei, *Clavis Philosophiæ Naturalis, De Mentis Humanæ Facultatibus*, § 76, p. 366. D'Alembert, *Mélanges*, t. v. p. 115. Cf. Scaliger, *De Subtilitate*, Ex. cix., where he observes that, in paralysis, heat is felt, after the power of apprehending gravity is gone.]

LECT. XXVII.

and cold, another for the dry and moist. Averroes[α] accerns a sense of titillation and a sense of hunger and thirst. Galen[β] also, I should observe, allowed a sense of heat and cold. Among modern philosophers, Cardan[γ] distinguishes four senses of touch or feeling; one, of the four primary tactile qualities of Aristotle, (that is, of cold and hot, and wet and dry); a second, of the light and heavy; a third, of pleasure and pain; and a fourth, of titillation. His antagonist, the elder Scaliger,[δ] distinguished as a sixth special sense the sexual appetite, in which he has been followed by Bacon,[ε] Buffon, Voltaire,[ζ] and others. From these historical notices you will see how marvellously incorrect is the statement,[η] that Locke was the first philosopher who originated this question, in allowing hunger and thirst to be the sensations of a sense different from tactile feeling. Hutcheson, in his work on the *Passions*,[θ] says, "the division of our external senses into five common classes is ridiculously imperfect. Some sensations, such as hunger and thirst, weariness and sickness, can be reduced to none of them; or if they are reduced to feelings, they are perceptions as different from the other ideas of touch, such as cold, heat, hardness, softness, as the ideas of taste or smell. Others have hinted at an external sense different from all of these." What that is, Hutcheson does not mention; and some of our Scotch philosophers have puzzled

Averroes. Galen. Cardan.

Scaliger.

Bacon, Buffon, Voltaire.

Locke.

Hutcheson.

α See Conimbricenses, *In De Anima*, lib. ii. c. xi. p. 327.—Ed.

β [Leidenfrost, *De Mente Humana*, c. ii. § 4, p. 16.]

γ *De Subtilitate*, lib. xiii. See *Reid's Works*, p. 867, note.—Ed.

δ *De Subtilitate*, Ex. cclxxxvi. § 3.—Ed.

ε [*Sylva Sylvarum*, cent. vii. 693. *Works*, edit. Montagu, iv. 361.]

ζ See *Reid's Works*, p. 121; and Poor, *Theoria Sensuum*, pars i. § 34, p. 38. Voltaire, *Dict. Philosophique*, art. *Sensation*, reduces this sense to that of Touch. Cf. *Traité de Métaphysique*, ch. iv. *Œuvres Complètes*, tom. vi. p. 651 (edit. 1817).—Ed.

η See *Lectures on Intellectual Philosophy*, by John Young, LL.D., p. 80.

θ Sect. i., third edition, p. 3, note.—Ed.

LECT. XXVII.

Adam Smith.

Kant.

themselves to conceive the meaning of his allusion. There is no doubt that he referred to the sixth sense of Scaliger. Adam Smith, in his posthumous *Essays*,[a] observes, that hunger and thirst are objects of feeling, not of touch; and that heat and cold are felt, not as pressing on the organ, but as in the organ. Kant[β] divides the whole bodily senses into two,—into a Vital Sense (*Sensus Vagus*), and an Organic Sense (*Sensus Fixus*). To the former class belong the sensations of heat and cold, shuddering, quaking, &c. The latter is divided into the five senses, of Touch Proper, Sight, Hearing, Taste, and Smell.

Kant's division general in Germany.

Brown.

This division has now become general in Germany, the Vital Sense receiving from various authors various synonyms, as *cœnæsthesis, common feeling, vital feeling*, and *sense of feeling, sensu latiori*, &c.; and the sensations attributed to it are heat and cold, shuddering, feeling of health, hunger and thirst, visceral sensations, &c. This division is, likewise, adopted by Dr Brown. He divides our sensations into those which are less definite, and into those which are more definite; and these, his two classes, correspond precisely to the *sensus vagus* and *sensus fixus* of the German philosophers.[γ]

Touch to be divided from sensible feeling. 1. From the analogy of the special senses.

The propriety of throwing out of the sense of Touch those sensations which afford us indications only of the subjective condition of the body, in other words, of dividing touch from sensible feeling, is apparent. In the first place, this is manifest on the analogy of the other special senses. These, as we have seen, are

α *Of the External Senses*, p. 262, (edit. 1800.)—ED.

β *Anthropologie*, § 15.—ED. [Previously to Kant, whose *Anthropologie* was first published in 1798, Leidenfrost, in his *De Mente Humana* (1793), c. ii. § 2, p. 14, distinguished the Vital Sense from the Organic Senses. See also Hübner's *Dissertation* (1794). Cf. Gruithuisen, *Anthropologie*, § 475 p. 364 (edit. 1810).]

γ Lectures xvii. xviii.—ED.

LECT. XXVII.

divided into two classes, according as perception proper or sensation proper predominates; the senses of Sight and Hearing pertaining to the first, those of Smell and Taste to the second. Here each is decidedly either perceptive or sensitive. But in Touch, under the vulgar attribution of qualities, perception and sensation both find their maximum. At the finger-points, this sense would give us objective knowledge of the outer world, with the least possible alloy of subjective feeling; in hunger and thirst, &c., on the contrary, it would afford us a subjective feeling of our own state, with the least possible addition of objective knowledge. On this ground, therefore, we ought to attribute to different senses perceptions and sensations so different in degree.

2. From the different quality of the perceptions and sensations themselves.

But, in the second place, it is not merely in the opposite degree of these two counter-elements that this distinction is founded, but likewise on the different quality of the groups of the perceptions and sensations themselves. There is nothing similar between these different groups, except the negative circumstance that there is no special organ to which positively to refer them; and, therefore, they are exclusively slumped together under that sense which is not obtrusively marked out and isolated by the mechanism of a peculiar instrument.

Special Sense of Touch,—its sphere and organic seat.

Limiting, therefore, the special sense of Touch to that of objective information, it is sufficient to say that this sense has its seat at the extremity of the nerves which terminate in the skin; its principal organs are the finger-points, the toes, the lips, and the tongue. Of these, the first is the most perfect. At the tips of the fingers, a tender skin covers the nervous papillæ, and here the nail serves not only as a protecting shield to the organ, but, likewise, by affording an opposition to the body which makes an impression on the finger-ends, it renders more distinct our perception of the nature of its surface. Through the great mobility of the fingers, of the wrist, and of the shoulder-joint, we are able with one, and still more effectually, with both hands, to manipulate an object on all sides, and, thereby, to attain a knowledge of its figure. We likewise owe to the sense of Touch a perception of those conformations of a body, according to which we call it rough or smooth, hard or soft, sharp or blunt. The repose or motion of a body is also perceived through the touch.

Proper organ of Touch requires, as condition of its exercise, the movement of the voluntary muscles.

To obviate misunderstanding, I should, however, notice that the proper organ of Touch,—the nervous papillæ,—requires, as the condition of its exercise, the movement of the voluntary muscles. This condition, however, ought not to be viewed as a part of the organ itself. This being understood, the perception of the weight of a body will not fall under this sense, as the nerves lying under the epidermis or scurf skin have little or no share in this knowledge. We owe it almost exclusively to the consciousness we have of the exertion of the muscles, requisite to lift with the hand a heavy body from the ground, or when it is laid on the shoulders or head, to keep our own body erect, and to carry the burthen from one place to another.

Two counter-questions regarding sphere of Sight.

I next proceed to consider two counter-questions, which are still agitated by philosophers. The first is, —Does Sight afford us an original knowledge of extension, or do we not owe this exclusively to Touch? The second is,—Does Touch afford us an original knowledge of extension, or do we not owe this exclusively to Sight?

LECT. XXVII.

Both questions are still undetermined; and, consequently, the vulgar belief is also unestablished, that we obtain a knowledge of extension originally both from sight and touch.

1. Does Vision afford us a primary knowledge of extension? or do we not owe this exclusively to Touch?

I commence, then, with the first,—Does Vision afford us a primary knowledge of extension, or do we not owe this knowledge exclusively to Touch? But, before entering on its discussion, it is proper to state to you, by preamble, what kind of extension it is that those would vindicate to sight, who answer this question in the affirmative. The whole primary objects of sight, then, are colours, and extensions, and forms or figures of extension. And here you will observe, it is not all kind of extension and form that is attributed to sight. It is not figured extension in all the three dimensions, but only extension as involved in plane figures; that is, only length and breadth.

Colour the proper object of Sight. This generally admitted.

It has generally been admitted by philosophers, after Aristotle, that colour is the proper object of sight, and that extension and figure, common to sight and touch, are only accidentally its objects, because supposed in the perception of colour.

Berkeley the first to deny that extension object of Sight.

The first philosopher, with whom I am acquainted, who doubted or denied that vision is conversant with extension, was Berkeley; but the clear expression of his opinion is contained in his *Defence of the Theory of Vision*, an extremely rare tract which has escaped the knowledge of all his editors and biographers, and is, consequently, not to be found in any of the editions of his collected works. It was almost certainly, therefore, wholly unknown to Condillac, who is the next philosopher who maintained the same opinion. This, however, he did not do either very explicitly or with-

Condillac.

out change; for the new doctrine which he hazards in his earlier work, in his later he again tacitly replaces by the old.[a] After its surrender by Condillac, the opinion was, however, supported, as I find, by Laboulinière.[β] Mr Stewart maintains that extension is not an object of sight. "I formerly," he says, "had occasion to mention several instances of very intimate associations formed between two ideas which have no necessary connection with each other. One of the most remarkable is, that which exists in every person's mind between the notions of *colour* and *extension*. The former of these words expresses (at least in the sense in which we commonly employ it) a sensation in the mind, the latter denotes a quality of an external object; so that there is, in fact, no more connection between the two notions than between those of pain and of solidity; and yet, in consequence of our always perceiving extension at the same time at which the sensation of colour is excited in the mind, we find it impossible to think of that sensation without conceiv-

Laboulinière.

Stewart.

a The order of Condillac's opinions is the reverse of that stated in the text. In his earliest work, the *Origine des Connoissances Humaines*, part i. sect. vi., he combats Berkeley's theory of vision, and maintains that extension exterior to the eye is discernible by sight. Subsequently, in the *Traité des Sensations*, part i. ch. xi., part iii. ch. iv. v., he asserts that the eye is incapable of perceiving extension beyond itself, and that this idea is originally due solely to the sense of touch. This opinion he again repeats in *l'Art de Penser*, part i. ch. xi. But neither Condillac nor Berkeley goes so far as to say that colour, regarded as an affection of the visual organism, is apprehended as absolutely unextended, as a mathematical point. Nor is this the question in dispute. But granting, as Condillac in his later view expressly asserts, that colour, as a visual sensation, necessarily occupies space, do we, by means of that sensation, acquire also the proper idea of extension, as composed of parts exterior to each other? In other words, does the sensation of different colours, which is necessary to the distinction of parts at all, necessarily suggest different and contiguous localities? This question is explicitly answered in the negative by Condillac, and in the affirmative by Sir W. Hamilton. Cf. *The Theory of Vision vindicated and explained*. London, 1733. See especially, §§ 41, 42, 44, 45, 46.—ED.

β See *Reid's Works*, p. 868.—ED.

LECT. XXVII.

ing extension along with it."[α] But before and after Stewart, a doctrine, virtually the same, is maintained by the Hartleian school; who assert, as a consequence of their universal principle of association, that the perception of colour suggests the notion of extension.[β]

Hartleian School.

Brown.

Then comes Dr Brown, who, in his *Lectures*, after having repeatedly asserted, that it is, and always has been, the universal opinion of philosophers, that the superficial extension of length and breadth becomes known to us by sight originally, proceeds, as he says, for the first time, to controvert this opinion;[γ] though it is wholly impossible that he could have been ignorant that the same had been done, at least by Condillac and Stewart. Brown himself, however, was to be treated somewhat in the fashion in which he treats his predecessors. Some twenty years ago, there were published the *Lectures on Intellectual Philosophy*, by the late John Young, LL.D., Professor of Philosophy in Belfast College; a work which certainly shows considerable shrewdness and ingenuity. This unfortunate speculator seems, however, to have been fated, in almost every instance, to be anticipated by Brown; and, as far as I have looked into these Lectures, I have been amused with the never-failing preamble,—of the astonishment, the satisfaction, and so forth, which the author expresses on finding, on the publication of Brown's *Lectures*, that the opinions which he himself, as he says, had always held and taught, were those also which had obtained the countenance of so distinguished a philosopher. The coincidence is, however,

John Young.

α *Elements of the Philosophy of the Human Mind*, vol. i. chap. v. part ii. § 1. *Works*, vol. ii. p. 306. [Cf. *Ibid.*, Note P.—Ed.]

β See Priestley, *Hartley's Theory*, Prop. 20. Belsham, *Elements of the Philosophy of the Mind*, p. 85. James Mill, *Analysis of the Human Mind*, vol. i. pp. 72, 73. Ed.

γ Lecture xxviii.—Ed.

too systematic and precise to be the effect of accident; and the identity of opinion between the two doctors can only, (plagiarism apart), be explained by borrowing from the hypothesis of a Pre-established Harmony between their minds.[α] Of course, they are both at one on the problem under consideration.[β]

But to return to Brown, by whom the argument against the common doctrine is most fully stated. He says:— Brown quoted.

"The universal opinion of philosophers is, that it is not colour merely which it (the simple original sensation of vision) involves, but extension also,—that there is a visible figure, as well as a tangible figure,—and that the visible figure involves, in our instant original perception, superficial length and breadth, as the tangible figure, which we learn to see, involves length, breadth, and thickness.

"That it is impossible for us, at present, to separate, in the sensation of vision, the colour from the extension, I admit; though not more completely impossible, than it is for us to look on the thousand feet of a meadow, and to perceive only the small inch of greenness on our retina; and the one impossibility, as much as the other, I conceive to arise only from intimate association, subsequent to the original sensations of sight. Nor do I deny, that a certain part of the retina,—which, being limited, must therefore have figure,—is affected by the rays of light that fall on it, as a certain breadth of nervous expanse is affected in all the other organs. I contend only, that the perception of this limited figure of the portion of the retina

α I now find, and have elsewhere stated, that the similarity between these philosophers arises from their borrowing, I may say stealing, from the same source,—De Tracy. See *Dissertations on Reid*, Note D, p. 868.

β See Young, *Lectures on Intellectual Philosophy*, p. 116.

LECT. XXVII. affected, does not enter into the sensation itself, more than, in our sensations of any other species, there is a perception of the nervous breadth affected.

"The immediate perception of visible figure has been assumed as indisputable, rather than attempted to be proved,—as before the time of Berkeley, the immediate visual perception of distance, and of the three dimensions of matter, was supposed, in like manner, to be without any need of proof;—and it is, therefore, impossible to refer to arguments on the subject. I presume, however, that the reasons which have led to this belief, of the immediate perception of a figure termed visible, as distinguished from that tangible figure, which we learn to see, are the following two,—the only reasons which I can even imagine,—that it is absolutely impossible, in our present sensations of sight, to separate colour from extension,—and that there are, in fact, a certain length and breadth of the retina, on which the light falls."[a]

Summary of Brown's argument. He then goes on to argue, at a far greater length than can be quoted, that the mere circumstance of a certain definite space, viz., the extended retina, being affected by certain sensations, does not necessarily involve the notion of extension. Indeed in all those cases in which it is supposed, that a certain diffusion of sensations excites the notion of extension, it seems to be taken for granted that the being knows already, that he has an extended body, over which these sensations are thus diffused. Nothing but the sense of touch, however, and nothing but those kinds of touch which imply the idea of continued resistance, can give us any notion of body at all. All mental affections which are regarded merely as feelings of the mind, and

[a] Lect. xxix, p. 185 (edit. 1830).—Ed.

which do not give us a conception of their external causes, can never be known to arise from anything which is extended or solid. So far, however, is the mere sensation of colour from being able to produce this, that touch itself, as felt in many of its modifications, could give us no idea of it. That the sensation of colour is quite unfit to give us any idea of extension, merely by its being diffused over a certain expanse of the retina, seems to be corroborated by what we experience in the other senses, even after we are perfectly acquainted with the notion of extension. In hearing, for instance, a certain quantity of the tympanum of the ear must be affected by the pulsations of the air; yet it gives us no idea of the dimensions of the part affected. The same may, in general, be said of taste and smell.

The perception of extension necessarily given in the perception of colours.

Now in all their elaborate argumentation on this subject, these philosophers seem never yet to have seen the real difficulty of their doctrine. It can easily be shown that the perception of colour involves the perception of extension. It is admitted that we have by sight a perception of colours, consequently, a perception of the difference of colours. But a perception of the distinction of colours necessarily involves the perception of a discriminating line; for if one colour be laid beside or upon another, we only distinguish them as different by perceiving that they limit each other, which limitation necessarily affords a breadthless line, —a line of demarcation. One colour laid upon another, in fact, gives a line returning upon itself, that is, a figure. But a line and a figure are modifications of extension. The perception of extension, therefore, is necessarily given in the perception of colours.

LECTURE XXVIII.

THE PRESENTATIVE FACULTY.—I. PERCEPTION.—RELATIONS OF SIGHT AND TOUCH TO EXTENSION.

LECT. XXVIII.

Recapitulation.

In my last Lecture, after showing you that the vulgar distribution of the Senses into five, stands in need of correction, and stating what that correction is, I proceeded to the consideration of some of the more important philosophical problems, which arise out of the relation of the senses to the elementary objects of Perception.

I then stated to you two counter-problems in relation to the genealogy of our empirical knowledge of extension; and as, on the one hand, some philosophers maintain that we do not perceive extension by the eye, but obtain this notion through touch, so, on the other, there are philosophers who hold that we do not perceive extension through the touch, but exclusively by the eye. The consideration of these counter-questions will, it is evident, involve a consideration of the common doctrine intermediate between these extreme opinions,—that we derive our knowledge of extension from both senses. I keep aloof from this discussion the opinion, that space, under which extension is included, is not an empirical or adventitious notion at all, but a native form of thought; for admitting this, still if space be also a necessary form of the external world, we shall also have an empirical perception of it by our senses, and the question, therefore, equally remains,—Through what sense, or senses, have we this perception? LECT. XXVIII.

In relation to the first problem, I stated that the position which denies to visual perception all cognisance of extension, was maintained by Condillac, by Laboulinière, by Stewart, by the followers of Hartley (Priestley, Belsham, Mill, &c.), and by Brown,—to say nothing of several recent authors in this country, and in America. I do not think it necessary to state to you the long process of reasoning on which, especially by Brown, this paradox has been grounded. It is sufficient to say, that there is no reason whatsoever adduced in its support, which carries with it the smallest weight. The whole argumentation in reply to the objections supposed by its defenders, is in reply to objections which no one, I conceive, who understood his case, would ever dream of advancing; while the only objection which it was incumbent on the advocates of the paradox to have answered, is passed over in total silence.

Proof that Sight is cognisant of extension.

This objection is stated in three words. All parties are, of course, at one in regard to the fact that we see colour. Those who hold that we see extension, admit that we see it only as coloured; and those who deny us any vision of extension, make colour the exclusive object of sight. In regard to this first position, all are, therefore, agreed. Nor are they less harmonious in reference to the second;—that the power of perceiving colour involves the power of perceiving the differences of colours. By sight we, therefore, perceive colour, and discriminate one colour, that is, one coloured body,—one sensation of colour, from another. This is admitted. A third position will also be denied by none, that the colours discriminated in vision, are, or may be, placed side by side in immediate juxtaposi-

LECT. XXVIII.

tion; or, one may limit another by being superinduced partially over it. A fourth position is equally indisputable,—that the contrasted colours, thus bounding each other, will form by their meeting a visible line, and that, if the superinduced colour be surrounded by the other, this line will return upon itself, and thus constitute the outline of a visible figure.

These four positions command a peremptory assent; they are all self-evident. But their admission at once explodes the paradox under discussion. And thus:—A line is extension in one dimension,—length; a figure is extension in two,—length and breadth. Therefore, the vision of a line is a vision of extension in length; the vision of a figure, the vision of extension in length and breadth. This is an immediate demonstration of the impossibility of the opinion in question; and it is curious that the ingenuity which suggested to its supporters the petty and recondite objections they have so operosely combated, should not have shown them this gigantic difficulty, which lay obtrusively before them.

Extension cannot be represented to the mind except as coloured.

So far, in fact, is the doctrine which divorces the perceptions of colour and extension from being true, that we cannot even represent extension to the mind except as coloured. When we come to the consideration of the Representative Faculty,—Imagination,—I shall endeavour to show you, (what has not been observed by psychologists,) that in the representation,—in the imagination, of sensible objects, we always represent them in the organ of Sense through which we originally perceived them. Thus, we cannot imagine any particular odour but in the nose; nor any sound but in the ear; nor any taste but in the mouth; and if we would represent any pain we have ever felt, this can only be done through the local nerves. In like

Sensible objects represented, in Imagination, in the organ of Sense, through which we originally perceived them.

manner, when we imagine any modification of light, we do so in the eye; and it is a curious confirmation of this, as is well known to physiologists, that when not only the external apparatus of the eye, which is a mere mechanical instrument, but the real organ of sight,—the optic nerves and their thalami, have become diseased, the patient loses, in proportion to the extent of the morbid affection, either wholly or in part, the faculty of recalling visible phænomena to his mind.

Vision, the sense by pre-eminence competent to the perception of extension.

I mention this at present in order to show, that Vision is not only a sense competent to the perception of extension, but the sense κατ' ἐξοχήν, if not exclusively, so competent,—and this in the following manner:— You either now know, or will hereafter learn, that no notion, whether native and general, or adventitious and generalised, can be represented in imagination, except in a concrete or singular example. For instance, you cannot imagine a triangle which is not either an equilateral, or an isosceles, or a scalene,—in short, some individual form of a triangle; nay more, you cannot imagine it, except either large or small, on paper or on a board, of wood or of iron, white or black or green; in short, except under all the special determinations which give it, in thought, as in existence, singularity or individuality. The same happens too with extension. Space I admit to be a native form of thought,—not an adventitious notion. We cannot but think it. Yet I cannot actually represent space in imagination, stript of all individualising attributes. In this act, I can easily annihilate all corporeal existence,—I can imagine empty space. But there are two attributes of which I cannot divest it, that is, shape and colour. This may sound almost ridiculous at first statement, but if you attend to the phænomenon, you will soon be satisfied of its truth. And

LECT. XXVIII.

Space or Extension cannot be represented in Imagination without shape.

first as to shape. Your minds are not infinite, and cannot, therefore, positively conceive infinite space. Infinite space is only conceived negatively,—only by conceiving it inconceivable; in other words, it cannot be conceived at all. But if we do our utmost to realise this notion of infinite extension by a positive act of imagination, how do we proceed? Why, we think out from a centre, and endeavour to carry the circumference of the sphere to infinity. But by no one effort of imagination can we accomplish this; and as we cannot do it at once by one infinite act, it would require an eternity of successive finite efforts,—an endless series of imaginings beyond imaginings, to equalise the thought with its object. The very attempt is contradictory. But when we leave off, has the imagined space a shape? It has: for it is finite; and a finite, that is, a bounded, space constitutes a figure. What, then, is this figure? It is spherical,—necessarily spherical; for as the effort of imagining space is an effort outwards from a centre, the space represented in imagination is necessarily circular. If there be no shape, there has been no positive imagination; and for any other shape than the orbicular no reason can be assigned. Such is the figure of space in a free act of phantasy.

This, however, will be admitted without scruple; for if real space, as it is well described by St Augustin, be a sphere whose centre is everywhere, and whose circumference is nowhere,[a] imagined space may be allowed to be a sphere whose circumference is

[a] The editors have not been able to discover this passage in St Augustin. As quoted in the text, with reference to space, it closely resembles the words of Pascal, *Pensées*, partie I. art. iv. (vol. ii. p. 64, edit. Faugère): "Tout ce monde visible n'est qu'un trait imperceptible dans l'ample sein de la nature. Nulle idée n'en approche. Nous avons beau enfler nos conceptions au delà des espaces imaginables: nous n'enfantons que des atomes, au

represented at any distance from its centre. But will its colour be as easily allowed? In explanation of this, you will observe that under colour I of course include black as well as white; the transparent as well as the opaque,—in short, any modification of light or darkness. This being understood, I maintain that it is impossible to imagine figure, extension, space, except as coloured in some determinate mode. You may represent it under any, but you must represent it under some, modification of light,—colour. Make the experiment, and you will find I am correct. But I anticipate an objection. The non-perception of colour, or the inability of discriminating colours, is a case of not unfrequent occurrence, though the subjects of this deficiency are, at the same time, not otherwise defective in vision. In cases of this description, there is, however, necessarily a discrimination of light and shade, and the colours that to us appear in all "the sevenfold radiance of effulgent light," to them appear only as different gradations of clare-obscure. Were this not the case, there could be no vision. Such persons, therefore, have still two great contrasts of colour,—black and white, and an indefinite number of intermediate gradations, in which to represent space to their imaginations. Nor is there any difficulty in the case of the blind, the absolutely blind,—the blind from birth. Blindness is the non-perception of colour; the

Nor without colour.

Objection obviated.

prix de la réalité des choses. C'est une sphère infinie, dont le centre est partout, la circonférence nulle part." But the expression is more usually cited as a definition of the Deity. In this relation it has been attributed to the mythical Hermes Trismegistus (see Alexander Alesius, *Summa Theol.* pars i. qu. vii. memb. 1), and to Empedocles (see Vincentius Bellovacensis, *Speculum Historiale*, lib. ii. c. 1; *Speculum Naturale*, lib. i. c. 4). It was a favourite expression with the mystics of the middle ages. See Müller, *Christian Doctrine of Sin*, vol. ii. p. 134 (Eng transl.) Some interesting historical notices of this expression will be found in a learned note in M. Havet's edition of Pascal's *Pensées*, p. 3.—Ed.

LECT. XXVIII.

non-perception of colour is simple darkness. The space, therefore, represented by the blind, if represented at all, will be represented black. Some modification of ideal light or darkness is thus the condition of the imagination of space. This of itself powerfully supports the doctrine, that vision is conversant with extension as its object. But if the opinion I have stated be correct, that an act of imagination is only realised through some organ of sense, the impossibility of representing space out of all relation to light and colour at once establishes the eye as the appropriate sense of extension and figure.

D'Alembert quoted in support of the view now given of the relation of Sight to extension.

In corroboration of the general view I have taken of the relation of Sight to extension, I may translate to you a passage by a distinguished mathematician and philosopher, who, in writing it, probably had in his eye the paradoxical speculation of Condillac. "It is certain," says D'Alembert,[a] "that sight alone, and independently of touch, affords us the idea of extension; for extension is the necessary object of vision, and we should see nothing if we did not see it extended. I even believe that sight must give us the notion of extension more readily than touch, because sight makes us remark more promptly and perfectly than touch, that contiguity, and, at the same time, that distinction of parts in which extension consists. Moreover, vision alone gives us the idea of the colour of objects. Let us suppose now parts of space differently coloured, and presented to our eyes; the difference of colours will necessarily cause us to observe the boundaries or limits which separate two neighbouring colours, and, consequently, will give us an idea of figure; for we conceive a figure when we conceive a limitation or boundary on all sides."

[a] *Mélanges*, t. v. p. 109.—ED.

I am confident, therefore, that we may safely establish the conclusion, that Sight is a sense principally conversant with extension; whether it be the only sense thus conversant, remains to be considered.

2. Does Touch afford us an original knowledge of extension, or do we owe this exclusively to Sight?

I proceed, therefore, to the second of the counter-problems,—to inquire whether Sight be exclusively the sense which affords us a knowledge of extension, or whether it does this only conjunctly with Touch. As some philosophers have denied to vision all perception of extension and figure, and given this solely to touch, so others have equally refused this perception to touch, and accorded it exclusively to vision.

The affirmative of the latter question maintained by Platner.

This doctrine is maintained among others by Platner,—a man no less celebrated as an acute philosopher, than as a learned physician, and an elegant scholar. I shall endeavour to render his philosophical German into intelligible English, and translate some of the preliminary sentences with which he introduces a curious observation made by him on a blind subject.

Platner quoted.

"It is very true, as my acute antagonist observes, that the gloomy extension which imagination presents to us as an actual object, is by no means the pure *a priori* representation of space. It is very true, that this is only an empirical or adventitious image, which itself supposes the pure or *a priori* notion of space, (or of extension), in other words, the necessity to think everything as extended. But I did not wish to explain the origin of this mental condition or form of thought objectively, through the sense of sight,—but only to say this much:—that empirical space, empirical extension, is dependent on the sense of sight,—that, allowing space or extension, as a form of thought, to be in us, were there even nothing correspondent to it out of us, still the unknown external things must operate upon us, and, in fact, through the sense of sight, do operate

LECT. XXVIII. upon us, if this unconscious form is to be brought into consciousness."

And after some other observations he goes on : "In regard to the visionless representation of space or extension,—the attentive observation of a person born blind, which I formerly instituted, in the year 1785, and, again, in relation to the point in question, have continued for three whole weeks,—this observation, I say, has convinced me, that the sense of touch, by itself, is altogether incompetent to afford us the representation of extension and space, and is not even cognisant of local exteriority, (*oertliches Auseinanderseyn*), in a word, that a man deprived of sight has absolutely no perception of an outer world, beyond the existence of something effective, different from his own feeling of passivity, and in general only of the numerical diversity,—shall I say of impressions, or of things? In fact, to those born blind, time serves instead of space. Vicinity and distance means in their mouths nothing more than the shorter or longer time, the smaller or greater number of feelings, which they find necessary to attain from some one feeling to some other. That a person blind from birth employs the language of vision,—that may occasion considerable error, and did, indeed, at the commencement of my observations, lead me wrong; but, in point of fact, he knows nothing of things as existing out of each other; and, (this in particular I have very clearly remarked), if objects, and the parts of his body touched by them, did not make different kinds of impression on his nerves of sensation, he would take everything external for one and the same. In his own body, he absolutely did not discriminate head and foot at all by their distance, but merely by the difference of the feelings, (and his perception of such difference was incredibly fine), which he experienced from the one and from the other; and, moreover, through time. In like manner, in external bodies, he distinguished their figure, merely by the varieties of impressed feelings; inasmuch, for example, as the cube, by its angles, affected his feeling differently from the sphere. No one can conceive how deceptive is the use of language accommodated to vision. When my acute antagonist appeals to Cheselden's case, which proves directly the reverse of what it is adduced to refute, he does not consider that the first visual impressions which one born blind receives after couching, do not constitute vision. For the very reason, that space and extension are empirically only possible through a perception of sight,—for that very reason, must such a patient, after his eyes are freed from the cataract, first learn to live in space; if he could do this previously, then would not the distant seem to him near,—the separate would not appear to him as one. These are the grounds which make it impossible for me to believe empirical space in a blind person; and from these I infer, that this form of sensibility, as Mr Kant calls it, and which, in a certain signification, may very properly be styled a pure representation, cannot come into consciousness otherwise than through the medium of our visual perception; without, however, denying that it is something merely subjective, or affirming that sight affords anything similar to this kind of representation. The example of blind geometers would likewise argue nothing against me, even if the geometers had been born blind; and this they were not, if, even in their early infancy, they had seen a single extended object." [a]

[a] *Philosophische Aphorismen*, vol. I. § 705, p. 439 *et seq.*, edit. 1793.—Ed.

LECT. XXVIII.

Phænomena that favour Platner's doctrine.

To what Platner has here stated I would add, from personal experiment, and observation upon others, that if any one who is not blind will go into a room of an unusual shape, wholly unknown to him, and into which no ray of light is allowed to penetrate, he may grope about for hours,—he may touch and manipulate every side and corner of it; still, notwithstanding every endeavour,—notwithstanding all the previous subsidiary notions he brings to the task, he will be unable to form any correct idea of the room. In like manner, a blindfolded person will make the most curious mistakes in regard to the figure of objects presented to him, if these are of any considerable circumference. But if the sense of touch in such favourable circumstances can effect so little, how much less could it afford us any knowledge of forms, if the assistance which it here brings with it from our visual conceptions, were wholly wanting?

Supported also by Cheselden's case of couching.

This view is, I think, strongly confirmed by the famous case of a young gentleman, blind from birth, couched by Cheselden;—a case remarkable for being perhaps, of those cured, that in which the cataract was most perfect, (it only allowed of a distinction of light and darkness); and, at the same time, in which the phænomena have been most distinctly described. In this latter respect, it is, however, very deficient; and it is saying but little in favour of the philosophical acumen of medical men, that the narrative of this case, with all its faults, is, to the present moment, the one most to be relied on.[a]

Now I contend, (though I am aware I have high authority against me), that if a blind man had been able to form a conception of a square or globe by mere

[a] See Nunneley, *On the Organs of Vision*, p. 31, (1858), for a recent case of couching, with careful observations, which confirm, in all essential particulars, the conclusions of Cheselden.—ED.

touch, he would, on first perceiving them by sight, be able to discriminate them from each other;[a] for this supposes only that he had acquired the primary notions of a straight and of a curved line. Again, if touch afforded us the notion of space or extension in general, the patient, on obtaining sight, would certainly be able to conceive the possibility of space or extension beyond the actual boundary of his vision. But of both of these Cheselden's patient was found incapable. As it is a celebrated case, I shall quote to you a few passages in illustration: you will find it at large in the *Philosophical Transactions* for the year 1728.

Cheselden quoted.

"Though we say of this gentleman, that he was blind," observes Mr Cheselden, "as we do of all people who have ripe cataracts; yet they are never so blind from that cause but that they can discern day from night; and for the most part, in a strong light, distinguish black, white, and scarlet; but they cannot perceive the shape of anything; for the light by which these perceptions are made, being let in obliquely through the aqueous humour, or the anterior surface of the crystalline, (by which the rays cannot be brought into a focus upon the retina,) they can discern in no other manner than a sound eye can through a glass of broken jelly, where a great variety of surfaces so differently refract the light, that the several distinct pencils of rays cannot be collected by the eye into their proper foci; wherefore the shape of an object in such a case cannot be at all discerned, though the colour may; and thus it was with this young gentleman, who, though he knew those colours asunder in a good light, yet when he

a On this question, see Locke, II. 9; and Sir W. Hamilton's note, *Essay on the Human Understanding*, *Reid's Works*, p. 137.—Ed.

LECT. XXVIII. saw them after he was couched, the faint ideas he had of them before were not sufficient for him to know them by afterwards; and therefore he did not think them the same which he had before known by those names."

. . . .

"When he first saw, he was so far from making any judgment about distances, that he thought all objects whatever touched his eyes (as he expressed it) as what he felt did his skin; and thought no objects so agreeable as those which were smooth and regular, though he could form no judgment of their shape, or guess what it was in any object that was pleasing to him. He knew not the shape of anything, nor any one thing from another, however different in shape or magnitude; but upon being told what things were, whose form he before knew from feeling, he would carefully observe, that he might know them again: but having too many objects to learn at once, he forgot many of them; and (as he said) at first learned to know, and again forgot a thousand things in a day. One particular only (though it may appear trifling) I will relate: Having often forgot which was the cat, and which the dog, he was ashamed to ask; but catching the cat (which he knew by feeling) he was observed to look at her steadfastly, and then setting her down, said, 'So, puss! I shall know you another time.'"

. . . .

"We thought he soon knew what pictures represented which were showed to him, but we found afterwards we were mistaken; for about two months after he was couched, he discovered at once they represented solid bodies, when, to that time, he considered them only as parti-coloured plains, or surfaces diversified with variety of paints; but even then he was no less surprised, expecting the pictures would feel like the things they represented, and was amazed when he found those parts, which by their light and shadow appeared now round and uneven, felt only flat like the rest; and asked which was the lying sense, feeling or seeing."[a]

The Author professes no decided opinion on the question.

The whole of this matter is still enveloped in great uncertainty, and I should be sorry either to dogmatise myself, or to advise you to form any decided opinion. Without, however, going the length of Platner, in denying the possibility of a geometer blind from birth, we may allow this, and yet vindicate exclusively to sight the power of affording us our empirical notions of space. The explanation of this supposes, however, an acquaintance with the doctrine of pure or *a priori* space, as a form of thought; it must, therefore, for the present be deferred.

How do we obtain our knowledge of Visual Distance?

The last question on which I shall touch, and with which I shall conclude the consideration of Perception in general, is,—How do we obtain our knowledge of Visual Distance? Is this original or acquired?

Visual distance, before Berkeley, regarded as an original perception.

With regard to the method by which we judge of distance, it was formerly supposed to depend upon an original law of the constitution, and to be independent of any knowledge gained through the medium of the external senses. This opinion was attacked by Berkeley in his *New Theory of Vision*, one of the finest examples, as Dr Smith justly observes, of philosophical analysis to be found in our own or in any other language; and in which it appears most clearly demonstrated, that our whole information on this subject is

a See Adam Smith's *Essays on Philosophical Subjects*. [Pp. 294, 295, 296, edit. 1800. Cf. *Reid's Works*, p. 137, note.—Ed.]

acquired by experience and association. This conclusion is supported by many circumstances of frequent occurrence, in which we fall into the greatest mistakes with respect to the distance of objects, when we form our judgment solely from the visible impression made upon the retina, without attending to the other circumstances which ordinarily direct us in forming our conclusions. It also obtains confirmation from the case of Cheselden, which I have already quoted. It clearly appears that, in the first instance, the patient had no correct ideas of distance; and we are expressly told that he supposed all objects to touch the eye, until he learned to correct his visible, by means of his tangible, impressions, and thus gradually to acquire more correct notions of the situation of surrounding bodies with respect to his own person.

Circumstances which assist us in forming our judgment respecting visual distance derived, 1. On certain states of the eye.

On the hypothesis that our ideas of distance are acquired, it remains for us to investigate the circumstances which assist us in forming our judgment respecting them. We shall find that they may be arranged under two heads, some of them depending upon certain states of the eye itself, and others upon various accidents that occur in the appearance of the objects. With respect to distances that are so short as to require the adjustment of the eye in order to obtain distinct vision, it appears that a certain voluntary effort is necessary to produce the desired effect: this effort, whatever may be its nature, causes a corresponding sensation, the amount of which we learn by experience to appreciate; and thus, through the medium of association, we acquire the power of estimating the distance with sufficient accuracy.

When objects are placed at only a moderate distance, but not such as to require the adjustment of the eye, in directing the two eyes to the object we incline them inwards; as is the case likewise with very short distances: so that what are termed the axes of the eyes, if produced, would make an angle at the object, the angle varying inversely as the distance. Here, as in the former case, we have certain perceptions excited by the muscular efforts necessary to produce a proper inclination of the axes, and these we learn to associate with certain distances. As a proof that this is the mode by which we judge of those distances where the optic axes form an appreciable angle, when the eyes are both directed to the same object, while the effort of adjustment is not perceptible,—it has been remarked, that persons who are deprived of the sight of one eye, are incapable of forming a correct judgment in this case.

2. On certain conditions of the object.

When we are required to judge of still greater distances, where the object is so remote as that the axes of the two eyes are parallel, we are no longer able to form our opinion from any sensation in the eye itself. In this case, we have recourse to a variety of circumstances connected with the appearance of the object; for example, its apparent size, the distinctness with which it is seen, the vividness of its colours, the number of intervening objects, and other similar accidents, all of which obviously depend upon previous experience, and which we are in the habit of associating with different distances, without, in each particular case, investigating the cause on which our judgment is founded.

Berkeley's proof thrown into doubt by the analogy of the lower animals.

The conclusions of science seem in this case to be decisive; and yet the whole question is thrown into doubt by the analogy of the lower animals. If in man the perception of distance be not original but acquired,

LECT. XXVIII.

the perception of distance must be also acquired by them. But as this is not the case in regard to animals, this confirms the reasoning of those who would explain the perception of distance in man, as an original, not as an acquired knowledge. That the Berkeleian doctrine is opposed by the analogy of the lower animals, is admitted by one of its most intelligent supporters, —Dr Adam Smith.[a]

Adam Smith quoted.

"That, antecedent to all experience," says Smith, "the young of at least the greater part of animals possess some instinctive perception of this kind, seems abundantly evident. The hen never feeds her young by dropping the food into their bills, as the linnet and the thrush feed theirs. Almost as soon as her chickens are hatched, she does not feed them, but carries them to the field to feed, where they walk about at their ease, it would seem, and appear to have the most distinct perception of all the tangible objects which surround them. We may often see them, accordingly, by the straightest road, run to and pick up any little grains which she shows them, even at the distance of several yards; and they no sooner come into the light than they seem to understand this language of Vision as well as they ever do afterwards. The young of the partridge and the grouse seem to have, at the same early period, the most distinct perceptions of the same kind. The young partridge, almost as soon as it comes from the shell, runs about among long grass and corn; the young grouse among long heath; and would both most essentially hurt themselves if they had not the most acute as well as distinct perception of the tangible objects which not only surround them but press upon them on all sides. This is the case, too, with the

[a] See *Essays—Of the External Senses*, p. 299-301, edit. 1800.—ED.

young of the goose, of the duck, and, so far as I have been able to observe, with those of at least the greater part of the birds which make their nests upon the ground, with the greater part of those which are ranked by Linnæus in the orders of the hen and the goose, and of many of those long-shanked and wading birds which he places in the order that he distinguishes by the name of Grallæ.

. . . . . .

"It seems difficult to suppose that man is the only animal of which the young are not endowed with some instinctive perception of this kind. The young of the human species, however, continue so long in a state of entire dependency, they must be so long carried about in the arms of their mothers or of their nurses, that such an instinctive perception may seem less necessary to them than to any other race of animals. Before it could be of any use to them, observation and experience may, by the known principle of the association of ideas, have sufficiently connected in their young minds each visible object with the corresponding tangible one which it is fitted to represent. Nature, it may be said, never bestows upon any animal any faculty which is not either necessary or useful, and an instinct of this kind would be altogether useless to an animal which must necessarily acquire the knowledge which the instinct is given to supply, long before that instinct could be of any use to it. Children, however, appear at so very early a period to know the distance, the shape, and magnitude of the different tangible objects which are presented to them, that I am disposed to believe that even they may have some instinctive perception of this kind; though possibly in a much weaker degree than the greater part of other animals.

LECT. XXVIII.

A child that is scarcely a month old, stretches out its hands to feel any little plaything that is presented to it. It distinguishes its nurse, and the other people who are much about it, from strangers. It clings to the former, and turns away from the latter. Hold a small looking-glass before a child of not more than two or three months old, and it will stretch out its little arms behind the glass, in order to feel the child which it sees, and which it imagines is at the back of the glass. It is deceived, no doubt; but even this sort of deception sufficiently demonstrates that it has a tolerably distinct apprehension of the ordinary perspective of Vision, which it cannot well have learnt from observation and experience."

LECTURE XXIX.

THE PRESENTATIVE FACULTY.—II. SELF-CONSCIOUSNESS.

LECT. XXIX.

Recapitulation. Principal points of difference between the Author's doctrine of Perception, and that of Reid and Stewart.

HAVING, in our last Lecture, concluded the consideration of External Perception, I may now briefly recapitulate certain results of the discussion, and state in what principal respects the doctrine I would maintain, differs from that of Reid and Stewart, whom I suppose always to hold, in reality, the system of an Intuitive Perception.

1. In regard to the relation of the external object to the senses.

In the first place,—in regard to the relation of the external object to the senses. The general doctrine on this subject is thus given by Reid: "A law of our nature regarding perception is, that we perceive no object, unless some impression is made upon the organ of sense, either by the immediate application of the object, or by some medium which passes between the object and the organ. In two of our senses, viz. Touch and Taste, there must be an immediate application of the object to the organ. In the other three, the object is perceived at a distance, but still by means of a medium, by which some impression is made upon the organ."[a]

Now this, I showed you, is incorrect. The only object ever perceived is the object in immediate contact, —in immediate relation, with the organ. What Reid,

a *Intellectual Powers*, Essay II. ch. II. [*Works*, p. 247.—ED.]

LECT. XXIX.

and philosophers in general, call the distant object, is wholly unknown to Perception; by reasoning we may connect the object perceived with certain antecedents, —certain causes, but these, as the results of an inference, cannot be the objects of perception. The only objects of perception are in all the senses equally immediate. Thus the object of my vision at present is not the paper or letters at a foot from my eye, but the rays of light reflected from these upon the retina. The object of your hearing is not the vibrations of my larynx, nor the vibrations of the intervening air; but the vibrations determined thereby in the cavity of the internal ear, and in immediate contact with the auditory nerves. In both senses, the external object perceived is the last effect of a series of unperceived causes. But to call these unperceived causes the *object* of perception, and to call the perceived effect,—the real object, only the *medium* of perception, is either a gross error or an unwarrantable abuse of language.

In all the senses, the external object in contact with the organ.

My conclusion is, therefore, that, in all the senses, the external object is in contact with the organ, and thus, in a certain signification, all the senses are only modifications of Touch. This is the simple fact, and any other statement of it is either the effect or the cause of misconception.

2. In regard to the number and consecution of the elementary phænomena.

In the second place,—in relation to the number and consecution of the elementary phænomena,—it is, and must be, admitted, on all hands, that perception must be preceded by an impression of the external object on the sense; in other words, that the material reality and the organ must be brought into contact, previous to, and as the condition of, an act of this faculty. On this point there can be no dispute. But the case is different in regard to the two following. It is asserted by philosophers in general:—1°. That the impression made on the organ must be propagated to the brain, before a cognition of the object takes place in the mind, —in other words, that an organic action must precede and determine the intellectual action; and 2°. That Sensation Proper precedes Perception Proper. In regard to the former assertion,—if by this were only meant, that the mind does not perceive external objects out of relation to its bodily organs, and that the relation of the object to the organism, as the condition of perception, must, therefore, in the order of nature, be viewed as prior to the cognition of that relation,—no objection could be made to the statement. But if it be intended, as it seems to be, that the organic affection precedes in the order of time the intellectual cognition,—of this we have no proof whatever. The fact as stated would be inconsistent with the doctrine of an intuitive perception; for if the organic affection were chronologically prior to the act of knowledge, the immediate perception of an object different from our bodily senses would be impossible, and the external world would thus be represented only in the subjective affections of our own organism. It is, therefore, more correct to hold, that the corporeal movement and the mental perception are simultaneous; and in place of holding that the intellectual action commences after the bodily has terminated,—in place of holding that the mind is connected with the body only at the central extremity of the nervous system, it is more simple and philosophical to suppose that it is united with the nervous system in its whole extent. The mode of this union is of course inconceivable: but the latter hypothesis of union is not more inconceivable than the former; and, while it has the testimony of consciousness in its

Common doctrine of philosophers regarding the organic impression.

In what respect inaccurate.

LECT. XXIX.

favour, it is otherwise not obnoxious to many serious objections to which the other is exposed.

Relation of Sensation proper to Perception proper.

In regard to the latter assertion,—viz., that a perception proper is always preceded by a sensation proper,—this, though maintained by Reid and Stewart, is even more manifestly erroneous than the former assertion, touching the precedence of an organic to a mental action. In summing up Reid's doctrine of Perception, Mr Stewart says, "To what does the statement of Reid amount? Merely to this; that the mind is so formed, that certain impressions produced on our organs of sense by external objects, are followed by correspondent sensations; and that these sensations, (which have no more resemblance to the qualities of matter, than the words of a language have to the things they denote), are followed by a perception of the existence and qualities of the bodies by which the impressions are made."[α] You will find in Reid's own works expressions which, if taken literally, would make us believe that he held perception to be a mere inference from sensation. Thus, "Observing that the agreeable sensation is raised when the rose is near, and ceases when it is removed, I am led, by my nature, to conclude some quality to be in the rose, which is the cause of this sensation. This quality in the rose is the object perceived; and that act of my mind, by which I have the conviction and belief of this quality, is what in this case I call perception."[β] I have, however, had frequent occasion to show you that we must not always interpret Reid's expressions very rigorously; and we are often obliged to save his philosophy from the consequences of his own loose and ambiguous lan-

α *Elements*, vol. i. ch. ii. § 2. *Works*, vol. ii. p. 111.

β *Intell. Powers*, Essay ii. ch. xvi. *Works*, p. 310.

guage. In the present instance, if Reid were taken at his word, his perception would be only an instinctive belief, consequent on a sensation, that there is some unknown external quality the cause of the sensation. Be this, however, as it may, there is no more ground for holding that sensation precedes perception, than for holding that perception precedes sensation. In fact, both exist only as they coexist. They do not indeed always coexist in the same degree of intensity, but they are equally original; and it is only by an act, not of the easiest abstraction, that we are able to discriminate them scientifically from each other.[α]

The faculty of Self-consciousness.

So much for the first of the two faculties by which we acquire knowledge,—the faculty of External Perception. The second of these faculties is Self-consciousness, which has likewise received, among others, the name of Internal or Reflex Perception. This faculty will not occupy us long, as the principal questions regarding its nature and operation have been already considered, in treating of Consciousness in general.[β]

Self-consciousness a branch of the Presentative Faculty.

I formerly showed you that it is impossible to distinguish Perception, or the other Special Faculties, from Consciousness,—in other words, to reduce Consciousness itself to a special faculty; and that the attempt to do so by the Scottish philosophers is self-contradictory.[γ] I stated to you, however, that though it be incompetent to establish a faculty for the immediate knowledge of the external world, and a faculty for the immediate knowledge of the internal, as two ultimate powers, exclusive of each other, and not merely subordinate forms of a higher immediate know-

α Compare *Reid's Works*, Note D*, p. 882 *et seq.*—Ed.

β See above, vol. I. lect. xl. *et seq.* —Ed.

γ See above, vol. I. lect. xliii. p. 224 *et seq.*—Ed.

ledge, under which they are comprehended or carried up into one,—I stated, I say, that though the immediate knowledges of matter and of mind are still only modifications of consciousness, yet that their discrimination, as subaltern faculties, is both allowable and convenient. Accordingly, in the scheme which I gave you of the distribution of Consciousness into its special modes,—I distinguished a faculty of External, and a faculty of Internal, Apprehension, constituting together a more general modification of consciousness, which I called the Acquisitive or Presentative or Receptive Faculty.

Philosophers less divided in their opinions touching Self-consciousness than in regard to Perception.

In regard to Self-consciousness,—the faculty of Internal Experience,—philosophers have been far more harmonious than in regard to External Perception. In fact, their differences touching this faculty originate rather in the ambiguities of language, and the different meanings attached to the same form of expression, than in any fundamental opposition of opinion in regard to its reality and nature. It is admitted equally by all to exist, and to exist as a source of knowledge; and the supposed differences of philosophers in this respect, are, as I shall show you, mere errors in the historical statement of their opinions.

Self-consciousness contrasted with Perception. Their fundamental forms.

The sphere and character of this faculty of acquisition, will be best illustrated by contrasting it with the other. Perception is the power by which we are made aware of the phænomena of the external world; Self-consciousness the power by which we apprehend the phænomena of the internal. The objects of the former are all presented to us in Space and Time; space and time are thus the two conditions,—the two fundamental forms, of external perception. The objects of the latter are all apprehended by us in Time and in Self;

time and self are thus the two conditions,—the two fundamental forms, of Internal Perception or Self-consciousness. Time is thus a form or condition common to both faculties; while space is a form peculiar to the one, self a form peculiar to the other. What I mean by the form or condition of a faculty, is that frame,—that setting, (if I may so speak), out of which no object can be known. Thus we only know, through Self-consciousness, the phænomena of the internal world, as modifications of the indivisible ego or conscious unit; we only know, through Perception, the phænomena of the external world, under space, or as modifications of the extended and divisible non-ego or known plurality. That the forms are native, not adventitious, to the mind, is involved in their necessity. What I cannot but think, must be *a priori*, or original to thought; it cannot be engendered by experience upon custom. But this is not a subject the discussion of which concerns us at present.

What meant by the form of a faculty.

It may be asked, if self or ego be the form of Self-consciousness, why is the not-self, the non-ego, not in like manner called the form of Perception? To this I reply, that the not-self is only a negation, and, though it discriminates the objects of the external cognition from those of the internal, it does not afford to the former any positive bond of union among themselves. This, on the contrary, is supplied to them by the form of space, out of which they can neither be perceived, nor imagined by the mind;—space, therefore, as the positive condition under which the non-ego is necessarily known and imagined, and through which it receives its unity in consciousness, is properly said to afford the condition or form of External Perception.

Objection obviated.

But a more important question may be started. If

LECT. XXIX.

If space be a necessary form of thought; is the mind itself extended?

space,—if extension, be a necessary form of thought, this, it may be argued, proves that the mind itself is extended. . The reasoning here proceeds upon the assumption, that the qualities of the subject knowing must be similar to the qualities of the object known. This, as I have already stated,[a] is a mere philosophical crotchet,—an assumption without a shadow even of probability in its favour. That the mind has the power of perceiving extended objects, is no ground for holding that it is itself extended. Still less can it be maintained, that because it has ideally a native or necessary conception of space, it must really occupy space. Nothing can be more absurd. On this doctrine, to exist as extended is supposed necessary in order to think extension. But if this analogy hold good, the sphere of ideal space which the mind can imagine, ought to be limited to the sphere of real space which the mind actually fills. This is not, however, the case; for though the mind be not absolutely unlimited in its power of conceiving space, still the compass of thought may be viewed as infinite in this respect, as contrasted with the petty point of extension, which the advocates of the doctrine in question allow it to occupy in its corporeal domicile.

The sphere of Self-consciousness.

The faculty of Self-consciousness affords us a knowledge of the phænomena of our minds. It is the source of internal experience. You will, therefore, observe, that, like External Perception, it only furnishes us with facts; and that the use we make of these facts, —that is, what we find in them, what we deduce from them,—belongs to a different process of intelligence. Self-consciousness affords the materials equally to all systems of philosophy; all equally admit it, and all elaborate the materials which this faculty supplies,

a See above, vol. ii. lect. xxv. p. 120 et seq.—Ed.

according to their fashion. And here I may merely notice, by the way, what, in treating of the Regulative Faculty, will fall to be regularly discussed, that these facts, these materials, may be considered in two ways. We may employ either Induction alone, or also Analysis. If we merely consider the phænomena which Self-consciousness reveals, in relation to each other,—merely compare them together, and generalise the qualities which they display in common, and thus arrange them into classes or groups governed by the same laws, we perform the process of Induction. By this process we obtain what is general, but not what is necessary. For example, having observed that external objects presented in perception are extended, we generalise the notion of extension or space. We have thus explained the possibility of a conception of space, but only of space as a general and contingent notion; for if we hold that this notion exists in the mind only as the result of such a process, we must hold it to be *a posteriori* or adventitious, and, therefore, contingent. Such is the process of Induction, or of Simple Observation. The other process, that of Analysis or Criticism, does not rest satisfied with this comparison and generalisation, which it, however, supposes. It proposes not merely to find what is general in the phænomena, but what is necessary and universal. It, accordingly, takes mental phænomena, and, by abstraction, throws aside all that it is able to detach, without annihilating the phænomena altogether,—in short, it analyses thought into its essential or necessary, and its accidental or contingent, elements.

Two modes of dealing with the phænomena given in Self-consciousness,—viz. either by Induction alone, or by Induction and Analysis together.

Thus, from Observation and Induction, we discover what experience affords as its general result; from Analysis and Criticism, we discover what experience

The sphere of Critical Analysis.

LECT. XXIX.

supposes as its necessary condition. You will notice, that the critical analysis of which I now speak, is limited to the objects of our internal observation; for in the phænomena of mind alone can we be conscious of absolute necessity. All necessity is, in fact, to us subjective; for a thing is conceived impossible only as we are unable to construe it in thought. Whatever does not violate the laws of thought, is, therefore, not to us impossible, however firmly we may believe that it will not occur. For example, we hold it absolutely impossible, that a thing can begin to be without a cause. Why? Simply because the mind cannot realise to itself the conception of absolute commencement. That a stone should ascend into the air, we firmly believe will never happen; but we find no difficulty in conceiving it possible. Why? Merely because gravitation is only a fact generalised by induction and observation; and its negation, therefore, violates no law of thought. When we talk, therefore, of the *necessity* of any external phænomenon, the expression is improper, if the necessity be only an inference of induction, and not involved in any canon of intelligence. For induction proves to us only what is, not what must be,—the actual, not the necessary.

All necessity to us subjective.

Historical notice of the employment of the Inductive and Critical Methods in philosophy. Locke.

The two processes of Induction or Observation, and of Analysis or Criticism, have been variously employed by different philosophers. Locke, for instance, limited himself to the former, overlooking altogether the latter. He, accordingly, discovered nothing necessary, or *a priori*, in the phænomena of our internal experience. To him all axioms are only generalisations of experience. In this respect he was greatly excelled by Descartes and Leibnitz. The latter, indeed, was the philosopher who clearly enunciated the principle, that

Descartes.

Leibnitz,—the first to enunciate

LECT. XXIX.

the phænomenon of necessity, in our cognitions, could not be explained on the ground of experience. "All the examples," he says, "which confirm a general truth, how numerous soever, would not suffice to establish the universal necessity of this same truth; for it does not follow, that what has hitherto occurred will always occur in future."[a] "If Locke," he adds, "had sufficiently considered the difference between truths which are necessary or demonstrative, and those which we infer from induction alone, he would have perceived that necessary truths could only be proved from principles which command our assent by their intuitive evidence; inasmuch as our senses can inform us only of what is, not of what must necessarily be." Leibnitz, however, was not himself fully aware of the import of the principle,—at least he failed in carrying it out to its most important applications; and though he triumphantly demonstrated, in opposition to Locke, the *a priori* character of many of those cognitions which Locke had derived from experience, yet he left to Kant the honour of having been the first who fully applied the critical analysis in the philosophy of mind.

Necessity as the criterion of truth native to the mind.

Kant,—the first who fully applied this criterion.

The faculty of Self-consciousness corresponds with the Reflection of Locke. Now there is an interesting question concerning this faculty,—whether the philosophy of Locke has been misapprehended and misrepresented by Condillac, and other of his French disciples, as Mr Stewart maintains; or, whether Mr Stewart has not himself attempted to vindicate the

Has the philosophy of Locke been misrepresented by Condillac, and other of his French disciples?

a *Nouveaux Essais*, Avant-propos, p. 5 (edit. Raspe).—Ed. [Cf. liv. i. c. i. § 5, p. 36; liv. ii. c. xvii. § 1, p. 116. *Letter to Burnet of Kemney* (1706), *Opera*, t. vi. p. 274 (edit. Dutens). *Letter to Bierling* (1710), *Opera*, t. v. p. 358. *Theodicée* (1710), i. § 2, p. 480 (Erdmann), or *Opera*, t. i. p. 65 (Dutens). *Monadologie* (1714), p. 707 (edit. Erdmann).]

LECT. XXIX.

tendency of Locke's philosophy on grounds which will not bear out his conclusions. Mr Stewart has canvassed this point at considerable length, both in his *Essays*[a] and in his *Dissertation on the Progress of Metaphysical, Ethical, and Political Philosophy.* In the latter, the point at issue is thus briefly stated:

Stewart quoted in vindication of Locke.

"The objections to which Locke's doctrine concerning the origin of our ideas, or, in other words, concerning the sources of our knowledge, are, in my judgment, liable, I have stated so fully in a former work, that I shall not touch on them here. It is quite sufficient, on the present occasion, to remark, how very unjustly this doctrine (imperfect, on the most favourable construction, as it undoubtedly is) has been confounded with those of Gassendi, of Condillac, of Diderot, and of Horne Tooke. The substance of all that is common in the conclusions of these last writers, cannot be better expressed than in the words of their master, Gassendi. 'All our knowledge,' he observes in a letter to Descartes, 'appears plainly to derive its origin from the senses; and although you deny the maxim, 'Quicquid est in intellectu praesse debere in sensu,' yet this maxim appears, nevertheless, to be true; since our knowledge is all ultimately obtained by an *influx* or *incursion* from things external; which knowledge afterwards undergoes various modifications by means of analogy, composition, division, amplification, extenuation, and other similar processes, which it is unnecessary to enumerate.' This doctrine of Gassendi's coincides exactly with that ascribed to Locke by Diderot and by Horne Tooke; and it differs only verbally from the more concise statement of Condillac, that 'our ideas are nothing more than transformed sensations.' 'Every idea,' says

[a] *Works*, vol. v. part i., essay i., p. 55 *et seq.*—ED.

the first of these writers, 'must necessarily, when brought to its state of ultimate decomposition, resolve itself into a sensible representation or picture; and since everything in our understanding has been introduced there by the channel of sensation, whatever proceeds out of the understanding is either chimerical, or must be able, in returning by the same road, to re-attach itself to its sensible archetype. Hence an important rule in philosophy,—that every expression which cannot find an external and a sensible object, to which it can thus establish its affinity, is destitute of signification.' Such is the exposition given by Diderot, of what is regarded in France as Locke's great and capital discovery; and precisely to the same purpose we are told by Condorcet, that 'Locke was the first who proved that all our ideas are compounded of sensations.' If this were to be admitted as a fair account of Locke's opinion, it would follow that he has not advanced a single step beyond Gassendi and Hobbes; both of whom have repeatedly expressed themselves in nearly the same words with Diderot and Condorcet. But although it must be granted, in favour of their interpretation of his language, that various detached passages may be quoted from his work, which seem, on a superficial view, to justify their comments, yet of what weight, it may be asked, are these passages, when compared with the stress laid by the author on *Reflection*, as an original source of our ideas, altogether different from *Sensation*? 'The other fountain,' says Locke, 'from which experience furnisheth the understanding with ideas, is the perception of the operations of our own minds within us, as it is employed about the ideas it has got; which operations, when the soul comes to reflect on and consider, do furnish the understanding with another set of ideas, which could not be had from things without; and such are Perception, Thinking, Doubting, Believing, Reasoning, Knowing, Willing, and all the different actings of our own minds, which, we being conscious of, and observing in ourselves, do from these receive into our understandings ideas as distinct as we do from bodies affecting our senses. This source of ideas every man has wholly in himself; and though it be not sense, as having nothing to do with external objects, yet it is very like it, and might properly enough be called *Internal Sense*. But as I call the other Sensation, so I call this Reflection; the ideas it affords being such only as the mind gets by reflecting on its own operations within itself.'[α] Again, 'The understanding seems to me not to have the least glimmering of any ideas which it doth not receive from one of these two. External objects furnish the mind with the ideas of sensible qualities; and the mind furnishes the understanding with ideas of its own operations.'[β]

Stewart's vindication unsatisfactory.

On these observations I must remark, that they do not at all satisfy me; and I cannot but regard Locke and Gassendi as exactly upon a par, and both as deriving all our knowledge from experience. The French philosophers are, therefore, in my opinion, fully justified in their interpretation of Locke's philosophy; and Condillac must, I think, be viewed as having simplified the doctrine of his master, without doing the smallest violence to its spirit. In the first place, I cannot concur with Mr Stewart in allowing any weight to Locke's distinction of Reflection, or Self-consciousness, as a

Condillac justified in his simplification of Locke's doctrine.

α Locke, *Works*, vol. i. p. 78. [*Essay*, B. ii. c. i. § 4.—Ed.]

β *Ibid.*, vol. i. p. 79. [*Ess.*, B. ii. c. i. § 5.—Stewart, *Dissertation*, part ii. § i. *Works*, vol. i. p. 224 *et seq.* —Ed.]

The Reflection of Locke,—compatible with Sensualism.

second source of our knowledge. Such a source of experience no sensualist ever denied, because no sensualist ever denied that sense was cognisant of itself. It makes no difference, that Locke distinguished Reflection from Sense, "as having nothing to do with external objects," admitting, however, that "they are very like," and that Reflection "might properly enough be called Internal Sense,"[α] while Condillac makes it only a modification of sense. It is a matter of no importance, that we do not call Self-consciousness by the name of *Sense*, if we allow that it is only conversant about the contingent. Now no interpretation of Locke can ever pretend to find in his Reflection a revelation to him of aught native or necessary to the mind, beyond the capability to act and suffer in certain manners,—a capability which no philosophy ever dreamt of denying. And if this be the case, it follows that the formal reduction, by Condillac, of Reflection to Sensation, is only a consequent following out of the principles of the doctrine itself.

Fundamental error of Stewart in regard to the philosophy of Gassendi.

Of how little import is the distinction of Reflection from Sensation, in the philosophy of Locke, is equally shown in the philosophy of Gassendi; in regard to which I must correct a fundamental error of Mr Stewart. I had formerly occasion to point out to you the unaccountable mistake of this very learned philosopher, in relation to Locke's use of the term Reflection,[β] which, both in his *Essays* and his *Dissertation*, he states was a word first employed by Locke in its psychological signification.[γ] Nothing, I stated, could

α *Essay*, B. ii. c. i. § 4.—Ed.

β See above, vol. i. lect. xiii. p. 234.—Ed.

γ Lee on Locke, makes apparently the same mistake. [See *Anti-Scepticism: or, Notes upon each Chapter of Mr Locke's Essay concerning Humane Understanding*, by Henry Lee, B.D., Preface, p. 7; London, 1702.—Ed.]

LECT. XXIX.

be more incorrect. When adopted by Locke, it was a word of universal currency, in a similar sense, in every contemporary system of philosophy, and had been so employed for at least a thousand years previously. This being understood, Mr Stewart's mistake in regard to Gassendi is less surprising. "The word *Reflection*," says Mr Stewart, "expresses the peculiar and characteristical doctrine, by which his system is distinguished from that of the Gassendists and Hobbists. All this, however, serves only to prove still more clearly, how widely remote his real opinion on this subject was from that commonly ascribed to him by the French and German commentators. For my own part, I do not think, notwithstanding some casual expressions which may seem to favour the contrary supposition, that Locke would have hesitated for a moment to admit with Cudworth and Price, that the *Understanding* is itself a source of new ideas. That it is by *Reflection*, (which, according to his own definition, means merely the exercise of the *Understanding* on the internal phenomena), that we get our ideas of Memory, Imagination, Reasoning, and of all other intellectual powers, Mr Locke has again and again told us; and from this principle it is so obvious an inference, that all the simple ideas which are necessarily implied in our intellectual operations, are ultimately to be referred to the same source, that we cannot reasonably suppose a philosopher of Locke's sagacity to admit the former proposition, and to withhold his assent to the latter."[a]

Gassendi, though a Sensationalist, admitted Reflection as a source of knowledge.

The inference which, in the latter part of this quotation, Mr Stewart speaks of, is not so obvious as he supposes, seeing that it was not till Leibnitz that the character of necessity was enounced, and clearly

[a] *Dissertation*, part II. § I., foot-note, *Works*, vol. I. p. 230.—ED.

enounced, as the criterion by which to discriminate the native from the adventitious cognitions of the mind. This is, indeed, shown by the example of Gassendi himself, who is justly represented by Mr Stewart as a Sensationalist of the purest water; but wholly misrepresented by him, as distinguished from Locke by his negation of any faculty corresponding to Locke's Reflection. So far is this from being correct,—Gassendi not only allowed a faculty of Self-consciousness analogous to the Reflection of Locke, he actually held such a faculty, and even attributed to it far higher functions than did the English philosopher; nay, what is more, held it under the very name of Reflection.[a] In fact, from the French philosopher, Locke borrowed this, as he did the principal part of his whole philosophy; and it is saying but little either for the patriotism or intelligence of their countrymen, that the works of Gassendi and Descartes should have been so long eclipsed in France by those of Locke, who was in truth only a follower of the one, and a mistaken refuter of the other. In respect to Gassendi, there are reasons that explain this neglect apart from any want of merit in himself; for he is a thinker fully equal to Locke in independence and vigour of intellect, and, with the exception of Leibnitz, he is, of all the great philosophers of modern times, the most varied and profound in learning.

And did not assimilate Reflection to Sense.

Now, in regard to the point at issue, so far is Gassendi from assimilating Reflection to Sense, as Locke virtually, if not expressly, does, and for which assimilation he has been principally lauded by those of his followers who analysed every mental process into Sensation,—so far, I say, is Gassendi from doing this, that

a See above, vol. i. lect. xiii. p. 234.—Ed.

LECT. XXIX.

he places Sense and Reflection at the opposite mental poles, making the former a mental function wholly dependent upon the bodily organism; the latter, an energy of intellect wholly inorganic and abstract from matter. The cognitive phænomena of mind Gassendi reduces to three general classes or faculties:—1°. Sense, 2°. Phantasy (or Imagination), and 3°. Intellect. The two former are, however, virtually one, inasmuch as Phantasy, on his doctrine, is only cognisant about the forms, which it receives from Sense, and is, equally with Sense, dependent on a corporeal organ. Intellect, on the contrary, he holds, is not so dependent, and that its functions are, therefore, of a kind superior to those of an organic faculty. These functions or faculties of Intellect he reduces to three. "The first," he says, (and I literally translate his words in order that I may show you how flagrantly he has been misrepresented), "is Intellectual Apprehension,—that is, the apprehension of things which are beyond the reach of Sense, and which, consequently, leaving no trace in the brain, are also beyond the ken of Imagination. Such, especially, is spiritual or incorporeal nature, as, for example, the Deity. For although in speaking of God, we say that He is incorporeal, yet in attempting to to realise Him to Phantasy, we only imagine something with the attributes of body. It must not, however, be supposed that this is all; for, besides and above the corporeal form which we thus imagine, there is, at the same time, another conception, which that form contributes, as it were, to veil and obscure. This conception is not confined to the narrow limits of Phantasy, (præter Phantasiæ cancellos est); it is proper to Intellect; and, therefore, such an apprehension ought not to be called an *imagination*, but an *intelligence*

His division of the cognitive phænomena of mind.

Intellect, according to Gassendi, has three functions,—1. Intellectual Apprehension.

or *intellection*, (non *imaginatio*, sed *intelligentia* vel *intellectio*, dici oportet)."[a] In his doctrine of Intellect, Gassendi takes, indeed, far higher ground than Locke; and it is a total reversal of his doctrine, when it is stated, that he allowed to the mind no different, no higher, apprehensions than the derivative images of sense. He says, indeed, and he says truly, that if we attempt to figure out the Deity in imagination, we cannot depict Him in that faculty, except under sensible forms—as, for example, under the form of a venerable old man. But does he not condemn this attempt as derogatory; and does he not allow us an intellectual conception of the Divinity, superior to the grovelling conditions of Phantasy? The Cartesians, however, were too well disposed to overlook the limits under which Gassendi had advanced his doctrine,—that the senses are the source of all our knowledge; and Mr Stewart has adopted, from the Port Royal *Logic*, a statement of Gassendi's opinion, which is, to say the least of it, partial and incomplete.

2. Reflection. 3. Reasoning.

The second function which Gassendi assigns to Intellect, is Reflection, and the third is Reasoning. It is with the former of these that we are at present concerned. Mr Stewart, you have seen, distinguishes the philosophy of Locke from that of his predecessor in this,—that the former introduced Reflection or Self-consciousness as a source of knowledge, which was overlooked or disallowed by the latter. Mr Stewart is thus wrong in the fact of Gassendi's rejection of any source of knowledge of the name and nature of Locke's Reflection. So far is this from being the case, that Gassendi attributes far more to this faculty than

a *Physica*, Sect. III, Memb. Post., p. 451.—Ed.
lib. ix. c. 3; *Opera*, Lugd. 1658, t. ii.

LECT. XXIX.

Locke; for he not only makes it an original source of knowledge, but founds upon the nature of its action a proof of the immateriality of mind. "To the second operation," he says, "belongs the Attention or Reflection of the Intellect upon its proper acts,—an operation by which it understands that it understands, and thinks that it thinks, (qua se intelligere intelligit, cogitatve se cogitare.)" "We have formerly," he adds, "shown that it is above the power of Phantasy to imagine that it imagines, because, being of a corporeal nature, it cannot act upon itself; in fact, it is as absurd to say that I imagine myself to imagine, as that I see myself to see." He then goes on to show, that the knowledge we obtain of all our mental operations and affections, is by this reflection of Intellect; that it is necessarily of an inorganic or purely spiritual character; that it is peculiar to man, and distinguishes him from the brutes; and that it aids us in the recognition of disembodied substances, in the confession of a God, and in according to Him the veneration which we owe Him.

The mere admission of a faculty of Self-consciousness, of no import in determining the anti-sensual character of a philosophy.

From what I have now said, you will see, that the mere admission of a faculty of Self-consciousness, as a source of knowledge, is of no import in determining the rational,—the anti-sensual, character of a philosophy; and that even those philosophers who discriminated it the most strongly from Sense, might still maintain that experience is not only the occasion, but the source, of all our knowledge. Such philosophers were Gassendi and Locke. On this faculty I do not think it necessary to dwell longer; and, in our next Lecture, I shall proceed to consider the Conservative Faculty,—Memory, properly so called.

LECTURE XXX.

THE CONSERVATIVE FACULTY.—MEMORY PROPER.

LECT. XXX.

Elementary phænomena may be distinct, while they depend on each other for their realisation.

I COMMENCED and concluded, in my last Lecture, the consideration of the second source of knowledge,—the faculty of Self-consciousness or Internal Perception. Through the powers of External and Internal Perception we are enabled to acquire information,—experience: but this acquisition is not of itself independent and complete; it supposes that we are also able to retain the knowledge acquired, for we cannot be said to get what we are unable to keep. The faculty of Acquisition is, therefore, only realised through another faculty,—the faculty of Retention or Conservation. Here, we have another example of what I have already frequently had occasion to suggest to your observation,—we have two faculties, two elementary phænomena, evidently distinct, and yet each depending on the other for its realisation. Without a power of acquisition, a power of conservation could not be exerted; and without the latter, the former would be frustrated, for we should lose as fast as we acquired. But as the faculty of Acquisition would be useless without the faculty of Retention, so the faculty of Retention would be useless without the faculties of Reproduction and Representation. That the mind retained, beyond the sphere of consciousness, a treasury of knowledge, would be of no avail, did it not possess the power of bringing out, and of displaying, in other words, of reproducing, and representing, this know-

This general principle illustrated by the phænomena of Acquisition, Retention, Reproduction, and Representation.

LECT. XXX.

ledge in consciousness. But because the faculty of Conservation would be fruitless without the ulterior faculties of Reproduction and Representation, we are not to confound these faculties, or to view the act of mind which is their joint result, as a simple and elementary phænomenon. Though mutually dependent on each other, the faculties of Conservation, Reproduction, and Representation are governed by different laws; and, in different individuals, are found greatly varying in their comparative vigour. The intimate connection of these three faculties, or elementary activities, is the cause, however, why they have not been distinguished in the analysis of philosophers; and why their distinction is not precisely marked in ordinary language. In ordinary language we have indeed words which, without excluding the other faculties, denote one of these more emphatically. Thus in the term *Memory*, the Conservative Faculty,—the phænomenon of Retention, is the central notion, with which, however, those of Reproduction and Representation are associated. In the term *Recollection*, again, the phænomenon of Reproduction is the principal notion, accompanied, however, by those of Retention and Representation, as its subordinates. This being the case, it is evident what must be our course in regard to the employment of common language. We must either abandon it altogether, or take the term that more proximately expresses our analysis, and, by definition, limit and specify its signification. Thus, in the Conservative Faculty, we may either content ourselves with the scientific terms of *Conservation* and *Retention* alone, or we may moreover use as a synonym the vulgar term *Memory*, determining its application, in our mouths, by a preliminary definition. And that the word *Memory* principally and properly denotes the

Hence these three faculties not distinguished by philosophers; nor in ordinary language.

Ordinary use of the terms Memory and Recollection.

power the mind possesses of retaining hold of the knowledge it has acquired, is generally admitted by philologers, and is not denied by philosophers. Of the latter, some have expressly avowed this. Of these I shall quote to you only two or three, which happen to occur the first to my recollection. Plato considers Memory simply as the faculty of Conservation, (ἡ μνήμη σωτηρία αἰσθήσεως.)[a] Aristotle distinguishes Memory, (μνήμη), as the faculty of Conservation from Reminiscence, (ἀνάμνησις), the faculty of Reproduction.[β] St Augustin, who is not only the most illustrious of the Christian fathers, but one of the profoundest thinkers of antiquity, finely contrasts Memory with Recollection or Reminiscence, in one of the most eloquent and philosophical chapters of his *Confessions*[γ]:—"Hæc omnia recipit *recolenda*, cum opus est, et *retractanda* grandis memoriæ recessus. Et nescio qui secreti atque ineffabiles sinus ejus; quæ omnia suis quæque foribus intrant ad eam, et reponuntur in ea. Nec ipsa tamen intrant, sed rerum sensarum imagines illic præsto sunt, cogitationi *reminiscenti* eas." The same distinction is likewise precisely taken by one of the acutest of modern philosophers, the elder Scaliger.[δ] "*Memoriam* voco hujusce cognitionis *conservationem*. *Reminiscentiam* dico, *repetitionem* disciplinæ, quæ e memoria delapsa fuerat." This is from his commentary on Aristotle's *History of Animals*; the following is from his *De Subtilitate*[ε]:—"Quid *Memoria*? Vis animæ communis ad *retinendum* tam rerum imagines, *i.e.* phantasmata, quam notiones universales; easque, vel simplices, vel complexas. Quid *Recordatio*? Opera

Memory properly denotes the power of Retention.

Acknowledged by Plato.

Aristotle.

St Augustin.

Julius Cæsar Scaliger.

α *Philebus*, [p. 34.—ED.]

β *De Memoria et Reminiscentia*, [c. 2, § 25. Cf. Conimbricenses, *In De Mem. et Rem.*, c. vii. p. 10.—ED.]

γ Lib. x. c. 8.—ED.

[δ *Aristotelis Historia de Animalibus, Julio Cæsare Scaligero Interprete*, Tolosæ 1619, p. 30.]

[ε Exercit. cccvii. § 28.]

LECT. XXX.

intellectus, species recolentis. Quid *Reminiscentia?* Disquisitio tectarum specierum; amotio importunarum, digestio obturbatarum." The father suggests the son, and the following occurs in the *Secunda Scaligerana*, which is one of the two collections we have of the table-talk of Joseph Scaliger. The one from which I quote was made by the brothers Vassan, whom the Dictator of Letters, from friendship to their learned uncles, (the Messrs Pithou), had received into his house, when pursuing their studies in the University of Leyden; and *Secunda Scaligerana* is made up of the notes they had taken of the conversations he had with them, and others in their presence. Scaliger, speaking of himself, is made to say: "I have not a good memory, but a good reminiscence; proper names do not easily recur to me, but when I think on them I find them out."[α] It is sufficient for our purpose that the distinction is here taken between the Retentive Power,—Memory, and the Reproductive Power,—Reminiscence. Scaliger's memory could hardly be called bad, though his reminiscence might be better; and these elements in conjunction go to constitute a good memory, in the comprehensive sense of the expression. I say the retentive faculty of that man is surely not to be despised, who was able to commit to memory Homer in twenty-one days, and the whole Greek poets in three months,[β] and who, taking him all in all, was the most learned man the world has ever seen. I might adduce many other authorities to the same effect; but this, I think, is sufficient to warrant me in using the term *Memory* exclusively to denote the faculty pos-

Joseph Scaliger.

α Tom. ii. p. 552.—ED.

β See Heinsius, *In Josephi Scaligeri Obitum; Funebris Oratio*, (1609), p. 15. His words are:—"Uno et viginti diebus Homerum, reliquos intra quartum mensum poetas, caeteros autem intra biennium scriptores perlustravit." See below, lect. xxxi. p. 224.—ED.

sessed by the mind of preserving what has once been present to consciousness, so that it may again be recalled and represented in consciousness.[α] So much for the verbal consideration.

Memory,—what.

By Memory or Retention, you will see, is only meant the condition of Reproduction; and it is, therefore, evident that it is only by an extension of the term that it can be called a faculty, that is, an active power. It is more a passive resistance than an energy, and ought, therefore, perhaps to receive rather the appellation of a capacity.[β] But the nature of this capacity or faculty we must now proceed to consider.

The fact of retention admitted.

In the first place, then, I presume that the fact of retention is admitted. We are conscious of certain cognitions as acquired, and we are conscious of these cognitions as resuscitated. That, in the interval, when out of consciousness, these cognitions do continue to subsist in the mind, is certainly an hypothesis, because whatever is out of consciousness can only be assumed; but it is an hypothesis which we are not only warranted, but necessitated, by the phænomena, to establish. I recollect, indeed, that one philosopher has proposed another hypothesis.

The hypothesis of Avicenna regarding retention.

Avicenna, the celebrated Arabian philosopher and physician, denies to the human mind the conservation of its acquired knowledge; and he explains the process of recollection by an irradiation of divine light, through which the recovered cognition is infused into the intellect.[γ] Assum-

α Suabedissen makes Memory equivalent to Retention; see his *Grundzüge der Lehre von dem Menschen*, p. 107. So Fries, Schmid. [Cf. Leibnitz, *Nouv. Ess.*, liv. i. c. i. § 5; liv. ii. c. xix. § 1. Conimbricenses, *In De Mem. et Rem.*, c. i. p. 2.] [Fracastorius, *De Intellectione*, lib. i., *Opera*, f. 126 (ed. 1584).—Ed.]

β See Suabedissen, as above.

γ See Conimbricenses, *In De Memoria et Reminiscentia*, [c. i. p. 2, edit. 1631. Cf. the same, *In De Anima*, lib. iii. c. v. qu. ii. art. ii. p. 430.—Ed.]

LECT. XXX.

ing, however, that the knowledge we have acquired is retained in and by the human mind, we must, of course, attribute to the mind a power of thus retaining it. The fact of memory is thus established.

Retention admits of explanation.

But if it cannot be denied, that the knowledge we have acquired by Perception and Self-consciousness, does actually continue, though out of consciousness, to endure; can we, in the second place, find any ground on which to explain the possibility of this endurance? I think we can, and shall adduce such an explanation, founded on the general analogies of our mental nature.

Similitudes suggested in illustration of the faculty of Retention. Cicero.

Before, however, commencing this, I may notice some of the similitudes which have been suggested by philosophers, as illustrative of this faculty. It has been compared to a storehouse,—Cicero calls it "*thesaurus omnium rerum*,"[α]—provided with cells or pigeon-holes in which its furniture is laid up and arranged.[β] It has been likened to a tablet on which characters were written or impressed.[γ] But of all these sensible resem-

Gassendi.

blances, none is so ingenious as that of Gassendi[δ] to the folds in a piece of paper or cloth; though I do not recollect to have seen it ever noticed. A sheet of paper, or cloth, is capable of receiving innumerable folds, and the folds in which it has been oftenest laid, it takes afterwards of itself. "Concipi charta valeat plicarum innumerabilium, inconfusarumque, et juxta suos ordines, suasque series repetendarum capax. Scilicet ubi unam seriem subtilissimarum induxerimus, superinducere licet alias, quæ primam quidem refrin-

α *De Oratore*, i. 5.—Ed.

β Cf. Plato, *Theætetus*, p. 197.—Ed.

γ Cf. Plato, *Theætetus*, p. 191. Arist., *De Anima*, iii. 4. Boethius, *De Consol. Phil.*, lib. v. metr. 4.—Ed.

δ *Physica*, Sect. iii., Membr. Post., lib. viii. c. 3. *Opera*, Lugd. 1658, vol. ii. p. 406.—Ed. [Cf. Descartes, *Œuvres*, t. ix. p. 167 (ed. Cousin).] [St. Hilaire, *Psychologie d'Aristote*, Préf. p. 18 *et seq.*—Ed.]

gant transversum, et in omnem obliquitatem; sed ita tamen, ut dum novæ plicæ, plicarumque series superinducuntur, priores omnes non modo remaneant, verum etiam possint facili negotio excitari, redire, apparere, quatenus una plica arrepta cæteræ, quæ in eadem serie quadam, quasi sponte sequuntur."

LECT. XXX.

All these resemblances, if intended as more than metaphors, are unphilosophical. We do not even obtain any insight into the nature of Memory from any of the physiological hypotheses which have been stated; indeed all of them are too contemptible even for serious criticism. "The mind affords us, however, in itself, the very explanation which we vainly seek in any collateral influences. The phænomenon of retention is, indeed, so natural, on the ground of the self-energy of mind, that we have no need to suppose any special faculty for memory; the conservation of the action of the mind being involved in the very conception of its power of self-activity.

These resemblances of use simply as metaphors.

The phænomenon of retention naturally arises from the self-energy of mind.

"Let us consider how knowledge is acquired by the mind. Knowledge is not acquired by a mere passive affection, but through the exertion of spontaneous activity on the part of the knowing subject; for though this activity be not exerted without some external excitation, still this excitation is only the occasion on which the mind develops its self-energy. But this energy being once determined, it is natural that it should persist, until again annihilated by other causes. This would in fact be the case, were the mind merely passive in the impression it receives; for it is a universal law of nature, that every effect endures as long as it is not modified or opposed by any other effect. But the mental activity, the act of knowledge, of which I now speak, is more than this; it is an

This specially shown. Knowledge acquired by the spontaneous activity of mind.

LECT. XXX.

energy of the self-active power of a subject one and indivisible: consequently, a part of the ego must be detached or annihilated, if a cognition once existent be again extinguished. Hence it is, that the problem most difficult of solution is not, how a mental activity endures, but how it ever vanishes. For, as we must here maintain not merely the possible continuance of certain energies, but the impossibility of the non-continuance of any one, we, consequently, stand in apparent contradiction to what experience shows us; showing us, as it does, our internal activities in a ceaseless vicissitude of manifestation and disappearance. This apparent contradiction, therefore, demands solution. If it be impossible, that an energy of mind which has once been should be abolished, without a laceration of the vital unity of the mind as a subject one and indivisible;—on this supposition, the question arises, How can the facts of our self-consciousness be brought to harmonise with this statement, seeing that consciousness proves to us, that cognitions once clear and vivid are forgotten; that feelings, wishes, desires, in a word, every act or modification, of which we are at one time aware, are at another vanished; and that our internal existence seems daily to assume a new and different aspect?

The problem most difficult of solution is not, how a mental activity endures, but how it ever vanishes.

"The solution of this problem is to be sought for in the theory of obscure or latent modifications, [that is, mental activities, real but beyond the sphere of consciousness, which I formerly explained.][a] The disappearance of internal energies from the view of internal perception, does not warrant the conclusion, that they no longer exist; for we are not always conscious of all the mental energies whose existence cannot be dis-

The difficulty removed by the principle of latent modifications. The obscuration of a mental activity arises from the weakening of the degree in

a See above, vol. I. lect. xviii. p. 338 et seq.—Ed.

allowed. Only the more vivid changes sufficiently affect our consciousness to become objects of its apprehension: we, consequently, are only conscious of the more prominent series of changes in our internal state; the others remain for the most part latent. Thus we take note of our memory only in its influence on our consciousness; and, in general, do not consider that the immense proportion of our intellectual possessions consists of our delitescent cognitions. All the cognitions which we possess, or have possessed, still remain to us,—the whole complement of all our knowledge still lies in our memory; but as new acquisitions are continually pressing in upon the old, and continually taking place along with them among the modifications of the ego, the old cognitions, unless from time to time refreshed and brought forward, are driven back, and become gradually fainter and more obscure. This obscuration is not, however, to be conceived as an obliteration, or as a total annihilation. The obscuration, the delitescence of mental activities, is explained by the weakening of the degree in which they affect our self-consciousness or internal sense. An activity becomes obscure, because it is no longer able adequately to affect this. To explain, therefore, the disappearance of our mental activities, it is only requisite to explain their weakening or enfeeblement,—which may be attempted in the following way:—Every mental activity belongs to the one vital activity of mind in general; it is, therefore, indivisibly bound up with it, and can neither be torn from, nor abolished in, it. But the mind is only capable, at any one moment, of exerting a certain quantity or degree of force. This quantity must, therefore, be divided among the different activities, so that each has only a

LECT. XXX.

which it affects self-consciousness.

The distribution of mental force explains the weakening of our activities, and the phenomenon of Forgetfulness.

LECT. XXX.

part; and the sum of force belonging to all the several activities taken together, is equal to the quantity or degree of force belonging to the vital activity of mind in general. Thus, in proportion to the greater number of activities in the mind, the less will be the proportion of force which will accrue to each; the feebler, therefore, each will be, and the fainter the vivacity with which it can affect self-consciousness. This weakening of vivacity can, in consequence of the indefinite increase in the number of our mental activities, caused by the ceaseless excitation of the mind to new knowledge, be carried to an indefinite tenuity, without the activities, therefore, ceasing altogether to be. Thus it is quite natural that the great proportion of our mental cognitions should have waxed too feeble to affect our internal perception with the competent intensity; it is quite natural that they should have become obscure or delitescent. In these circumstances it is to be supposed that every new cognition, every newly-excited activity, should be in the greatest vivacity, and should draw to itself the greatest amount of force: this force will, in the same proportion, be withdrawn from the other earlier cognitions; and it is they, consequently, which must undergo the fate of obscuration. Thus is explained the phænomenon of Forgetfulness or Oblivion. And here, by the way, it should perhaps be noticed, that forgetfulness is not to be limited merely to our cognitions; it applies equally to the feelings and desires.

And the phænomenon of Distraction and Attention.

"The same principle illustrates, and is illustrated by, the phænomenon of Distraction and Attention. If a great number of activities are equally excited at once, the disposable amount of mental force is equally distributed among this multitude, so that each activity only attains a low degree of vivacity; the state of mind which results from this is Distraction. Attention is the state the converse of this; that is, the state in which the vital activity of mind is, voluntarily or involuntarily, concentrated, say, in a single activity; in consequence of which concentration this activity waxes stronger, and, therefore, clearer. On this theory, the proposition with which I started,—that all mental activities, all acts of knowledge, which have been once excited, persist,—becomes intelligible; we never wholly lose them, but they become obscure. This obscuration can be conceived in every infinite degree, between incipient latescence and irrecoverable latency. The obscure cognition may exist simply out of consciousness, so that it can be recalled by a common act of reminiscence. Again, it may be impossible to recover it by an act of voluntary recollection; but some association may revivify it enough to make it flash after a long oblivion into consciousness. Further, it may be obscured so far that it can only be resuscitated by some morbid affection of the system; or, finally, it may be absolutely lost for us in this life, and destined only for our reminiscence in the life to come.

Two observations regarding Memory, that arise out of the preceding theory.

1. The law of retention extends over all the phenomena of mind alike.

"That this doctrine admits of an immediate application to the faculty of Retention, or Memory Proper, has been already signified. And in further explanation of this faculty, I would annex two observations, which arise out of the preceding theory. The first is, that retention, that memory, does not belong alone to the cognitive faculties, but that the same law extends, in like manner, over all the three primary classes of the mental phænomena. It is not ideas, notions, cognitions only, but feelings and conations, which are held

LECT. XXX.

fast, and which can, therefore, be again awakened."[*] This fact of the conservation of our practical modifications is not indeed denied; but pyschologists usually so represent the matter, as if, when feelings or conations are retained in the mind, this takes place only through the medium of the memory; meaning by this, that we must, first of all, have had notions of these affections, which notions being preserved, they, when recalled to mind, do again awaken the modification they represent. From the theory I have detailed to you, it must be seen that there is no need of this intermediation of notions, but that we immediately retain feelings, volitions, and desires, no less than notions and cognitions; inasmuch as all the three classes of fundamental phænomena arise equally out of the vital manifestations of the same one and indivisible subject.

2. The various attempts to explain memory by physiological hypotheses are unnecessary.

"The second result of this theory is, that the various attempts to explain memory by physiological hypotheses are as unnecessary as they are untenable. This is not the place to discuss the general problem touching the relation of mind and body. But in proximate reference to memory, it may be satisfactory to show, that this faculty does not stand in need of such crude modes of explanation. It must be allowed, that no faculty affords a more tempting subject for materialistic conjecture. No other mental power betrays a greater dependence on corporeal conditions than memory. Not only in general does its vigorous or feeble activity essentially depend on the health and indisposition of the body, more especially of the nervous systems; but there is manifested a connection between certain functions of memory and certain parts of the

Memory greatly dependent on corporeal conditions.

* [Cf. Tetens, *Versuche über die menschliche Natur*, I. p. 56.]

LECT. XXX.

cerebral apparatus."[a] This connection, however, is such, as affords no countenance to any particular hypotheses at present in vogue. For example, after certain diseases, or certain affections of the brain, some partial loss of memory takes place. Perhaps the patient loses the whole of his stock of knowledge previous to the disease; the faculty of acquiring and retaining new information remaining entire. Perhaps he loses the memory of words, and preserves that of things. Perhaps he may retain the memory of nouns, and lose that of verbs, or *vice versa;* nay, what is still more marvellous, though it is not a very unfrequent occurrence, one language may be taken neatly out of his retention, without affecting his memory of others. "By such observations, the older psychologists were led to the various physiological hypotheses by which they hoped to account for the phænomena of retention,—as, for example, the hypothesis of permanent material impressions on the brain, or of permanent dispositions in the nervous fibres to repeat the same oscillatory movements,—of particular organs for the different functions of memory, —of particular parts of the brain as the repositories of the various classes of ideas,—or even of a particular fibre, as the instrument of every several notion. But all these hypotheses betray only an ignorance of the proper object of philosophy, and of the true nature of the thinking principle. They are at best but useless; for if the unity and self-activity of mind be not denied, it is manifest, that the mental activities, which have been once determined, must persist, and these corporeal explanations are superfluous. Nor can it be argued, that the limitations to which the Retentive, or rather

Physiological hypotheses of the older psychologists, regarding memory.

[a] H. Schmid, *Versuch einer Metaphysik der inneren Natur*, [p. 231-236; translated with occasional brief interpolations.—Ed.]

LECT. XXX.

the Reproductive, Faculty is subjected in its energies, in consequence of its bodily relations, prove the absolute dependence of memory on organisation, and legitimate the explanation of this faculty by corporeal agencies; for the incompetency of this inference can be shown from the contradiction in which it stands to the general laws of mind, which, howbeit conditioned by bodily relations, still ever preserves its self-activity and independence."[a]

Two qualities requisite to a good memory, viz. Retention and Reproduction.

There is perhaps no mental power in which such extreme differences appear, in different individuals, as in memory. To a good memory there are certainly two qualities requisite,—1°. The capacity of Retention, and 2°. The faculty of Reproduction. But the former quality appears to be that by which these marvellous contrasts are principally determined. I should only fatigue you, were I to enumerate the prodigious feats of retention, which are proved to have been actually performed. Of these, I shall only select the one which, upon the whole, appears to me the most extraordinary, both by reason of its own singularity, and because I am able to afford it some testimony, in confirmation of the veracity of the illustrious scholar by whom it is narrated, and which has most groundlessly been suspected by his learned editor. The story I am about to detail to you is told by Muretus, in the first chapter of the third book of his incomparable work, the *Variæ Lectiones*.[β]

[a] H. Schmid, *Versuch einer Metaphysik*, [p. 235-236.—Ed.]

[β] *Opera*, edit. Ruhnken., tom. ii. p. 55.—Ed. Muretus is one of the most distinguished philologers and critics of modern times; and from himself to Cicero, a period of sixteen centuries, there is to be found no one who equalled him in Latin eloquence. Besides numerous editions of his several treatises, his works have been republished in a collected form six several times; and the editor of the edition before the one at present [1837] in the course of publication, by Professor Frotscher of Leipsig, was Ruhnkenius, perhaps the greatest scholar of the eighteenth century.

After noticing the boast of Hippias, in Plato, that he could repeat, upon hearing once, to the amount of five hundred words, he observes that this was nothing as compared with the power of retention possessed by Seneca the rhetorician. In his *Declamations*, Seneca, complaining of the inroads of old age upon his faculties of mind and body, mentions, in regard to the tenacity of his now failing memory, that he had been able to repeat two thousand names read to him, in the order in which they had been spoken; and that, on one occasion, when at his studies, two hundred unconnected verses having been pronounced by the different pupils of his preceptor, he repeated them in a reversed order, that is, proceeding from the last to the first uttered. After quoting the passage from Seneca, of which I have given you the substance, Muretus remarks, that this statement had always appeared to him marvellous, and almost incredible, until he himself had been witness of a fact to which he never could otherwise have afforded credit. The sum of this statement is, that at Padua there dwelt, in his neighbourhood, a young man, a Corsican by birth, and of a good family in that island, who had come thither for the cultivation of civil law, in which he was a diligent and distinguished student. He was a frequent visitor at the house and gardens of Muretus, who having heard that he possessed a remarkable art, or faculty of memory, took occasion, though incredulous in regard to reports, of requesting from him a specimen of his power. He at once agreed; and having adjourned with a considerable party of distinguished auditors into a saloon, Muretus began to dictate words, Latin, Greek, barbarous, significant and non-significant, disjoined and connected, until he wearied himself, the young man

The remarkable case of retention, narrated by Muretus.

LECT. XXX.

who wrote them down, and the audience who were present;—"we were all," he says, "marvellously tired." The Corsican alone was the one of the whole company alert and fresh, and continually desired Muretus for more words; who declared he would be more than satisfied, if he could repeat the half of what had been taken down, and at length he ceased. The young man, with his gaze fixed upon the ground, stood silent for a brief season, and then, says Muretus, "vidi facinus mirificissimum. Having begun to speak, he absolutely repeated the whole words, in the same order in which they had been delivered, without the slightest hesitation; then, commencing from the last, he repeated them backwards till he came to the first. Then again, so that he spoke the first, the third, the fifth, and so on; did this in any order that was asked, and all without the smallest error. Having subsequently become familiarly acquainted with him, I have had other and frequent experience of his power. He assured me, (and he had nothing of the boaster in him,) that he could recite, in the manner I have mentioned, to the amount of thirty-six thousand words. And what is more wonderful, they all so adhered to the mind that, after a year's interval, he could repeat them without trouble. I know, from having tried him, he could do so after a considerable time, (post multos dies). Nor was this all. Franciscus Molinus, a patrician of Venice, was resident with me, a young man ardently devoted to literature, who, as he had but a wretched memory, besought the Corsican to instruct him in the art. The hint of his desire was enough, and a daily course of instruction commenced, and with such success that the pupil could, in about a week or ten days, easily repeat to the extent of five hundred words or more, in any order that was prescribed." "This," adds Muretus, "I should hardly venture to record, fearing the suspicion of falsehood, had not the matter been very recent, for a year has not elapsed, and had I not as fellow-witnesses, Nicolaus the son of Petrus Lippomanus, Lazarus the son of Francis Mocenicus, Joannes the son of Nicolaus Malipetrus, George the son of Laurence Contarenus—all Venetian nobles, worthy and distinguished young men, besides other innumerable witnesses. The Corsican stated that he received the art from a Frenchman who was his domestic tutor." Muretus terminates the narrative by alleging sundry examples of a similar faculty, possessed in antiquity by Cyrus, Simonides, and Apollonius Tyanæus.

Ruhnkenius unduly sceptical in regard to this case.

Now, on this history, Ruhnkenius has the following note, in reference to the silence of Muretus in regard to the name of the Corsican: "Ego nomen hominis tam mirabilis, citius quam patriam requisiissem. Idque pertinebat ad fidem narrationi faciendam." This scepticism is, I think, out of place. It would perhaps have been warranted, had Muretus not done far more than was necessary to establish the authenticity of the story; and, after the testimonies to whom he appeals, the omission of the Corsican's name is a matter of little import. But I am surprised that one confirmatory circumstance has escaped so learned a scholar as Ruhnkenius, seeing that it occurs in the works of a man with whose writings no one was more familiar. Muretus and Paulus Manutius were correspondents, and Manutius, you must know, was a Venetian. Now, in the letters of Manutius to Muretus, at the date of the occurrence in question, there is frequent mention made of Molino, in whom Manutius seems to have felt much interest; and, on one occasion, there is an allusion,

LECT. XXX.

(which I cannot at the moment recover, so as to give you the precise expressions), to Molino's cultivation of the Art of Memory, and to his instructor.[α] This, if it were wanted, corroborates the narrative of Muretus, whose trustworthiness, I admit, was not quite as transcendent as his genius.[β]

α See *Pauli Manutii Epistolæ*, vol. i. lib. iii. ep. xiii. p. 154 (edit. Krause, 1720): "Molino, parum abest, quin vehementer invideam; quid ni? *artem Memoriæ* tenenti. Verumtamen impedit amor, a quo abesse solet invidia: etiam ea spes, quod ille, quo eum bono *alienus homo* impertivit, civi suo, homini amantissimo, certe numquam denegabit." Cf. vol. iii. *Notæ ad Epistolas*, p. 1138.—Ed.

β "As Sophocles says that memory is the queen of things, and because the nurse of poetry herself is a daughter of Mnemosyne, I shall mention here another once world-renowned Corsican of Calvi—Giulio Guidi, in the year 1581, the wonder of Padua, on account of his unfortunate memory. He could repeat thirty-six thousand names after once hearing them. People called him *Guidi della gran memoria*. But he produced nothing; his memory had killed all his creative faculty. Pico von Mirandola, who lived before him, produced; but he died young. It is with the precious gift of memory, as with all other gifts—they are a curse of the gods when they give too much."—Gregorovius, *Wanderings in Corsica*, vol. ii. book vi. chap. vi. p. 31 (Constable's edition). [A case similar to that narrated by Muretus is given by Joseph Scaliger in the *Secunda Scaligerana*, v. *Mémoire*, t. ii. p. 450-451, edit. 1740.—Ed.]

LECTURE XXXI.

THE REPRODUCTIVE FACULTY.—LAWS OF ASSOCIATION.

LECT. XXXI.

Recapitulation.

In my last Lecture, I entered on the consideration of that faculty of mind by which we keep possession of the knowledge acquired by the two faculties of External Perception, and Self-consciousness; and I endeavoured to explain to you a theory of the manner in which the fact of retention may be accounted for, in conformity to the nature of mind, considered as a self-active and indivisible subject. At the conclusion of the Lecture, I gave you, *instar omnium*, one memorable example of the prodigious differences which exist between mind and mind in the capacity of retention. Before passing from the faculty of Memory, considered simply as the power of conservation, I may notice two opposite doctrines, that have been maintained, in regard to the relation of this faculty to the higher powers of mind. One of these doctrines holds, that a great development of memory is incompatible with a high degree of intelligence; the other, that a high degree of intelligence supposes such a development of memory as its condition.

Two opposite doctrines maintained in regard to the relation of Memory to the higher powers of mind.

The former of these opinions is one very extensively prevalent, not only among philosophers, but among mankind in general, and the words,—*Beati memoria, expectantes judicium*,—have been applied to express the supposed incompatibility of great memory and

1. That a great power of memory is incompatible with a high degree of intelligence.

LECT. XXXI.

sound judgment.[a] There seems, however, no valid ground for this belief. If an extraordinary power of retention is frequently not accompanied with a corresponding power of intelligence, it is a natural, but not a very logical, procedure to jump to the conclusion, that a great memory is inconsistent with a sound judgment.

This opinion refuted by facts.—Examples of high intelligence and great memory.

The opinion is refuted by the slightest induction; for we immediately find, that many of the individuals who towered above their fellows in intellectual superiority, were almost equally distinguished for the capacity of their memory. I recently quoted to you a passage from the *Scaligerana*, in which

Joseph Scaliger.

Joseph Scaliger is made to say that he had not a good memory, but a good reminiscence; and he immediately adds, "never or rarely are judgment and a great memory found in conjunction." Of this opinion Scaliger himself affords the most illustrious refutation. During his lifetime, he was hailed as the Dictator of the Republic of Letters, and posterity has ratified the decision of his contemporaries, in crowning him as the prince of philologers and critics. But to elevate a man to such an eminence, it is evident, that the most consummate genius and ability were conditions. And

His great powers of memory testified to by Casaubon.

what were the powers of Scaliger, let Isaac Casaubon,[β] among a hundred other witnesses, inform us; and Casaubon was a scholar second only to Scaliger himself in erudition. "Nihil est quod discere quisquam vellet, quod ille (Scaliger) docere non posset: Nihil legerat, (quid autem ille non legerat?), quod non statim meminisset; nihil tam obscurum aut abolitum in

a [Niethammer, *Der Streit des Philanthropinismus und Humanismus*, p. 294.] [Ausserdem sey es eine selbst Sprichwörtlich gewordene Erfahrung, (beati memoria exspectant judicium), dass vorherrschende *Gedächtnissfertigkeit* der *Urtheilskraft* Abbruch thue.—Ed.]

β [*Præfatio in Opuscula Jos. Justi Scaligeri.*]

ullo vetere scriptore Græco, Latino, vel Hebræo, de quo interrogatus non statim responderet. Historias omnium populorum, omnium ætatum, successiones imperiorum, res ecclesiæ veteris in numerato habebat: animalium, plantarum, metallorum, omniumque rerum naturalium, proprietates, differentias, et appellationes, qua veteres, qua recentes, tenebat accurate. Locorum situs, provinciarum fines et varias pro temporibus illarum divisiones ad unguem callebat; nullam disciplinarum, scientiarumve graviorum reliquerat intactam; linguas tam multas tam exacte sciebat, ut, vel si hoc unum per totum vitæ spatium egisset, digna res miraculo potuerit videri."

Grotius, Pascal. Leibnitz, Euler. Donellus. Muratori. Ben Jonson.

For intellectual power of the highest order, none were distinguished above Grotius and Pascal; and Grotius[α] and Pascal[β] forgot nothing they had ever read or thought. Leibnitz[γ] and Euler[δ] were not less celebrated for their intelligence than for their memory, and both could repeat the whole of the *Æneid*. Donellus[ε] knew the *Corpus Juris* by heart, and yet he was one of the profoundest and most original speculators in jurisprudence. Muratori,[ζ] though not a genius of the very highest order, was still a man of great ability and judgment; and so powerful was his retention, that in making quotations, he had only to read his passages, put the books in their place, and then to write out from memory the words. Ben Jonson[η] tells

α *Grotii Manes Vindicati* (1727), pars post., p. 685.—ED.

β *Pensées*, Préface (ed. Renouard). Cf. *Stewart's Works*, vol. ii. p. 378-379, and relative foot-note.—ED.

γ Fontenelle, *Eloge de M. Leibnitz*,—*Leibn. Op.*, p. xx. (ed. Dutens).—ED.

δ [Blunde, *Versuch einer Systematischen Behandlung der empirischen Psychologie*, i. 356.]

ε Teissier, *Eloges des Hommes Savans*, t. iv. p. 146.—ED.

ζ [Blunde, *Versuch*, &c., as above.] [*Vita di Muratori*, c. xi. p. 236.—ED.]

η *Timber; or, Discoveries made upon Men and Matter* (*Works*, edit. Gifford, vol. ix. p. 169).—ED.

LECT. XXXI.

us that he could repeat all he had ever written, and whole books that he had read. Themistocles[α] could call by their names the twenty thousand citizens of Athens; Cyrus[β] is reported to have known the name of every soldier in his army. Hortensius, after Cicero, the greatest orator of Rome, after sitting a whole day at a public sale, correctly enunciated from memory all the things sold, their prices, and the names of the purchasers.[γ] Niebuhr,[δ] the historian of Rome, was not less distinguished for his memory than for his acuteness. In his youth, he was employed in one of the public offices of Denmark; part of a book of accounts having been destroyed, he restored it from his recollection. Sir James Mackintosh was, likewise, remarkable for his power of memory. An instance I can give you, which I witnessed myself. In a conversation I had with him, we happened to touch upon an author whom I mentioned in my last Lecture,—Muretus; and Sir James recited from his oration in praise of the massacre of St Bartholomew some considerable passages. Mr Dugald Stewart, and the late Dr Gregory, are, likewise, examples of great talent united with great memory.

Themistocles. Cyrus. Hortensius. Niebuhr. Sir James Mackintosh. Dugald Stewart. Dr Gregory.

2. That a high degree of intelligence supposes great power of memory.

But if there be no ground for the vulgar opinion, that a strong faculty of retention is incompatible with intellectual capacity in general, the converse opinion is not better founded, which has been maintained, among others, by Hoffbauer.[ε] This doctrine does not, however, deserve an articulate refutation; for the

α Cicero, *De Senectute*, c. vii. Val. Maximus, viii. 7.—Ed.

β Pliny, *Nat. Hist.*, vii. 24. Quintilian, *Orat.*, xi. 2.—Ed.

γ Seneca, (M.) *Controv.*, Pref.—Ed.

δ See *Life of Niebuhr*, vol. ii. p. 412-413, where a similar anecdote is mentioned, but not exactly as stated in the text. See also vol. i. c. vii. p. 298.—Ed.

ε [See Blunde, *Versuch einer systematischen Behandlung der empirischen Psychologie*, i. 357, where Hoffbauer is referred to.] [See Hoffbauer, *Naturlehre der Seele in Briefen*, p. 151-183.—Ed.]

common experience of every one sufficiently proves, that intelligence and memory hold no necessary proportion to each other. On this subject I may refer you to Mr Stewart's excellent chapter on Memory in the first volume of his *Elements*.[a]

The Reproductive Faculty. This name inappropriate; the limitation in which it is here employed.

I now pass to the next faculty in order,—the faculty which I have called the Reproductive. I am not satisfied with this name; for it does not precisely of itself mark what I wish to be expressed,—viz., the process by which what is lying dormant in memory is awakened, as contradistinguished from the representation in consciousness of it as awakened. The two processes certainly suppose each other; for we cannot awaken a cognition without its being represented,—the representation being, in fact, only its state of waking; nor can a latent thought or affection be represented, unless certain conditions be fulfilled, by which it is called out of obscurity into the light of consciousness. The two processes are relative and correlative, but not more identical than hill and valley. I am not satisfied, I say, with the term *reproduction* for the process by which the dormant thought or affection is aroused; for it does not clearly denote what it is intended to express. Perhaps the *Resuscitative Faculty* would have been better; and the term *reproduction* might have been employed to comprehend the whole process, made up of the correlative acts of retention, resuscitation, and representation. Be this, however, as it may, I shall at present continue to employ the term, in the limited meaning I have already assigned.

Interest excited by the phænomena of Reproduction.

The phænomenon of Reproduction is one of the most wonderful in the whole compass of psychology; and it is one in the explanation of which philosophy

a Chap. vi. *Works*, ii. 343.—Ed.

LECT. XXXI.

The Schoolmen.

has been more successful than in almost any other. The scholastic psychologists seem to have regarded the succession in the train of thought, or, as they called it, the excitation of the species, with peculiar wonder, as one of the most inscrutable mysteries of nature;[a] and

Aristotle's analysis of the phenomenon, — nearly perfect.

yet, what is curious, Aristotle has left almost as complete an analysis of the laws by which this phenomenon is regulated, as has yet been accomplished. It required, however, a considerable progress in the inductive philosophy of mind, before this analysis of Aristotle could be appreciated at its proper value; and, in fact, it was only after modern philosophers had rediscovered the principal laws of Association, that it was found that these laws had been more completely

Julius Cæsar Scaliger.

given two thousand years before. Joseph Scaliger, speaking of his father, whose philosophical acuteness I have more than once had occasion to commemorate, says, "My father declared, that of the causes of three things in particular he was wholly ignorant,—of the interval of fevers, of the ebb and flow of the sea, and of reminiscence."[β] The excitation of the species is de-

Poncius.

clared by Poncius[γ] to be "one of the most difficult secrets of nature" (ex difficilioribus naturæ arcanis);

Oviedo.

and Oviedo,[δ] a Jesuit schoolman, says, "therein lies the very greatest mystery of all philosophy, (maximum totius philosophiæ sacramentum), never to be competently explained by human ingenuity;" "and this because we can neither discover the cause which, for example, in the recitation of an oration, excites the species in the order in which they are excited, nor the reason why often, when wishing to recollect a matter,

a See *Reid's Works*, p. 889.—Ed.

β [*Prima Scaligerana*, v. "Causa,"] [t. ii. p. 46, edit. 1740.—Ed.]

γ [Poncius, *Cursus Philosophicus*, *De Anima*, Disp. lxiii. qu. iii. concl. 3.]

δ [Franciscus de Oviedo, *Cursus Philosophicus*, *De Anima*, Cont. v. punct. iv. n. 13.] [Cf. *Reid's Works*, Note D**, p. 889.—Ed.]

we do not, whereas when not wishing to recollect it, we sometimes do. Hence the same Poncius says, that for the excitation of the species we must either recur at once to God, or to some sufficient cause, which, however, he does not specify."[α]

Reproduction,—what.

The faculty of Reproduction is governed by the laws which regulate the Association of the mental train; or, to speak more correctly, reproduction is nothing but the result of these laws. Every one is conscious of a ceaseless succession or train of thoughts, one thought suggesting another, which again is the cause of exciting a third, and so on. In what manner, it may be asked, does the presence of any thought determine the introduction of another? Is the train subject to laws, and if so, by what laws is it regulated?

The train of thought subject to laws. This illustrated by Hobbes.

That the elements of the mental train are not isolated, but that each thought forms a link of a continuous and uninterrupted chain, is well illustrated by Hobbes. "In a company," he says, "in which the conversation turned upon the late civil war, what could be conceived more impertinent than for a person to ask abruptly, what was the value of a Roman denarius? On a little reflection, however, I was easily able to trace the train of thought which suggested the question; for the original subject of discourse naturally introduced the history of the king, and of the treachery of those who surrendered his person to his enemies; this again introduced the treachery of Judas Iscariot, and the sum of money which he received for his reward."[β]

The expression train of thought includes the phænomena

But if thoughts, and feelings, and conations, (for you must observe, that the train is not limited to the phænomena of cognition only),[γ] do not arise of them-

α [Fr. Bonæ Spei, *Physica*, pars iv. *In de Anima*, disp. x. p. 94. Cf. Ancillon, *Essais Philos.* (*Nouv. Mél.*), t. ii. c. iii. p. 139.]

β *Leviathan*, part i. chap. iii.—Ed.

γ [Cf. Fries, *Anthropologie*, i. § 8, p. 29, edit. 1820; *Kritik*, i. § 33. H. Schmid, *Versuch einer Metaphysik der*

LECT. XXXI.

of Cognition, Feeling, and Conation.

Is there any law, besides that of simple connection, which regulates this train?

selves, but only in causal connection with preceding and subsequent modifications of mind, it remains to be asked and answered,—Do the links of this chain follow each other under any other condition than that of simple connection,—in other words, may any thought, feeling, or desire, be connected with any other? Or, is the succession regulated by other and special laws, according to which certain kinds of modification exclusively precede, and exclusively follow, each other? The slightest observation of the phænomenon shows, that the latter alternative is the case; and on this all philosophers are agreed. Nor do philosophers differ in regard to what kind of thoughts, (and under that term, you will remark, I at present include also *feelings* and *conations*), are associated together. They differ almost exclusively in regard to the subordinate question, of how these thoughts ought to be classified, and carried up into system. This, therefore, is the question to which I shall address myself; referring you for illustrations and examples of the fact and effects of Association, to the chapter on the subject in the first volume of Mr Stewart's *Elements*,[a] in which you will find its details treated with great elegance and ability.

The point on which philosophers differ; and question to be considered.

Conditions of Reproduction, as generalized by philosophers;—in all seven.

In my last Lecture, I explained to you how thoughts, once experienced, remain, though out of consciousness, still in possession of the mind; and I have now to show you, how these thoughts, retained in memory, may, without any excitation from without, be again retrieved by an excitation or awakening from other thoughts within. Philosophers, having observed, that

inneren *Natur*, pp. 236, 242. Eschenmayer, *Psychologie*, § 75, p. 69. F. A. Carus, *Psychologie*, i. p. 183. Stewart, *Elements*, I. c. v. *Works*, vol. ii. p. 257. Brown, *Philosophy of the Human Mind*, lect. xliv. p. 282 (edit. 1830). Dr J. Young, *Lectures on Intellectual Philosophy*, p. 281.] [For Aristotle, see *Reid's Works*, p. 892-893.—Ed.]

[a] Chap. v. *Works*, ii. 252.—Ed.

one thought determined another to arise, and that this determination only took place between thoughts which stood in certain relations to each other, set themselves to ascertain and classify the kinds of correlation under which this occurred, in order to generalise the laws by which the phænomenon of Reproduction was governed. Accordingly, it has been established, that thoughts are associated, that is, are able to excite each other;—1°, If coexistent, or immediately successive, in time; 2°, If their objects are conterminous or adjoining in space; 3°, If they hold the dependence to each other of cause and effect, or of mean and end, or of whole and part; 4°, If they stand in a relation either of contrast or of similarity; 5°, If they are the operations of the same power, or of different powers conversant about the same object; 6°, If their objects are the sign and the signified; or, 7°, Even if their objects are accidentally denoted by the same sound. These, as far as I recollect, are all the classes to which philosophers have attempted to reduce the principles of Mental Association. Aristotle recalled the laws of this connection to four, or rather to three,—Contiguity in time and space, Resemblance, and Contrariety.[a] He even seems to have thought they might all be carried up into the one law of Coexistence. Aristotle implicitly, St Augustin[β] explicitly,—what has never been observed,—reduces association to a single canon,—viz., Thoughts which have once coexisted in the mind are afterwards associated. This law, which I would call the law of Redintegration, was afterwards enounced by Malebranche,[γ] Wolf,[δ] and Bilfinger;[ε] but without

Aristotle reduces the laws of association to three; and implicitly to one canon.

St Augustin explicitly reduces these laws to one,—which the author calls the law of Redintegration.

Malebranche. Wolf. Bilfinger.

a *De Memoria et Reminiscentia*, c. ii. § 6.—Ed.

β *Confessiones*, lib. x. chap. xix.—Ed.

γ *Recherche de la Vérité*, liv. ii. c. v.—Ed.

δ *Psychologia Empirica*, § 230.—Ed.

ε See *Reid's Works*, p. 899.—Ed.

any reference to St Austin. Hume, who thinks himself the first philosopher who had ever attempted to generalise the laws of association, makes them three,—Resemblance, Contiguity in time and place, and Cause and Effect.[α] Gerard[β] and Beattie[γ] adopt, with little modification, the Aristotelic classification. Omitting a hundred others, whose opinions would be curious in a history of the doctrine, I shall notice only Stewart and Brown. Stewart,[δ] after disclaiming any attempt at a complete enumeration, mentions two classes of circumstances as useful to be observed. "The relations," he says, "upon which some of them are founded, are perfectly obvious to the mind; those which are the foundation of others, are discovered only in consequence of particular efforts of attention. Of the former kind are the relations of Resemblance and Analogy, of Contrariety, of Vicinity in time and place, and those which arise from accidental coincidences in the sound of different words. These, in general, connect our thoughts together, when they are suffered to take their natural course, and when we are conscious of little or no active exertion. Of the latter kind are the relations of Cause and Effect, of Means and End, of Premises and Conclusion; and those others which regulate the train of thought in the mind of the philosopher, when he is engaged in a particular investigation."

Hume. Gerard. Beattie. Stewart. Brown. Stewart quoted.

Brown[ε] divides the circumstances affecting association into primary and secondary. Under the primary laws of Suggestion, he includes Resemblance, Contrast, Contiguity in time and place,—a classification iden-

Brown's classification.

α *Enquiry concerning Human Understanding*, sect. iii.—ED.

β *Essay on Taste*, part iii. § i. pp. 167, 168, edit. 1759.—ED.

γ *Dissertations, Moral and Critical.*—*Of Imagination*, c. ii. § 1 *et seq.*, p. 78. Cf. pp. 9, 145.—ED.

δ *Elements*, vol. ii. c. v. part i. § 2. *Works*, vol. iii. p. 263.—ED.

ε *Philosophy of the Human Mind*, lects. xxxiv.-xxxvii.—ED.

tical with Aristotle's. By the secondary, he means the vivacity, the recentness, and the frequent repetition of our thoughts,—circumstances which, though they exert an influence on the recurrence of our thoughts, belong to a different order of causes from those we are at present considering.[α]

The laws enumerated admit of reduction to two; and these two again to one grand law.

Now all the laws which I have hitherto enumerated may be easily reduced to two,—the law of the Simultaneity, and the law of the Resemblance or Affinity, of Thought.[β] Under Simultaneity I include Immediate Consecution in time; to the other category of Affinity every other circumstance may be reduced. I shall take the several cases I have above enumerated, and having exemplified their influence as associating principles, I shall show how they are all only special modifications of the two laws of Simultaneity and Affinity; which two laws, I shall finally prove to you, are themselves only modifications of one supreme law,—the law of Redintegration.

The influence of the special laws, as associating principles, illustrated.

I. The law of Simultaneity.

The first law,—that of Simultaneity, or of Coexistence and Immediate Succession in time,—is too evident to require any illustration. "In passing along a road," as Mr Stewart[γ] observes, "which we have formerly travelled in the company of a friend, the particulars of the conversation in which we were then engaged, are frequently suggested to us by the objects we meet with. In such a scene, we recollect that a particular subject was started; and in passing the different houses, and plantations, and rivers, the arguments we were discussing when we last saw them, recur spontaneously to the memory. The connection

α See *Reid's Works*, p. 910.—Ed.

β See H. Schmid, *Versuch einer Metaphysik der inneren Natur*, p. 241. [Cf. Fries, *Anthropologie*, i. § 6, p. 29 (edit. 1820).]

γ *Elements*, vol. i. c. v. part i. § 1. *Works*, vol. ii. pp. 252, 253.—Ed.

LECT. XXXI. which is formed in the mind between the words of a language and the ideas they denote; the connection which is formed between the different words of a discourse we have committed to memory; the connection between the different notes of a piece of music in the mind of the musician, are all obvious instances of the same general law of our nature."

II. The law of Affinity.

The second law,—that of the Affinity of thoughts,—will be best illustrated by the cases of which it is the more general expression.

1. The case of resembling, analogous, or partially identical objects.

In the first place, in the case of resembling, or analogous, or partially identical objects, it will not be denied that these virtually suggest each other. The imagination of Alexander carries me to the imagination of Cæsar, Cæsar to Charlemagne, Charlemagne to Napoleon. The vision of a portrait suggests the image of the person portrayed. In a company one anecdote suggests another analogous. This principle is admirably illustrated from the mouth of Shakespeare's Merchant of Venice:—

"My wind, cooling my broth,
Would blow me to an ague, when I thought,
What harm a wind too great might do at sea.
I should not see the sandy hour-glass run,
But I should think of shallows and of flats,
And see my wealthy Andrew dock'd in sand,
Vailing her high top lower than her ribs,
To kiss her burial. Should I go to church,
And see the holy edifice of stone,
And not bethink me strait of dang'rous rocks?
Which touching but my gentle vessel's side,
Would scatter all the spices on the stream,
Enrobe the roaring waters with my silks;
And in a word,—but even now worth this,
And now worth nothing." [a]

That resembling, analogous, or partially identical objects stand in reciprocal affinity, is apparent; they are its strongest exemplifications. So far there is no difficulty.

[a] *Merchant of Venice*, act i. scene I.

2. The case of contrary or contrasted thoughts.

In the second place, thoughts standing to each other in the relation of contrariety or contrast, are mutually suggestive. Thus the thought of vice suggests the thought of virtue; and, in the mental world, the prince and the peasant, kings and beggars, are inseparable concomitants. On this principle are dependent those associations which constitute the charms of antithesis and wit. Thus the whole pathos of Milton's apostrophe to light, lies in the contrast of his own darkness to the resplendent object he addresses:

> "Hail, holy light, offspring of heaven first-born,
> . . . . . Thee I revisit safe,
> And feel thy sovran vital lamp; but thou
> Revisit'st not these eyes, that roll in vain
> To find thy piercing ray, and find no dawn." α

It is contrast that animates the Ode of Horace to Archytas:

> "Te maris et terræ, numeroque carentis arenæ
> Mensorem cohibent, Archyta,
> Pulveris exigui prope littus parva Matinum
> Munera; nec quidquam tibi prodest
> Aërias tentasse domos, animoque rotundum
> Percurrisse polum, morituro." β

The same contrast illuminates the stanza of Gray:

> "The boast of heraldry, the pomp of power,
> And all that beauty, all that wealth ere gave,
> Awaits alike the inevitable hour;
> The paths of glory lead but to the grave."

And in what else does the beauty of the following line consist, but in the contrast and connection of life and death; life being represented as but a wayfaring from grave to grave?—

> Τίς βίος;—ἐκ τύμβοιο θορὼν, ἐπὶ τύμβον ὁδεύω. γ

α *Paradise Lost*, book iii.—Ed.

β *Carm.*, i. xxviii.—Ed.

γ [Gregor. Nazianz. *Carm.*, xiv.]

LECT. XXXI.

Who can think of Marius sitting amid the ruins of Carthage, without thinking of the resemblance of the consul and the city,—without thinking of the difference between their past and present fortunes? And in the incomparable epigram of Molsa on the great Pompey, the effect is produced by the contrast of the life and death of the hero, and in the conversion of the very fact of his posthumous dishonour into a theme of the noblest panegyric.

> "Dux, Pharia quamvis jaceas inhumatus arena,
> Non ideo fati est sævior ira tui:
> Indignum fuerat tellus tibi victa sepulcrum;
> Non decuit cœlo, te nisi, Magne, tegi." [a]

Depends on the logical principle,—that the knowledge of contraries is one.

Thus that objects, though contrasted, are still akin,—still stand to each other in a relation of affinity, depends on their logical analogy. The axiom that the knowledge of contraries is one, proves that the thought of the one involves the thought of the other.[β]

3. The law of contiguity.

In the third place, objects contiguous in place are associated. You recollect the famous passage of Cicero in the first chapter of the fifth book *De Finibus*, of which the following is the conclusion:—"Tanta vis admonitionis est in locis, ut, non sine causa, ex his memoriæ deducta sit disciplina. . . . Id quidem infinitum in hac urbe; quocumque enim ingredimur, in aliquam historiam vestigium ponimus." But how do objects adjacent in place stand in affinity to each other? Simply because local contiguity binds up objects, otherwise unconnected, into a single object of perceptive thought.

4. The law

In the fourth place, thoughts of the whole and the

a [*Carmina Illustrium Poetarum Italorum*, t. vi. 369. Florentiæ, 1719.]

β [Alex. Aphrodisiensis (*In Top.* I. 18) makes Contrariety equivalent to Similarity, inasmuch as contraries &c., have common attributes.]

parts, of the thing and its properties, of the sign and the thing signified,—of these it is superfluous to illustrate either the reality of the influence, or to show that they are only so many forms of affinity; both are equally manifest. But in this case affinity is not the only principle of association; here simultaneity also occurs. One observation I may make to show, that what Mr Stewart promulgates as a distinct principle of association, is only a subordinate modification of the two great laws I have laid down,—I mean his association of objects, arising from accidental coincidences in the sound of the words by which they are denoted. Here the association between the objects or ideas is not immediate. One object or idea signified suggests its term signifying. But a complete or partial identity in sound suggests another word, and that word suggests the thing or thought it signifies. The two things or thoughts are thus associated, only mediately, through the association of their signs, and the several immediate associations are very simple examples of the general laws.

Of whole and parts, &c.

5. The law of cause and effect.

In the fifth place, thoughts of causes and effects reciprocally suggest each other. Thus the falling snow excites the imagination of an inundation; a shower of hail a thought of the destruction of the fruit; the sight of wine carries us back to the grapes, or the sight of the grapes carries us forward to the wine; and so forth. But cause and effect not only naturally but necessarily suggest each other; they stand in the closest affinity, and, therefore, whatever phænomena are subsumed under this relation, as indeed under all relations, are, consequently, also in affinity.

I have now, I think, gone through all the circumstances which philosophers have constituted into sepa-

All these separate laws thus resolved into two:—Simultaneity and Affinity: and these again are resolvable into the one grand law of Redintegration.

rate laws of Association; and shown that they easily resolve themselves into the two laws of Simultaneity and Affinity. I now proceed to show you that these two laws themselves are reducible to that one law, which I would call the law of Redintegration or Totality, which, as I already stated, I have found incidentally expressed by St Augustin.[α] This law may be thus enounced,—Those thoughts suggest each other which had previously constituted parts of the same entire or total act of cognition. Now to the same entire or total act belong, as integral or constituent parts, in the first place, those thoughts which arose at the same time, or in immediate consecution; and in the second, those thoughts which are bound up into one by their mutual affinity. Thus, therefore, the two laws of Simultaneity and Affinity are carried up into unity, in the higher law of Redintegration or Totality; and by this one law the whole phænomena of Association may be easily explained.[β]

α *Confessiones*, x. 19.—ED.

β For historical notices of the law of Redintegration, see *Reid's Works*, Note D**, p. 889 *et seq.* Compare with the doctrine of the text the author's theory of Association, as partially developed in Note D***, p. 910 *et seq.*—ED.

LECTURE XXXII.

THE REPRODUCTIVE FACULTY.—LAWS OF ASSOCIATION. SUGGESTION AND REMINISCENCE.

LECT. XXXII.

Recapitulation.

In our last Lecture we were occupied with the phænomena of Reproduction, as the result of the laws which govern the succession of our mental train. These laws, as they have been called, of the Association of our Thoughts, comprehend equally the whole phænomena of mind,—the Cognitions, the Feelings, the Desires. I enumerated to you the principal heads under which philosophers had classed the circumstances which constitute between thoughts a bond of association,—a principle of mutual suggestion; and showed you that these could all easily be reduced to two laws,—the law of Simultaneity, and the law of Affinity. By the former of these, objects coexistent or immediately consequent in time are associated; by the latter, things which stand in a mutual affinity to each other, either objectively and in themselves, or subjectively, through the modes under which the mind conceives them, are in like manner reciprocally suggestive. These two laws, I further showed you, might themselves be carried up into one supreme principle of Association, which I called the law of Redintegration or of Totality; and according to which thoughts or mental activities, having once formed parts of the same total thought or mental activity, tend ever after immediately to suggest each other. Out of this universal law every special

LECT. XXXII.

law of Association may easily be evolved, as they are all only so many modified expressions of this common principle,—so many applications of it to cases more or less particular.

No legitimate presumption against the truth of the law of Redintegration, if found inexplicable.

But this law being established by induction and generalisation, and affording an explanation of the various phænomena of Association, it may be asked, How is this law itself explained? On what principle of our intellectual nature is it founded? To this no answer can be legitimately demanded. It is enough for the natural philosopher to reduce the special laws of the attraction of distant bodies to the one principle of gravitation; and his theory is not invalidated, because he can give no account of how gravitation is itself determined. In all our explanations of the phænomena of mind and matter, we must always arrive at an ultimate fact or law, of which we are wholly unable to afford an ulterior explanation. We are, therefore, entitled to decline attempting any illustration of the ground on which the supreme fact or law of Association reposes; and if we do attempt such illustration, and fail in the endeavour, no presumption is, therefore, justly to be raised against the truth of the fact or principle itself.

Attempted illustration of the ground on which this law reposes, from the unity of the subject of the mental energies.

But an illustration of this great law is involved in the principle of the unity of the mental energies, as the activities of the subject one and indivisible, to which I have had occasion to refer.[a] "The various acts of mind must not be viewed as single,—as isolated, manifestations; they all belong to the one activity of the ego: and, consequently, if our various mental energies are only partial modifications of the same general activity, they must all be associated among themselves. Every mental energy,—every thought,

[a] See above, lect. xxx. vol. ii. p. 211.—Ed.

feeling, desire that is excited, excites at the same time all other previously existent activities, in a certain degree; it spreads its excitation over the whole activities of the mind, as the agitation of one place of a sheet of water expands itself, in wider and wider circles, over the whole surface of the fluid,[α] although, in proportion to its eccentricity, it is always becoming fainter, until it is at last not to be perceived. The force of every internal activity exists only in a certain limited degree; consequently, the excitation it determines has only likewise a certain limited power of expansion, and is continually losing in vigour in proportion to its eccentricity. Thus there are formed particular centres, particular spheres, of internal unity, within which the activities stand to each other in a closer relation of action and reaction; and this, in proportion as they more or less belong already to a single energy,—in proportion as they gravitate more or less proximately to the same centre of action. A plurality, a complement, of several activities forms, in a stricter sense, one whole activity for itself; an invigoration of any of its several activities is, therefore, an invigoration of the part of a whole activity; and as a part cannot be active for itself alone, there, consequently, results an invigoration of the whole, that is, of all the other parts of which it is composed. Thus the supreme law of association,—that activities excite each other in proportion as they have previously belonged, as parts, to one whole activity,—is explained from the still more universal principle of the unity of all our mental energies in general.[β]

"But, on the same principle we can also explain

α Cf. Pope, *Essay on Man*, iv. 363. —Ed.

β [Cf. Fries, *Anthropologie*, i. 29, § 6; *Kritik*, i. § 33.]

LECT. XXXII.

The laws of Simultaneity and Affinity, explicable on the same principle.

the two subaltern laws of Simultaneity and Affinity. The phænomena of mind are manifested under a twofold condition or form; for they are only revealed, 1°, As occurrences in time; and, 2°, As the energies or modifications of the ego, as their cause and subject. Time and self are thus the two forms of the internal world. By these two forms, therefore, every particular, every limited, unity of operation, must be controlled;—on them it must depend. And it is precisely these two forms that lie at the root of the two laws of Simultaneity and Affinity. Thus acts which are exerted at the same time, belong, by that very circumstance, to the same particular unity,—to the same definite sphere of mental energy; in other words, constitute through their simultaneity a single activity. Thus energies, however heterogeneous in themselves, if developed at once, belong to the same activity,—constitute a particular unity; and they will operate with a greater suggestive influence on each other, in proportion as they are more closely connected by the bond of time. On the other hand, the affinity of mental acts or modifications will be determined by their particular relations to the ego, as their cause or subject. As all the activities of mind obtain a unity in being all the energies of the same soul or active principle in general, so they are bound up into particular unities, inasmuch as they belong to some particular faculty,—resemble each other in the common ground of their manifestation. Thus cognitions, feelings, and volitions, severally awaken cognitions, feelings, and volitions; for they severally belong to the same faculty, and, through that identity, are themselves constituted into distinct unities: or again, a thought of the cause suggests a thought of the effect, a thought of the mean suggests a thought of the end, a thought of the part suggests a thought of the whole; for cause and effect, end and mean, whole and parts, have subjectively an indissoluble affinity, as they are all so many necessary forms or organisations of thought. In like manner, the notions of all resembling objects suggest each other, for they possess some common quality, through which they are in thought bound up in a single act of thought. Even the notions of opposite and contrasted objects mutually excite each other upon the same principle; for these are logically associated, inasmuch as, by the laws of thought, the notion of one opposite necessarily involves the notion of the other; and it is also a psychological law, that contrasted objects relieve each other. *Opposita, juxta posita, se invicem collustrant.* When the operations of different faculties are mutually suggestive, they are, likewise, internally connected by the nature of their action; for they are either conversant with the same object, and have thus been originally determined by the same affection from without, or they have originally been associated through some form of the mind itself: thus moral cognitions, moral feelings, and moral volitions, may suggest each other, through the common bond of morality; the moral principle in this case uniting the operations of the three fundamental powers into one general activity."[a]

Before leaving this subject, I must call your attention to a circumstance which I formerly incident-

[a] H. Schmid, *Versuch einer Metaph.* p. 242-4; [translated with occasional brief interpolations. Cf. *Reid's Works*, Notes D** and D***.—Ed.]

LECT. XXXII.

Thoughts, apparently unassociated, seem to follow each other immediately.

ally noticed.* It sometimes happens that thoughts seem to follow each other immediately, between which it is impossible to detect any bond of association. If this anomaly be insoluble, the whole theory of association is overthrown. Philosophers have accordingly set themselves to account for this phænomenon. To deny the fact of the phænomenon is impossible; it must, therefore, be explained on the hypothesis of association. Now, in their attempts at such an explanation, all philosophers agree in regard to the first step of the solution, but they differ in regard to the second. They agree in this,—that, admitting the apparent, the phænomenal, immediacy of the consecution of the two unassociated thoughts, they deny its reality. They all affirm, that there have actually intervened one or more thoughts, through the mediation of which, the suggestion in question has been effected, and on the assumption of which intermediation the theory of association remains intact. For example, let us suppose that A and C are thoughts, not on any law of association suggestive of each other, and that A and C appear to our consciousness as following each other immediately. In this case, I say, philosophers agree in supposing, that a thought B, associated with A and with C, and which consequently could be awakened by A, and could awaken C, has intervened. So far they are at one. But now comes their separation. It is asked, how can a thought be supposed to intervene, of which consciousness gives us no indication? In reply to this, two answers have been made. By one set of philosophers, among whom I may particularly specify Mr Stewart, it is said, that the immediate thought B, having been awakened by A, did rise into

Two modes of explication adopted by philosophers.

* See above, vol. I. lect. xviii. p. 351.—Ed.

consciousness, suggested C, and was instantly forgotten. This solution is apparently that exclusively known in Britain. Other philosophers, following the indication of Leibnitz, by whom the theory of obscure or latent activities was first explicitly promulgated, maintain that the intermediate thought never did rise into consciousness. They hold that A excited B, but that the excitement was not strong enough to rouse B from its state of latency, though strong enough to enable it obscurely to excite C, whose latency was less, and to afford it vivacity sufficient to rise into consciousness.

To be explained on the principle of latent modifications of mind.

Of these opinions, I have no hesitation in declaring for the latter. I formerly showed you an analysis of some of the most palpable and familiar phænomena of mind, which made the supposition of mental modifications latent, but not inert, one of absolute necessity. In particular, I proved this in regard to the phænomena of Perception.[a] But the fact of such latencies being established in one faculty, they afford an easy and philosophical explanation of the phænomena in all. In the present instance, if we admit, as admit we must, that activities can endure, and consequently can operate out of consciousness, the question is at once solved. On this doctrine, the whole theory of association obtains an easy and natural completion; as no definite line can be drawn between clear and obscure activities, which melt insensibly into each other; and both, being of the same nature, must be supposed to operate under the same laws. In illustration of the mediatory agency of latent thoughts in the process of suggestion, I formerly alluded to an analogous phænomenon under the laws of physical

a See above, vol. I. lect. xviii. p. 349.—Ed.

LECT. XXXII.

motion, which I may again recall to your remembrance. If a series of elastic balls, say of ivory, are placed in a straight line, and in mutual contact, and if the first be sharply struck, what happens? The intermediate balls remain at rest; the last alone is moved.

The counter solution untenable.

The other doctrine, which proceeds upon the hypothesis that we can be conscious of a thought and that thought be instantly forgotten, has everything against it, and nothing in its favour. In the first place, it does not, like the counter hypothesis of latent agencies, only apply a principle which is already proved to exist; it, on the contrary, lays its foundation in a fact which is not shown to be real. But in the second place, this fact is not only not shown to be real: it is improbable,—nay impossible; for it contradicts the whole analogy of the intellectual phænomena. The memory or retention of a thought is in proportion to its vivacity in consciousness; but that all trace of its existence so completely perished with its presence, that reproduction became impossible, even the instant after,—this assumption violates every probability, in gratuitously disallowing the established law of the proportion between consciousness and memory. But on this subject, having formerly spoken, it is needless now again to dwell.[a]

So much for the laws of Association,—the laws to which the faculty of Reproduction is subjected.

This faculty, I formerly mentioned, might be considered as operating, either spontaneously, without any interference of the will, or as modified in its action by the intervention of volition. In the one case, as in the other, the Reproductive Faculty acts in sub-

a See above, vol. I. lect. xviii. p. 353.—Ed.

LECT. XXXII.

servience to its own laws. In the former case, one thought is allowed to suggest another according to the greater general connection subsisting between them; in the latter, the act of volition, by concentrating attention upon a certain determinate class of associating circumstances, bestows on these circumstances an extraordinary vivacity, and, consequently, enables them to obtain the preponderance, and exclusively to determine the succession of the intellectual train. The former of these cases, where the Reproductive Faculty is left wholly to itself, may not improperly be called Spontaneous Suggestion, or Suggestion simply; the latter ought to obtain the name of Reminiscence or Recollection, (in Greek ἀνάμνησις.) The employment of these terms in these significations, corresponds with the meaning they obtain in common usage. Philosophers have not, however, always so applied them. But as I have not entered on a criticism of the analyses attempted by philosophers of the faculties, so I shall say nothing in illustration of their perversion of the terms by which they have denoted them.

The Reproductive Faculty divided into two: Spontaneous Suggestion, and Reminiscence.

Recollection or Reminiscence supposes two things. "First, it is necessary that the mind recognise the identity of two representations, and then it is necessary that the mind be conscious of something different from the first impression, in consequence of which it affirms to itself that it had formerly experienced this modification. It is passing marvellous, this conviction that we have of the identity of two representations; for they are only similar, not the same. Were they the same, it would be impossible to discriminate the thought reproduced from the thought originally

What Reminiscence involves.

LECT. XXXII.

St Augustin's analysis of this power,—detailed.

experienced."[α] This circumstance justly excited the admiration of St Augustin, and he asks how, if we had actually forgotten a thing, we could so categorically affirm,—it is not that, when some one named to us another; or, it is that, when it is itself presented. The question was worthy of his subtlety, and the answer does honour to his penetration. His principle is, that we cannot seek in our own memory for that of which we have no sort of recollection, "Quod omnino obliti fueramus amissum quærere non possumus."[β] We do not seek what has been our first reflective thought in infancy, the first reasoning we have performed, the first free act which raised us above the rank of automata. We are conscious that the attempt would be fruitless; and even if modifications thus lost should chance to recur to our mind, we should not be able to say with truth that we had recollected them, for we should have no criterion by which to recognise them, "Cujus nisi memor essem, si offeretur mihi, non invenirem, quia non agnoscerem." And what is the consequence he deduces? It is worthy of your attention.

Its conditions,—the law of totality.

From the moment, then, that we seek aught in our memory, we declare, by that very act, that we have not altogether forgotten it; we still hold of it, as it were, a part, and by this part, which we hold, we seek that which we do not hold, "Ergo non totum exciderat; sed ex parte qua tenebatur, alia quærebatur." And what is the secret motive which determines us to this research? It is that our memory feels, that it does not see together all that it was accustomed to see together, "Quia sentiebat se memoria non simul vol-

α Ancillon, *Essais Philosophiques*, ii. pp. 141-142.—Ed. [Cf. André, *Traité de l'Homme*, i. 277.]

β *Confessiones*, lib. x. cc. 18, 19.

vere quæ simul solebat." It feels with regret that it still only discovers a part of itself, and hence its disquietude to seek out what is missing, in order to reannex it to the whole; like to those reptiles, if the comparison may be permitted, whose members when cut asunder seek again to reunite, "Et quasi detruncata consuetudine claudicans, reddi quod deerat flagitabat." But when this detached portion of our memory at length presents itself,—the name, for example, of a person which had escaped us; how shall we proceed to reannex it to the other? We have only to allow nature to do her work. For if the name, being pronounced, goes of itself to reunite itself to the thought of the person, and to place itself, so to speak, upon his face, as upon its ordinary seat, we will say, without hesitation,—there it is. And if, on the contrary, it obstinately refuses to go there to place itself, in order to rejoin the thought to which we had else attached it, we will say peremptorily and at once,—no, it does not suit, "Non connectitur, quia non simul cum illo cogitari consuevit." But when it suits, where do we discover this luminous accordance which consummates our research? And where can we discover it, except in our memory itself,—in some back chamber, I mean, of that labyrinth where what we considered as lost had only gone astray, "Et unde adest, nisi ex ipsa memoria." And the proof of this is manifest. When the name presents itself to our mind, it appears neither novel nor strange, but old and familiar, like an ancient property of which we have recovered the title-deeds, "Non enim quasi novum credimus, sed recordantes approbamus."

Such is the doctrine of one of the profoundest thinkers of antiquity, and whose philosophical opinions, were they collected, arranged, and illustrated, would raise him to as high a rank among metaphysicians, as he already holds among theologians.

Defect in the analysis of Memory and Reproduction by psychologists,—in recognising only a consecutive order of association.

"Among psychologists, those who have written on Memory and Reproduction with the greatest detail and precision, have still failed in giving more than a meagre outline of these operations. They have taken account only of the notions which suggest each other, with a distinct and palpable notoriety. They have viewed the associations only in the order in which language is competent to express them; and as language, which renders them still more palpable and distinct, can only express them in a consecutive order,—can only express them one after another, they have been led to suppose that thoughts only awaken in succession. Thus, a series of ideas mutually associated, resembles, on the doctrine of philosophers, a chain in which every link draws up that which follows; and it is by means of these links that intelligence labours through, in the act of reminiscence, to the end which it proposes to attain.[a]

"There are some, indeed, among them, who are ready to acknowledge, that every actual circumstance is associated to several fundamental notions, and, consequently, to several chains, between which the mind may choose; they admit even that every link is attached to several others, so that the whole forms a kind of trellis,—a kind of net-work, which the mind may traverse in every direction, but still always in a single direction at once,—always in a succession similar to that of speech. This manner of explaining reminiscence is founded solely on this,—that, content

a Cf. *Reid's Works*, p. 906, note †.—Ed.

to have observed all that is distinctly manifest in the phænomenon, they have paid no attention to the under play of the latescent activities,—paid no attention to all that custom conceals, and conceals the more effectually in proportion as it is more completely blended with the natural agencies of mind.

Element in the phænomena, which the common theory fails to explain,—the movement of thought from one order of subjects to another.

"Thus their theory, true in itself, and departing from a well-established principle,—the Association of Ideas, explains in a satisfactory manner a portion of the phænomena of Reminiscence; but it is incomplete, for it is unable to account for the prompt, easy, and varied operation of this faculty, or for all the marvels it performs. On the doctrine of the philosophers, we can explain how a scholar repeats, without hesitation, a lesson he has learned, for all the words are associated in his mind according to the order in which he has studied them; how he demonstrates a geometrical theorem, the parts of which are connected together in the same manner; these and similar reminiscences of simple successions present no difficulties which the common doctrine cannot resolve. But it is impossible, on this doctrine, to explain the rapid and certain movement of thought, which, with a marvellous facility, passes from one order of subjects to another, only to return again to the first; which advances, retrogrades, deviates, and reverts, sometimes marking all the points on its route, again clearing, as if in play, immense intervals; which runs over now in a manifest order, now in a seeming irregularity, all the notions relative to an object, often relative to several, between which no connection could be suspected; and this without hesitation, without uncertainty, without error, as the hand of a skilful musician expatiates over

LECT. XXXII.

the keys of the most complex organ. All this is inexplicable on the meagre and contracted theory on which the phænomena of reproduction have been thought explained."α

Conditions under which Reminiscence is determined to exertion.

1. Momentary circumstances the causes of our mental activity.

2. The determining circumstance must constitute a want.

"To form a correct notion of the phænomena of Reminiscence, it is requisite, that we consider under what conditions it is determined to exertion. In the first place, it is to be noted that, at every crisis of our existence, momentary circumstances are the causes which awaken our activity, and set our recollection at work to supply the necessaries of thought.β In the second place, it is as constituting a want, (and by *want* I mean the result either of an act of desire or of volition), that the determining circumstance tends principally to awaken the thoughts with which it is associated. This being the case, we should expect, that each circumstance which constitutes a want, should suggest, likewise, the notion of the object, or objects, proper to satisfy it; and this is what actually happens. It is, however, further to be observed, that it is not enough that the want suggests the idea of the object; for if that idea were alone, it would remain without effect, since it could not guide me in the procedure I should follow. It is necessary, at the same time, that to the idea of this object there should be associated the notion of the relation of this object to the want, of the place where I may find it, of the means by which I may procure it, and turn it to account, &c. For instance, I wish to make a quotation:—This want awakens in me the idea of the author in whom the passage is to be found, which I

α Cardaillac, [*Études Élémentaires de Philosophie*, t. ii. c. v. p. 124 et seq.—Ed.]

β [Saepe jam spatio obrutam
Levis exoletam memoriam revocat nota.
Seneca, *Œdipus*, v. 820.]

am desirous of citing; but this idea would be fruitless, unless there were conjoined, at the same time, the representation of the volume, of the place where I may obtain it, of the means I must employ, &c.

Conditions under which a want is effective to determine reminiscence.

"Hence I infer, in the first place, that a want does not awaken an idea of its object alone, but that it awakens it accompanied with a number, more or less considerable, of accessory notions, which form, as it were, its train or attendance. This train may vary according to the nature of the want which suggests the notion of an object; but the train can never fall wholly off, and it becomes more indissolubly attached to the object, in proportion as it has been more frequently called up in attendance.

1. Awakens the idea of its object along with certain accessory notions.

"I infer, in the second place, that this accompaniment of accessory notions, simultaneously suggested with the principal idea, is far from being as vividly and distinctly represented in consciousness as that idea itself; and when these accessories have once been completely blended with the habits of the mind, and its reproductive agency, they at length finally disappear, becoming fused, as it were, in the consciousness of the idea to which they are attached. Experience proves this double effect of the habits of reminiscence. If we observe our operations relative to the gratification of a want, we shall perceive that we are far from having a clear consciousness of the accessory notions; the consciousness of them is, as it were, obscured, and yet we cannot doubt that they are present to the mind, for it is they that direct our procedure in all its details.

2. These accessory notions less vividly presented in consciousness than the idea itself.

"We must, therefore, I think, admit that the thought of an object immediately suggested by a desire, is always accompanied by an escort more or less nume-

LECT. XXXII.

The accessory notions, the more influential on our conduct, as they are further withdrawn from consciousness.

rous of accessory thoughts, equally present to the mind, though, in general, unknown in themselves to consciousness; that these accessories are not without their influence in guiding the operations elicited by the principal notion; and, it may even be added, that they are so much the more calculated to exert an effect in the conduct of our procedure, in proportion as, having become more part and parcel of our habits of reproduction, the influences they exert are further withdrawn, in ordinary, from the ken of consciousness."[a]

Illustrated by the case of reading.

The same thing may be illustrated by what happens to us in the case of reading. Originally each word, each letter, was a separate object of consciousness. At length, the knowledge of letters and words and lines being, as it were, fused into our habits, we no longer have any distinct consciousness of them, as severally concurring to the result, of which alone we are conscious. But that each word and letter has its effect,—an effect which can at any moment become an object of consciousness,—is shown by the following experiment. If we look over a book for the occurrence of a particular name or word, we glance our eye over a page from top to bottom, and ascertain, almost in a moment, that it is or is not to be found therein. Here the mind is hardly conscious of a single word, but that of which it is in quest; but yet it is evident, that each other word and letter must have produced an obscure effect, and which effect the mind was ready to discriminate and strengthen, so as to call it into clear consciousness, whenever the effect was found to be that which the letters of the word sought for could determine. But, if the mind be not unaffected by the multitude of letters and words which it surveys,

[a] Cardaillac, [*Études Élément. de Philos.*, t. ii. c. v. p. 128 *et seq.*—ED.]

if it be able to ascertain whether the combination of letters constituting the word it seeks, be or be not actually among them, and all this without any distinct consciousness of all it tries and finds defective;—why may we not suppose,—why are we not bound to suppose, that the mind may, in like manner, overlook its book of memory, and search among its magazines of latescent cognitions for the notions of which it is in want, awakening these into consciousness, and allowing the others to remain in their obscurity?

Grounds for inferring that we have not yet compassed the faculty of Reminiscence in its whole extent.

"A more attentive consideration of the subject will show, that we have not yet divined the faculty of Reminiscence in its whole extent. Let us make a single reflection. Continually struck by relations of every kind, continually assailed by a crowd of perceptions and sensations of every variety, and, at the same time, occupied with a complement of thoughts; we experience at once, and we are more or less distinctly conscious of, a considerable number of wants,—wants, sometimes real, sometimes factitious or imaginary,—phænomena, however, all stamped with the same characters, and all stimulating us to act with more or less of energy. And as we choose among the different wants which we would satisfy, as well as among the different means of satisfying that want which we determine to prefer; and as the motives of this preference are taken either from among the principal ideas relative to each of these several wants, or from among the accessory ideas which habit has established into their necessary escorts;—in all these cases it is requisite, that all the circumstances should at once, and from the moment they have taken the character of wants, produce an effect, correspondent to that which, we have seen, is caused by each in par-

LECT. XXXII.

ticular. Hence we are compelled to conclude, that the complement of the circumstances by which we are thus affected, has the effect of rendering always present to us, and, consequently, of placing at our disposal, an immense number of thoughts; some of which certainly are distinctly recognised, being accompanied by a vivid consciousness, but the greater number of which, although remaining latent, are not the less effective in continually exercising their peculiar influence on our modes of judging and acting.*

"We might say, that each of these momentary circumstances is a kind of electric shock which is communicated to a certain portion,—to a certain limited sphere, of intelligence; and the sum of all these circumstances is equal to so many shocks which, given at once at so many different points, produce a general agitation. We may form some rude conception of this phænomenon by an analogy. We may compare it, in the former case, to those concentric circles which are presented to our observation on a smooth sheet of water, when its surface is agitated by throwing in a pebble; and, in the latter case, to the same surface when agitated by a number of pebbles thrown simultaneously at different points.

This farther shown from the relations of our thoughts among themselves, and with the determining circumstances of the moment.

"To obtain a clearer notion of this phænomenon, I may add some observations on the relation of our thoughts among themselves, and with the determining circumstances of the moment.

"1°, Among the thoughts, notions, or ideas which belong to the different groups, attached to the principal representations simultaneously awakened, there are

* [Cf. Wolf, *Psychologia Rationalis*, §§ 96, 97. Maynettus Maynetius, *In Arist. De Sensu et Sensili*, particula 78, pp. 155, 156, (Florence, 1555), and Simon Simonius, *ibid.*, p. 257.]

some reciprocally connected by relations proper to themselves; so that, in this whole complement of co-existent activities, these tend to excite each other to higher vigour, and, consequently, to obtain for themselves a kind of pre-eminence in the group or particular circle of activity to which they belong.

"2°, There are thoughts associated, whether as principals or accessories, to a greater number of determining circumstances, or to circumstances which recur more frequently. Hence they present themselves oftener than the others, they enter more completely into our habits, and take, in a more absolute manner, the character of customary or habitual notions. It hence results, that they are less obtrusive, though more energetic, in their influence, enacting, as they do, a principal part in almost all our deliberations; and exercising a stronger influence on our determinations.

"3°, Among this great crowd of thoughts, simultaneously excited, those which are connected with circumstances which more vividly affect us, assume not only the ascendant over others of the same description with themselves, but likewise predominate over all those which are dependent on circumstances of a feebler determining influence.

"From these three considerations we ought, therefore, to infer, that the thoughts connected with circumstances on which our attention is more specially concentrated, are those which prevail over the others; for the effect of attention is to render dominant and exclusive the object on which it is directed, and during the moment of attention, it is the circumstance to which we attend that necessarily obtains the ascendant.

LECT. XXXII.

General conclusions. Thoughts awakened not only in succession, but simultaneously.

"Thus, if we appreciate correctly the phænomena of Reproduction or Reminiscence, we shall recognise, as an incontestable fact, that our thoughts suggest each other, not one by one successively, as the order to which language is astricted might lead us to infer; but that the complement of circumstances under which we at every moment exist, awakens simultaneously a great number of thoughts; these it calls into the presence of the mind, either to place them at our disposal, if we find it requisite to employ them, or to make them co-operate in our deliberations by giving them, according to their nature and our habits, an influence, more or less active, on our judgments and consequent acts.

Of these notions only few the objects of clear consciousness.

"It is also to be observed, that in this great crowd of thoughts always present to the mind, there is only a small number of which we are distinctly conscious: and that in this small number we ought to distinguish those which, being clothed in language, oral or mental, become the objects of a more fixed attention; those which hold a closer relation to circumstances more impressive than others; or which receive a predominant character by the more vigorous attention we bestow on them. As to the others, although not the objects of clear consciousness, they are nevertheless present to the mind, there to perform a very important part as motive principles of determination; and the influence which they exert in this capacity is even the more powerful in proportion as it is less apparent, being more disguised by habit."[a]

[a] Cardaillac, [*Etudes Elément. de Philos.*, t. ii. c. v. p. 184 *et seq.*—Ed.]

LECTURE XXXIII.

THE REPRESENTATIVE FACULTY.—IMAGINATION.

LECT. XXXIII. Recapitulation.

In my last Lecture, I concluded the special consideration of the elementary process of calling up or resuscitating out of unconsciousness the mental modifications which the mind, by its Retentive Faculty, preserves from absolute extinction;—the process to which I gave the not unexceptionable name of the Reproductive, and which, as left to its spontaneous action, or as modified by the will, obtains the several denominations of Suggestion, or of Reminiscence. In the latter part of the Lecture, I was engaged in showing that the common doctrine in regard to Reproduction is altogether inadequate to the phænomena,—that it allows to the mind only the power of reproducing the minima of thought in succession, as in speech it can only enunciate these one after another; whereas, in the process of Suggestion and Reminiscence, thoughts are awakened simultaneously in multitudes, in so far as to be brought into the immediate presence of the mind; in other words, they all, like the letters of a writing which we glance over, produce their effect, but those only upon which the mind concentrates its attention are drawn out into the light and foreground of consciousness.

Having thus terminated the separate consideration of the two first of the three correlative processes of

LECT. XXXIII.

Retention, Reproduction, and Representation, I proceed to the special discussion of the last,—the Representative Faculty.

The Faculty of Representation,—what.

By the faculty of Representation, as I formerly mentioned, I mean strictly the power the mind has of holding up vividly before itself the thoughts which, by the act of Reproduction, it has recalled into consciousness. Though the processes of Representation and Reproduction cannot exist independently of each other, they are nevertheless not more to be confounded into one than those of Reproduction and Conservation. They are, indeed, discriminated by differences sufficiently decisive. Reproduction, as we have seen, operates, in part at least, out of consciousness. Representation, on the contrary, is only realised as it is realised in consciousness; the degree or vivacity of the representation being always in proportion to the degree or vivacity of our consciousness of its reality. Nor are the energies of Representation and Reproduction always exerted by the same individual in equal intensity, any more than the energies of Reproduction and Retention. Some minds are distinguished for a higher power of manifesting one of these phænomena; others, for manifesting another; and as it is not always the person who forgets nothing, who can most promptly recall what he retains, so neither is it always the person who recollects most easily and correctly, who can exhibit what he remembers in the most vivid colours. It is to be recollected, however, that Retention, Reproduction, and Representation, though not in different persons of the same relative vigour, are, however, in the same individuals, all strong or weak in reference to the same classes of objects. For example, if a man's memory be more peculiarly retentive of words, his verbal reminiscence and ima-

Representation and Reproduction not always exerted by the same individual in equal intensity; but all strong or weak in the same individuals in reference to the same classes of objects.

gination will, in like manner, be more particularly energetic.

I formerly observed, that philosophers not having carried their psychological analysis so far as the constituent or elementary processes, the faculties in their systems are only precarious unions of these processes, in binary or even trinary combination,—unions, consequently, in which hardly any two philosophers are at one. In common language, it is not of course to be expected that there should be found terms to express the result of an analysis, which had not even been performed by philosophers; and, accordingly, the term *Imagination* or *Phantasy*, which denotes most nearly the representative process, does this, however, not without an admixture of other processes, which it is of consequence for scientific precision that we should consider apart.

Philosophers have divided Imagination into Reproductive, (Conception), and Productive.

Philosophers have divided Imagination into two,—what they call the Reproductive and the Productive. By the former, they mean imagination considered as simply re-exhibiting, representing the objects presented by perception, that is, exhibiting them without addition, or retrenchment, or any change in the relations which they reciprocally held, when first made known to us through sense. This operation Mr Stewart[a] has discriminated as a separate faculty, and bestowed on it the name of Conception. This discrimination and nomenclature, I think unfortunate. The discrimination is unfortunate, because it is unphilosophical to distinguish, as a separate faculty, what is evidently only a special application of a common power. The nomenclature is unfortunate,

This discrimination unfortunate in itself, and in its nomenclature.

[a] *Elements*, vol. i. part i. c. 3. *Works*, vol. ii. p. 144. On Reid's use of the term Conception, see Sir W. Hamilton's Edition of his *Works*, p. 360, note †, and p. 407, note ‡.—Ed.

LECT. XXXIII.

for the term *Conception*, which means a taking up in bundles, or grasping into unity,—this term, I say, ought to have been left to denote, what it previously was, and only properly could be, applied to express, —the notions we have of classes of objects, in other words, what have been called our *general ideas*. Be this, however, as it may, it is evident, that the Reproductive Imagination, (or Conception, in the abusive language of the Scottish philosophers), is not a simple faculty. It comprises two processes:—first, an act of representation strictly so called; and, secondly, an act of reproduction, arbitrarily limited by certain contingent circumstances; and it is from the arbitrary limitation of this second constituent, that the faculty obtains the only title it can exhibit to an independent existence. Nor can the Productive Imagination establish a better claim to the distinction of a separate faculty than the Reproductive. The Productive or Creative Imagination is that which is usually signified by the term *Imagination* or *Fancy*, in ordinary language. Now, in the first place, it is to be observed, that the terms *productive* or *creative* are very improperly applied to Imagination, or the Representative Faculty of mind. It is admitted on all hands, that Imagination creates nothing, that is, produces nothing new; and the terms in question are, therefore, by the acknowledgment of those who employ them, only abusively applied to denote the operations of Fancy, in the new arrangement it makes of the old objects furnished to it by the senses. We have now, therefore, only to consider, whether, in this corrected meaning, Imagination, as a plastic energy, be a simple or a complex operation. And that it is a complex operation, I do not think it will be at all difficult to prove.

Imagination, as a plastic energy, is a complex operation.

In the view I take of the fundamental processes, the act of representation is merely the energy of the mind in holding up to its own contemplation what it is determined to represent. I distinguish, as essentially different, the representation, and the determination to represent. I exclude from the faculty of Representation all power of preference among the objects it holds up to view. This is the function of faculties wholly different from that of Representation, which, though active in representing, is wholly passive as to what it represents.

The act of representation,—what.

What, then, it may be asked, are the powers by which the Representative Faculty is determined to represent, and to represent this particular object, or this particular complement of objects, and not any other? These are two. The first of these is the Reproductive Faculty. This faculty is the great immediate source from which the Representative receives both the materials and the determination to represent; and the laws by which the Reproductive Faculty is governed, govern also the Representative. Accordingly, if there were no other laws in the arrangement and combination of thought than those of association, the Representative Faculty would be determined in its manifestations, and in the character of its manifestations, by the Reproductive Faculty alone; and, on this supposition, representation could no more be distinguished from reproduction than reproduction from association.

Two powers by which the Representative Faculty is determined to energy.

1. The Reproductive Faculty.

But there is another elementary process which we have not yet considered,—Comparison, or the Faculty of Relations, to which the representative act is likewise subject, and which plays a conspicuous part in determining in what combinations objects are repre-

2. The Faculty of Relations.

LECT. XXXIII.

sented. By the process of Comparison, the complex objects,—the congeries of phænomena called up by the Reproductive Faculty, undergo various operations. They are separated into parts, they are analysed into elements; and these parts and elements are again compounded in every various fashion. In all this the Representative Faculty co-operates. It, first of all, exhibits the phænomena as called up by the laws of ordinary association. In this it acts as handmaid to the Reproductive Faculty. It then exhibits the phænomena as variously elaborated by the analysis and synthesis of the Comparative Faculty, to which, in like manner, it performs the part of a subsidiary.

The Imagination of common language is equivalent to the processes of Representation and Comparison.

This being understood, you will easily perceive, that the Imagination of common language,—the Productive Imagination of philosophers,—is nothing but the Representative process *plus* the process to which I would give the name of the *Comparative*. In this compound operation, it is true that the representative act is the most conspicuous, perhaps the most essential, element. For, in the first place, it is a condition of the possibility of the act of comparison,—of the act of analytic synthesis,—that the material on which it operates, (that is, the objects reproduced in their natural connections), should be held up to its observation in a clear light, in order that it may take note of their various circumstances of relation; and, in the second, that the result of its own elaboration, that is, the new arrangements which it proposes, should be realised in a vivid act of representation. Thus it is, that, in the view both of the vulgar and of philosophers, the more obtrusive, though really the more subordinate, element in this compound process has been elevated into the principal constituent; whereas, the act of comparison,—the act of separation and reconstruction, has been regarded as identical with the act of representation.

The process of Representation the principal constituent of Imagination, as commonly understood.

Thus Imagination, in the common acceptation of the term, is not a simple but a compound faculty,—a faculty, however, in which representation,—the vivid exhibition of an object,—forms the principal constituent. If, therefore, we were obliged to find a common word for every elementary process of our analysis,—*Imagination* would be the term, which, with the least violence to its meaning, could be accommodated to express the Representative Faculty.

Imagination not limited to objects of sense.

By Imagination, thus limited, you are not to suppose that the faculty of representing mere objects of sense alone is meant. On the contrary, a vigorous power of representation is as indispensable a condition of success in the abstract sciences, as in the poetical and plastic arts; and it may, accordingly, be reasonably doubted whether Aristotle or Homer were possessed of the more powerful imagination. "We may, indeed, affirm, that there are as many different kinds of imagination as there are different kinds of intellectual activity. There is the imagination of abstraction, which represents to us certain phases of an object to the exclusion of others, and, at the same time, the sign by which the phases are united; the imagination of wit, which represents differences and contrasts, and the resemblances by which these are again combined; the imagination of judgment, which represents the various qualities of an object, and binds them together under the relations of substance, of attribute, of mode; the imagination of reason, which represents a principle in connection with its consequences, the effect in dependence on its cause;

LECT. XXXIII. the imagination of feeling, which represents the accessory images, kindred to some particular sentiment, and which thereby confer on it greater compass, depth, and intensity; the imagination of volition, which represents all the circumstances which concur to persuade or dissuade from a certain act of will; the imagination of the passions, which, according to the nature of the affection, represents all that is homogeneous or analogous; finally, the imagination of the poet, which represents whatever is new, or beautiful, or sublime,—whatever, in a word, it is determined to represent by any interest of art."[a] The term *imagination*, however, is less generally applied to the representations of the Comparative Faculty considered in the abstract, than to the representations of sensible objects, concretely modified by comparison. The two kinds of imagination are in fact not frequently combined. Accordingly, using the term in this its ordinary extent, that is, in its limitation to objects of sense, it is finely said by Mr Hume: "Nothing is more dangerous to reason than the flights of imagination, and nothing has been the occasion of more mistakes among philosophers. Men of bright fancies may, in this respect, be compared to those angels whom the Scriptures represent as covering their eyes with their wings."[β]

Three principal orders in which Imagination represents ideas.

Considering the Representative Faculty in subordination to its two determinants, the faculty of Reproduction and the faculty of Comparison or Elaboration, we may distinguish three principal orders in which Imagination represents ideas:—"1°, The Natural order; 2°, The Logical order; 3°, The Poetical order. The natural order is that in which we receive the impression of external objects, or the order according to which our

1. The natural order.

a Ancillon, *Essais Philosophiques*, ii. 151.

β *Treatise of Human Nature*, book i. part iv. § 7.—Ed.

thoughts spontaneously group themselves. The logical order consists in presenting what is universal, prior to what is contained under it as particular, or in presenting the particulars first, and then ascending to the universal which they constitute. The former is the order of deduction, the latter that of induction. These two orders have this in common, that they deliver to us notions in the dependence in which the antecedent explains the subsequent. The poetical order consists in seizing individual circumstances, and in grouping them in such a manner that the imagination shall represent them so as they might be offered by the sense. The natural order is involuntary; it is established independently of our concurrence. The logical order is a child of art, it is the result of our will; but it is conformed to the laws of intelligence, which tend always to recall the particular to the general, or the general to the particular. The poetical order is exclusively calculated on effect. Pindar would not be a lyric poet, if his thoughts and images followed each other in the common order, or in the logical order. The state of mind in which thought and feeling clothe themselves in lyric forms, is a state in which thoughts and feelings are associated in an extraordinary manner—in which they have, in fact, no other relation than that which groups and moves them around the dominant thought or feeling which forms the subject of the ode.

LECT. XXXIII.

2. The logical order.

3. The poetical order.

"Thoughts which follow each other only in the natural order, or as they are associated in the minds of men in general, form tedious conversations and tiresome books. Thoughts, on the other hand, whose connection is singular, capricious, extraordinary, are unpleasing; whether it be that they strike us as improbable, or that the effort which has been required

Associations tedious, unpleasing, and agreeable.

LECT. XXXIII.

to produce, supposes a corresponding effort to comprehend. Thoughts whose association is at once simple and new, and which, though not previously witnessed in conjunction, are yet approximated without a violent exertion,—such thoughts please universally, by affording the mind the pleasures of novelty and exercise at once.

Peculiar kinds of imagination determined by peculiar orders of association.

"A peculiar kind of imagination, determined by a peculiar order of association, is usually found in every period of life, in every sex, in every country, in every religion. A knowledge of men principally consists in a knowledge of the principles by which their thoughts are linked and represented. The study of this is of importance to the instructor, in order to direct the character and intellect of his pupils; to the statesman, that he may exert his influence on the public opinion and manners of a people; to the poet, that he may give truth and reality to his dramatic situations; to the orator, in order to convince and persuade; to the man of the world, if he would give interest to his conversation.

Difference between a cultivated and a vulgar mind.

"Authors who have made a successful study of this subject skim over a multitude of circumstances under which an occurrence has taken place; because they are aware that it is proper to reject what is only accessory to the object which they would present in prominence. A vulgar mind forgets and spares nothing; he is ignorant that conversation is always but a selection; that every story is subject to the laws of dramatic poetry,—*festinat ad eventum*; and that all which does not concur to the effect destroys or weakens it. The involuntary associations of their thoughts are imperative on minds of this description; they are held in thraldom to the order and circumstances in which their perceptions were originally obtained."[α] This has not, of course, escaped the notice of the greatest observer of human nature. Mrs Quickly, in reminding Falstaff of his promise of marriage, supplies a good example of this peculiarity. "Thou didst swear to me upon a parcel-gilt goblet, sitting in my Dolphin chamber, at the round table, by a sea-coal fire, upon Wednesday in Whitsun week, when the prince broke thy head for likening his father to a singing man of Windsor,"—and so forth. In Martinus Scriblerus, the coachman thus describes a scene in the Bear Garden: "He saw two men fight a prize; one was a fair man, a sergeant in the guards; the other black, a butcher; the sergeant had red breeches, the butcher blue; they fought upon a stage, about four o'clock, and the sergeant wounded the butcher in the leg."

Dreaming an effect of imagination, determined by association.

"Dreaming, Somnambulism, Reverie, are so many effects of imagination, determined by association,—at least states of mind in which these have a decisive influence. If an impression on the sense often commences a dream, it is by imagination and suggestion that it is developed and accomplished. Dreams have frequently a degree of vivacity which enables them to compete with the reality; and if the events which they represent to us were in accordance with the circumstances of time and place in which we stand, it would be almost impossible to distinguish a vivid dream from a sensible perception."[β] "If," says Pascal,[γ] "we dreamt every night the same thing, it would perhaps affect us as powerfully as the objects which we perceive every day. And if an artisan were certain of dreaming every night for twelve hours that he was king, I am convinced that he would be almost as

α Ancillon, *Essais Philos.*, ii. 152-156.—Ed.

β Ancillon, *Ess. Phil.*, ii. 159.—Ed.

γ *Pensées*, partie i. art. vi. § 20. Vol. ii. p. 102, (edit. Faugère.)—Ed.

happy as a king, who dreamt for twelve hours that he was an artisan. If we dreamt every night that we were pursued by enemies and harassed by horrible phantoms, we should suffer almost as much as if that were true, and we should stand in as great dread of sleep, as we should of waking, had we real cause to apprehend these misfortunes. . . . . It is only because dreams are different and inconsistent, that we can say, when we awake, that we have dreamt; for life is a dream a little less inconstant." Now the case which Pascal here hypothetically supposes, has

Case of dreaming mentioned by Abel.

actually happened. In a very curious German work, by Abel, entitled *A Collection of Remarkable Phænomena from Human Life*,[a] I find the following case, which I abridge:—A young man had a cataleptic attack, in consequence of which a singular effect was operated in his mental constitution. Some six minutes after falling asleep, he began to speak distinctly, and almost always of the same objects and concatenated events, so that he carried on from night to night the same history, or rather continued to play the same part. On wakening, he had no reminiscence whatever of his dreaming thoughts,—a circumstance, by the way, which distinguishes this as rather a case of somnambulism than of common dreaming. Be this, however, as it may, he played a double part in his existence. By day he was the poor apprentice of a merchant; by night he was a married man, the father of a family, a senator, and in affluent circumstances. If during his vision anything was said in regard to his waking state, he declared it unreal and a dream. This case, which is established on the best evidence, is, as far as I am aware, unique.

[a] *Sammlung und Erklärung merkwürdiger Erscheinungen aus dem menschlichen Leben* (1784), ii. p. 124 et seq.—Ed.

The influence of dreams upon our character is not without its interest. A particular tendency may be strengthened in a man solely by the repeated action of dreams. Dreams do not, however, as is commonly supposed, afford any appreciable indication of the character of individuals. It is not always the subjects that occupy us most, when awake, that form the matter of our dreams; and it is curious that the persons the dearest to us are precisely those about whom we dream most rarely.

Somnambulism.

Somnambulism is a phænomenon still more astonishing. In this singular state, a person performs a regular series of rational actions, and those frequently of the most difficult and delicate nature, and, what is still more marvellous, with a talent to which he could make no pretension when awake.[a] His memory and reminiscence supply him with recollections of words and things, which perhaps were never at his disposal in the ordinary state; he speaks more fluently a more refined language; and, if we are to credit what the evidence on which it rests hardly allows us to disbelieve, he has not only perceptions through other channels than the common organs of sense, but the sphere of his cognitions is amplified to an extent far beyond the limits to which sensible perception is confined. This subject is one of the most perplexing in the whole compass of philosophy; for, on the one hand, the phænomena are so marvellous that they cannot be believed, and yet, on the other, they are of so unambiguous and palpable a character, and the witnesses to their reality are so numerous, so intelligent, and so high above every suspicion of deceit, that it is equally impossible to deny credit to what is attested by such ample and unexceptionable evidence.

a Cf. Ancillon, *Essais Philos.*, ii. 161.—Ed.

LECT. XXXIII.

Reverie.

"The third state, that of Reverie or Castle-building, is a kind of waking dream, and does not differ from dreaming, except by the consciousness which accompanies it. In this state, the mind abandons itself without a choice of subject, without control over the mental train, to the involuntary associations of imagination. The mind is thus occupied without being properly active; it is active, at least, without effort. Young persons, women, the old, the unemployed, and the idle, are all disposed to reverie. There is a pleasure attached to its illusions, which renders it as seductive as it is dangerous. The mind, by indulgence in this dissipation, becomes enervated, it acquires the habit of a pleasing idleness, loses its activity, and at length even the power and the desire of action."[α]

The happiness and misery of the individual dependent on the character of his habitual associations.

"The happiness and misery of every individual of mankind depends almost exclusively on the particular character of his habitual associations, and the relative kind and intensity of his imagination. It is much less what we actually are, and what we actually possess, than what we imagine ourselves to be and have, that is decisive of our existence and fortune."[β] Apicius committed suicide to avoid starvation, when his fortune was reduced to somewhere, in English money, about £100,000. The Roman epicure imagined that he could not subsist on what, to men in general, would seem more than affluence.

The influence of imagination on human life.

"Imagination, by the attractive or repulsive pictures with which, according to our habits and associations, it fills the frame of our life, lends to reality a magical charm, or despoils it of all its pleasantness. The imaginary happy and the imaginary miserable are

α Ancillon, *Essais Philos.*, ii. 162.—Ed.

β Ancillon, *Essais Philos.*, ii. 163, 164.—Ed.

common in the world, but their happiness and misery are not the less real; everything depends on the mode in which they feel and estimate their condition. Fear, hope, the recollection of past pleasures, the torments of absence and of desire, the secret and almost resistless tendency of the mind towards certain objects, are the effects of association and imagination. At a distance, things seem to us radiant with a celestial beauty, or in the lurid aspect of deformity. Of a truth, in either case we are equally wrong. When the event which we dread, or which we desire, takes place, when we obtain, or when there is forced upon us, an object environed with a thousand hopes, or with a thousand fears, we soon discover that we have expected too much or too little; we thought it by anticipation infinite in good or evil, and we find it in reality not only finite but contracted. 'With the exception,' says Rousseau,[α] 'of the self-existent Being, there is nothing beautiful, but that which is not.' In the crisis whether of enjoyment or suffering, happiness is not so much happiness, nor misery so much misery, as we had anticipated. In the past, thanks to a beneficent Creator, our joys reappear as purer and more brilliant than they had been actually experienced; and sorrow loses not only its bitterness, but is changed even into a source of pleasing recollection."[β] "Suavis laborum est præteritorum memoria," says Cicero;[γ] while "hæc olim meminisse juvabit,"[δ] is, in the words of Virgil, the consolation of a present infliction. "In early youth, the present and the future are displayed in a factitious magnificence; for at this

α *Nouvelle Héloise*, part vi. liv. viii. —Ed.

β Ancillon, *Ess. Phil.*, ii. 164-5.—Ed.

γ *De Finibus*, ii. 32, translated from Euripides, (quoted by Macrobius, *Sat.*, vii. 2):—Ὡς ἡδύ τοι σωθέντα μεμνῆσθαι πόνων.—Ed.

δ *Æneid*, i. 203.—Ed.

LECT. XXXIII.

period of life imagination is in its spring and freshness, and a cruel experience has not yet exorcised its brilliant enchantments. Hence the fair picture of a golden age, which all nations concur in placing in the past; it is the dream of the youth of mankind."[α] In old age, again, where the future is dark and short, imagination carries us back to the re-enjoyment of a past existence. "The young," says Aristotle,[β] "live forwards in hope, the old live backwards in memory;" as Martial has well expressed it,

> "Hoc est
> Vivere bis, vita posse priore frui."[γ]

From all this, however, it appears that the present is the only time in which we never actually live; we live either in the future, or in the past. So long as we have a future to anticipate, we contemn the present; and when we can no longer look forward to a future, we revert and spend our existence in the past. In the words of Manilius:

> "Victuros agimus semper, nec vivimus unquam."[δ]

In the words of Pope:

> "Man never is, but always to be blest."[ε]

Imagination employs the organs of sense in the representations of sensible objects.

I shall terminate the consideration of Imagination Proper by a speculation concerning the organ which it employs in the representations of sensible objects. The organ which it thus employs seems to be no other than the organs themselves of Sense, on which the original impressions were made, and through which they were originally perceived. Experience has shown, that Imagination depends on no one part of the cerebral apparatus exclusively. There is no portion of

α Ancillon, *Essais Philos.*, ii. 164.—Ed.

β *Rhet.*, ii. cc. 12, 13.—Ed.

γ Lib. x. epigr. 23.—Ed.

δ *Astronomicon*, iv. 4.—Ed.

ε *Essay on Man*, i. 95.—Ed.

the brain which has not been destroyed by mollification, or induration, or external lesion, without the general faculty of Representation being injured. But experience equally proves, that the intracranial portion of any external organ of sense cannot be destroyed, without a certain partial abolition of the Imagination Proper. For example, there are many cases recorded by medical observers, of persons losing their sight, who have also lost the faculty of representing the images of visible objects. They no longer call up such objects by reminiscence, they no longer dream of them. Now in these cases, it is found that not merely the external instrument of sight,—the eye, has been disorganised, but that the disorganisation has extended to those parts of the brain which constitute the internal instrument of this sense,—that is, the optic nerves and thalami. If the latter,—the real organ of vision, remain sound, the eye alone being destroyed, the imagination of colours and forms remains as vigorous as when vision was entire. Similar cases are recorded in regard to the deaf. These facts, added to the observation of the internal phænomena which take place during our acts of representation, make it, I think, more than probable that there are as many organs of Imagination as there are organs of Sense. Thus I have a distinct consciousness, that, in the internal representation of visible objects, the same organs are at work which operate in the external perception of these; and the same holds good in an imagination of the objects of Hearing, Touch, Taste, and Smell.

Voluntary motions imitated in and by the imagination.

But not only sensible perceptions, voluntary motions likewise are imitated in and by the imagination. I can, in imagination, represent the action of speech, the play of the muscles of the countenance, the movement

LECT. XXXIII. of the limbs; and when I do this, I feel clearly that I awaken a kind of tension in the same nerves through which, by an act of will, I can determine an overt and voluntary motion of the muscles; nay, when the play of imagination is very lively, this external movement is actually determined. Thus we frequently see the countenances of persons under the influence of imagination undergo various changes; they gesticulate with their hands, they talk to themselves, and all this is in consequence only of the imagined activity going out into real activity. I should, therefore, be disposed to conclude, that, as in Perception the living organs of sense are from without determined to energy, so in Imagination they are determined to a similar energy by an influence from within.

LECTURE XXXIV.

THE ELABORATIVE FACULTY.—CLASSIFICATION.—ABSTRACTION.

LECT. XXXIV.

The Elaborative Faculty,—what and how designated.

THE faculties with which we have been hitherto engaged, may be regarded as subsidiary to that which we are now about to consider. This, to which I gave the name of the Elaborative Faculty,—the Faculty of Relations,—or Comparison,—constitutes what is properly denominated, Thought. It supposes always at least two terms, and its act results in a judgment, that is, an affirmation or negation of one of these terms of the other. You will recollect that, when treating of Consciousness in general, I stated to you, that consciousness necessarily involves a judgment; and as every act of mind is an act of consciousness, every act of mind, consequently, involves a judgment.[a] A consciousness is necessarily the consciousness of a determinate something; and we cannot be conscious of anything without virtually affirming its existence, that is, judging it to be. Consciousness is thus primarily a judgment or affirmation of existence. Again, consciousness is not merely the affirmation of naked existence, but the affirmation of a certain qualified or determinate existence. We are conscious that we exist only in and through our consciousness that we exist in this or that particular state,—

Every act of mind involves a judgment.

a See above, vol. I. p. 201.—ED. [Cf. Aristotle, *De Motione Animalium*, c. vi. ['Η φαντασία καὶ ἡ αἴσθησις . . . κριτικά.—ED.] *Post An.*, ii. c. ult. Galien-Arnoult, *Programme*, pp. 31, 103, 105. Reid, *Int. Powers*, Ess. vi.] [c. I. *Works*, p. 414.—ED.]

LECT. XXXIV. that we are so or so affected,—so or so active; and we are only conscious of this or that particular state of existence, inasmuch as we discriminate it as different from some other state of existence, of which we have been previously conscious and are now reminiscent; but such a discrimination supposes, in consciousness, the affirmation of the existence of one state of a specific character, and the negation of another. On this ground it was that I maintained, that consciousness necessarily involves, besides recollection, or rather a certain continuity of representation, also judgment or comparison; and, consequently, that, so far from comparison or judgment being a process always subsequent to the acquisition of knowledge, through perception and self-consciousness, it is involved as a condition of the acquisitive process itself. In point of fact, the various processes of Acquisition (Apprehension), Representation, and Comparison, are all mutually dependent. Comparison cannot judge without something to compare; we cannot originally acquire,—apprehend, we cannot subsequently represent our knowledge, without in either act attributing existence, and a certain kind of existence, both to the object known and to the subject knowing, that is, without enouncing certain judgments and performing certain acts of comparison; I say without performing certain acts of comparison, for taking the mere affirmation that a thing is,—this is tantamount to a negation that it is not, and necessarily supposes a comparison,—a collation, between existence and non-existence.

Defect in the analysis of this faculty by philosophers.

What I have now said may perhaps contribute to prepare you for what I am hereafter to say of the faculty or elementary process of Comparison,—a faculty which, in the analysis of philosophers, is exhibited only in part; and even that part is not preserved in its integrity. They take into account only a fragment of the process, and that fragment they again break down into a plurality of faculties. In opposition to the views hitherto promulgated in regard to Comparison, I will show that this faculty is at work in every, the simplest, act of mind; and that, from the primary affirmation of existence in an original act of consciousness to the judgment contained in the conclusion of an act of reasoning, every operation is only an evolution of the same elementary process,—that there is a difference in the complexity, none in the nature, of the act; in short, that the various products of Analysis and Synthesis, of Abstraction and Generalisation, are all merely the results of Comparison, and that the operations of Conception or Simple Apprehension, of Judgment, and of Reasoning, are all only acts of Comparison, in various applications and degrees. LECT. XXXIV.

What I have, therefore, to prove is, in the first place, that Comparison is supposed in every, the simplest, act of knowledge; in the second, that our factitiously simple, our factitiously complex, our abstract, and our generalised notions, are all merely so many products of Comparison; in the third, that Judgment, and, in the fourth, that Reasoning, is identical with Comparison. In doing this, I shall not formally distribute the discussion into these heads, but shall include the proof of what I have now advanced, while tracing Comparison from its simplest to its most complex operations. Positions to be established.

The first or most elementary act of Comparison, or of that mental process in which the relation of two terms is recognised and affirmed, is the judgment virtually pronounced, in an act of Perception, of the non-ego, or, in an act of Self-consciousness, of the ego. This is the primary affirmation of existence. The notion of Comparison as determined by objective conditions.

LECT. XXXIV.

The first act.

existence is one native to the mind. It is the primary condition of thought. The first act of experience awoke it, and the first act of consciousness was a subsumption of that of which we were conscious under this notion; in other words, the first act of consciousness was an affirmation of the existence of something. The first or simplest act of comparison is thus the discrimination of existence from non-existence; and the first or simplest judgment is the affirmation of existence, in other words, the denial of non-existence.[a]

Second.

But the something of which we are conscious, and of which we predicate existence, in the primary judgment, is twofold,—the ego and the non-ego. We are conscious of both, and affirm existence of both. But we do more; we do not merely affirm the existence of each out of relation to the other, but, in affirming their existence, we affirm their existence in duality, in difference, in mutual contrast; that is, we not only affirm the ego to exist, but deny it existing as the non-ego; we not only affirm the non-ego to exist, but deny it existing as the ego. The second act of comparison is thus the discrimination of the ego and the non-ego; and the second judgment is the affirmation, that each is not the other.

Third.

The third gradation in the act of comparison, is in the recognition of the multiplicity of the coexistent or successive phænomena, presented either to Perception or Self-consciousness, and the judgment in regard to their resemblance or dissimilarity.

Fourth.

The fourth is the comparison of the phænomena with the native notion of Substance, and the judgment is the grouping of these phænomena into different

a [Cf. Troxler, *Logik*, ii. 20 *et seq.* Reinhold, *Theorie des menschlichen Erkenntniss-vermögens und Metaphysik*, i. 290. Beneke, *Psychologische Skizzen*, i. 227 *et seq.* Cousin, *Cours de l'Histoire de la Philosophie*, (xviii Siècle) leçons xxiii., xxiv. Garnier, *Cours de Psychologie*, p. 67.]

LECT. XXXIV.

bundles, as the attributes of different subjects. In the external world, this relation constitutes the distinction of things; in the internal, the distinction of powers.

Fifth.

The fifth act of comparison is the collation of successive phænomena under the native notion of Causality, and the affirmation or negation of their mutual relation as cause and effect.

Comparison viewed as determined by the necessities of the thinking subject.

So far the process of comparison is determined merely by objective conditions; hitherto it has followed only in the footsteps of nature. In those, again, we are now to consider, the procedure is, in a certain sort, artificial, and determined by the necessities of the thinking subject itself. The mind is finite in its powers of comprehension; the objects, on the contrary, which are presented to it are, in proportion to its limited capacities, infinite in number. How then is this disproportion to be equalised? How can the infinity of nature be brought down to the finitude of man? This is done by means of Classification. Objects, though infinite in number, are not infinite in variety; they are all, in a certain sort, repetitions of the same common qualities, and the mind, though lost in the multitude of particulars,—individuals, can easily grasp the classes into which their resembling attributes enable us to assort these. This whole process of Classification is a mere act of Comparison, as the following deduction will show.

Classification shown to be an act of Comparison.

1. In regard to Complex or Collective notions.

In the first place, this may be shown in regard to the formation of Complex notions, with which, as the simplest species of classification, we may commence. By Complex or Collective notions, I mean merely the notion of a class formed by the repetition of the same constituent notion.[a] Such are the notions of *an army*,

[a] Cf. Locke, *Essay on the Human Understanding*, book ii. c. xii. § 5. Degerando, *Des Signes*, t. i. c. vii. p. 170.—Ed.

LECT. XXXIV.

a forest, a town, a number. These are names of classes, formed by the repetition of the notion of *a soldier*, of *a tree*, of *a house*, of *a unit*. You are not to confound, as has sometimes been done, the notion of *an army, a forest, a town, a number*, with the notions of *army, forest, town*, and *number*; the former, as I have said, are complex or collective, the latter are general or universal notions.

It is evident that a collective notion is the result of comparison. The repetition of the same constituent notion supposes that these notions were compared, their identity or absolute similarity affirmed.

In this, the simplest act of Classification, the mind is dependent on language.

In the whole process of classification, the mind is in a great measure dependent upon language for its success; and in this, the simplest of the acts of classification, it may be proper to show how language affords to mind the assistance it requires. Our complex notions being formed by the repetition of the same notion, it is evident that the difficulty we can experience in forming an adequate conception of a class of identical constituents, will be determined by the difficulty we have in conceiving a multitude. "But the comprehension of the mind is feeble and limited; it can embrace at once but a small number of objects. It would thus seem that an obstacle is raised to the extension of our complex ideas at the very outset of our combinations. But here language interposes, and supplies the mind with the force of which it is naturally destitute."[α] We have formerly seen that the mind cannot in one act embrace more than five or six, at the utmost seven, several units.[β] How then does it proceed? "When, by a first combination, we have obtained a complement of notions

[α] Degerando, *Des Signes*, t. i. c. vii. p. 165.

[β] See above, vol. i. lect. xiv. p. 254. —ED.

as complex as the mind can embrace, we give this complement a name. This being done, we regard the assemblage of units thus bound up under a collective name as itself a unit, and proceed, by a second combination, to accumulate these into a new complement of the same extent. To this new complement we give another name; and then again proceed to perform, on this more complex unit, the same operation we had performed on the first; and so we may go on rising from complement to complement to an indefinite extent. Thus, a merchant, having received a large unknown sum of money in crowns, counts out the pieces by fives, and having done this till he has reached twenty, he lays them together in a heap; around these, he assembles similar piles of coin, till they amount, let us say, to twenty; and he then puts the whole four hundred into a bag. In this manner he proceeds until he fills a number of bags, and placing the whole in his coffers, he will have a complex or collective notion of the quantity of crowns which he has received."[a] It is on this principle that arithmetic proceeds,—tens, hundreds, thousands, myriads, hundreds of thousands, millions, &c., are all so many factitious units which enable us to form notions, vague indeed, of what otherwise we could have obtained no conception at all. So much for complex or collective notions, formed without decomposition, —a process which I now go on to consider.

Decomposition twofold. 1. In the interests of the Fine Arts.

Our thought,—that is, the sum total of the perceptions and representations which occupy us at any given moment, is always, as I have frequently observed, compound. The composite objects of thoughts may be decomposed in two ways, and for the sake of two different interests. In the first place, we may

[a] Degerando, *Des Signes*, t. i. c. vii. p. 165, [slightly abridged.—Ed.]

LECT. XXXIV. decompose in order that we may recombine, influenced by the mere pleasure which this plastic operation affords us. This is poetical analysis and synthesis. On this process it is needless to dwell. It is evidently the work of comparison. For example, the minotaur, or chimæra, or centaur, or gryphon (hippogryph), or any other poetical combination of different animals, could only have been effected by an act in which the representations of these animals were compared, and in which certain parts of one were affirmed, compatible with certain parts of another. How, again, is the imagination of all ideal beauty or perfection formed? Simply by comparing the various beauties or excellencies of which we have had actual experience, and thus being enabled to pronounce in regard to their common and essential quality.

2. In the interest of Science.

In the second place, we may decompose in the interest of science; and as the poetical decomposition was principally accomplished by a separation of integral parts, so this is principally accomplished by an abstraction of constituent qualities. On this process it is necessary to be more particular.

Abstraction of the senses.

Suppose an unknown body is presented to my senses, and that it is capable of affecting each of these in a certain manner. "As furnished with five different organs, each of which serves to introduce a certain class of perceptions and representations into the mind, we naturally distribute all sensible objects into five species of qualities. The human body, if we may so speak, is thus itself a kind of abstractive machine. The senses cannot but abstract. If the eye did not abstract colours, it would see them confounded with odours and with tastes, and odours and tastes would necessarily become objects of sight."

"The abstraction of the senses is thus an operation

the most natural; it is even impossible for us not to perform it. Let us now see whether abstraction by the mind be more arduous than that of the senses."[α] We have formerly found that the comprehension of the mind is extremely limited; that it can only take cognisance of one object at a time, if that be known with full intensity; and that it can accord a simultaneous attention to a very small plurality of objects, and even that imperfectly. Thus it is that attention fixed on one object is tantamount to a withdrawal,—to an abstraction, of consciousness from every other. Abstraction is thus not a positive act of mind, as it is often erroneously described in philosophical treatises,—it is merely a negation to one or more objects, in consequence of its concentration on another.

Abstraction,—what.

This being the case, Abstraction is not only an easy and natural, but a necessary result. "In studying an object we neither exert all our faculties at once, nor at once apply them to all the qualities of an object. We know from experience that the effect of such a mode of procedure is confusion. On the contrary, we converge our attention on one alone of its qualities,—nay, contemplate this quality only in a single point of view, and retain it in that aspect until we have obtained a full and accurate conception of it. The human mind proceeds from the confused and complex to the distinct and constituent, always separating, always dividing, always simplifying; and this is the only mode in which, from the weakness of our faculties, we are able to apprehend and to represent with correctness."[β]

Abstraction,—a natural and necessary process.

α Laromiguière, [*Leçons de Philosophie*, partie II. leçon xi., t. ii. p. 310.—Ed.] Condillac, [*L'Art de Penser*, part I. c. viii.; *Cours*, t. iii. p. 295.—Ed.] [Cf. Fonseca, *Isagoge Philosophica*], [c. iv. p. 712, appended to his *Institut. Dialect.* (edit. 1604.)—Ed.]

β Laromiguière, *Leçons*, t. ii. p. 317.—Ed.

LECT. XXXIV.

Synthesis necessary after analysis.

"It is true, indeed, that after having decomposed everything, we must, as it were, return on our steps by recomposing everything anew; for unless we do so, our knowledge would not be conformable to the reality and relations of nature. The simple qualities of body have not each a proper and independent existence; the ultimate faculties of mind are not so many distinct and independent existences. On either side, there is a being one and the same; on that side, at once extended, solid, coloured, &c.; on this, at once capable of thought, feeling, desire, &c.

"But although all, or the greater number of, our cognitions comprehend different fasciculi of notions, it is necessary to commence by the acquisition of these notions one by one, through a successive application of our attention to the different attributes of objects. The abstraction of the intellect is thus as natural as that of the senses. It is even imposed upon us by the very constitution of our mind."[a]

The expression, abstraction of the senses.

"I am aware that the expression, *abstraction of the senses*, is incorrect; for it is the mind always which acts, be it through the medium of the senses. The impropriety of the expression is not, however, one which is in danger of leading into error; and it serves to point out the important fact, that abstraction is not always performed in the same manner. In Perception,—in the presence of physical objects, the intellect abstracts colours by the eyes, sounds by the ear, &c. In Representation, and when the external object is absent, the mind operates on its reproduced cognitions, and looks at them successively in their different points of view."[β]

"However abstraction be performed, the result is notions which are simple, or which approximate to

a Laromiguière, *Leçons*, t. ii. p. 312.—Ed.

β Laromiguière, *Leçons*, t. ii. p. 314 slightly abridged.—Ed.

simplicity; and if we apply it with consistency and order to the different qualities of objects, we shall attain at length to a knowledge of these qualities and of their mutual dependencies; that is, to a knowledge of objects as they really are. In this case, abstraction becomes analysis, which is the method to which we owe all our cognitions."[a]

The process of abstraction is familiar to the most uncultivated minds; and its uses are shown equally in the mechanical arts as in the philosophical sciences. "A carpenter," says Kames,[β] speaking of the great utility of abstraction, "considers a log of wood with regard to hardness, firmness, colour, and texture; a philosopher, neglecting these properties, makes the log undergo a chemical analysis, and examines its taste, its smell, and component principles; the geometrician confines his reasoning to the figure, the length, breadth, and thickness; in general, every artist, abstracting from all other properties, confines his observations to those which have a more immediate connection with his profession."

Abstraction the work of comparison.

But is Abstraction, or rather, is exclusive attention, the work of Comparison? This is evident. The application of attention to a particular object, or quality of an object, supposes an act of will,—a choice or preference, and this again supposes comparison and judgment. But this may be made more manifest from a view of the act of Generalisation, on which we are about to enter.

Generalisation. Idea abstract and individual.

The notion of the figure of the desk before me is an abstract idea,—an idea that makes part of the total notion of that body, and on which I have concentrated my attention, in order to consider it exclu-

a Laromiguière, *Leçons*, t. ii. p. 345. —Ed.

β *Elements of Criticism*, Appendix, § 40; vol. ii. p. 533, ed. 1758.—Ed.

LECT. XXXIV.

sively. This idea is abstract, but it is at the same time individual; it represents the figure of this particular desk, and not the figure of any other body. But had we only individual abstract notions, what would be our knowledge? We should be cognisant only of qualities viewed apart from their subjects; (and of separate phænomena there exist none in nature); and as these qualities are also separate from each other, we should have no knowledge of their mutual relations.[a]

Abstract General notions,—what and how formed.

It is necessary, therefore, that we should form Abstract General notions. This is done when, comparing a number of objects, we seize on their resemblances; when we concentrate our attention on these points of similarity, thus abstracting the mind from a consideration of their differences; and when we give a name to our notion of that circumstance in which they all agree. The general notion is thus one which makes us know a quality, property, power, action, relation; in short, any point of view under which we recognise a plurality of objects as a unity. It makes us aware of a quality, a point of view, common to many things. It is a notion of resemblance; hence the reason why general names or terms, the signs of general notions, have been called *terms of resemblance*, (*termini similitudinis*). In this process of generalisation, we do not stop short at a first generalisation. By a first generalisation we have obtained a number of classes of resembling individuals. But these classes we can compare together, observe their similarities, abstract from their differences, and bestow on their common circumstance a common name. On these second classes we can again perform the same

[a] We should also be overwhelmed with their number.—*Jotting.*

operation, and thus ascending the scale of general notions, throwing out of view always a greater number of differences, and seizing always on fewer similarities in the formation of our classes, we arrive at length at the limit of our ascent in the notion of *being* or *existence*. Thus placed on the summit of the scale of classes, we descend by a process the reverse of that by which we have ascended; we divide and subdivide the classes, by introducing always more and more characters, and laying always fewer differences aside; the notions become more and more composite, until we at length arrive at the individual.

Twofold quantity in notions.—Extension and Comprehension.

I may here notice that there is a twofold kind of quantity to be considered in notions. It is evident, that in proportion as the class is high, it will, in the first place, contain under it a greater number of classes, and, in the second, will include the smallest complement of attributes. Thus *being* or *existence* contains under it every class; and yet when we say that a thing exists, we say the very least of it that is possible. On the other hand, an individual, though it contain nothing but itself, involves the largest amount of predication. For example, when I say,—this is Richard, I not only affirm of the subject every class from existence down to man, but likewise a number of circumstances proper to Richard as an individual. Now, the former of these quantities, the external, is called the *Extension* of a notion, (*quantitas ambitus*); the latter, the internal quantity, is called its *Comprehension* or *Intension*, (*quantitas complexus*). The extension of a notion is, likewise, styled its *circuit*, *region*, *domain*, or *sphere* (*sphæra*), also its *breadth* (πλάτος). On the other hand, the comprehension of a notion is, likewise, called its *depth* (βάθος).

Their designations.

LECT. XXXIV.

Their law.

These names we owe to the Greek logicians.[α] The internal and external quantities are in the inverse ratio of each other. The greater the extension, the less the comprehension; the greater the comprehension, the less the extension.[β]

α [See Ammonius, *In Categ.*, f. 33. Gr., f. 29. Lat. Brandis, *Scholia in Arist.*, p. 45.] [Αἱ κατηγορίαι καὶ πλάτος ἔχουσι καὶ βάθος, βάθος μὲν τὴν εἰς τὰ μερικώτερα αὐτῶν πρόοδον, πλάτος δὲ τὴν εἰς τὰ πλάγια μετάστασιν, οἷον ἵνα βάθος μὲν λάβῃς οὕτω τὴν οὐσίαν καὶ τὸ σῶμα καὶ τὸ ἔμψυχον καὶ τὸ ζῷον καὶ οὕτως ἐφεξῆς, πλάτος δέ, ὅταν διέλῃς τὴν οὐσίαν εἰς σῶμα καὶ ἀσώματον.—Ed.]

β [Cf. *Port Royal Logic*, part i. c. vi. p. 74. Eugenios,] [Λογική, b. i. c. iv. p. 194 *et seq.*—Ed.]

LECTURE XXXV.

THE ELABORATIVE FACULTY.—GENERALISATION.—NOMINALISM AND CONCEPTUALISM.

LECT. XXXV.

Recapitulation.

I ENTERED, in my last Lecture, on the discussion of that great cognitive power which I called the Elaborative Faculty,—the Faculty of Relations,—the Discursive Faculty,—Comparison, or Judgment; and which corresponds to what the Greek philosophers understood by διάνοια, when opposed, as a special faculty, to νοῦς. I showed you, that, though a comparison,—a judgment, involved the supposition of two relative terms, still it was an original operation, in fact involved in consciousness, and a condition of every energy of thought. But, besides the primary judgments of existence,—of the existence of the ego and non-ego, and of their existence in contrast to, and in exclusion of, each other,—I showed that this process is involved in perception, external and internal; inasmuch as the recognitions,—that the objects presented to us by the Acquisitive Faculty are many and complex, that one quality is different from another, and that different bundles of qualities are the properties of different things or subjects,—are all so many acts of Comparison or Judgment.

This being done, I pointed out that a series of operations were to be referred to this faculty, which, by philosophers, had been made the functions of specific powers. Of these operations I enumerated :—

LECT. XXXV. 1°, Composition or Synthesis; 2°, Abstraction, Decomposition or Analysis; 3°, Generalisation; 4°, Judgment; and, 5°, Reasoning.

The first of these,—Composition or Synthesis,—which is shown in the formation of Complex or Collective notions, I stated to you was the result of an act of comparison. For a complex notion, (I gave you as examples *an army, a forest, a town*), being only the repetition of notions absolutely similar, this similarity could be ascertained only by comparison. In speaking of this process, I explained the support afforded in it to the mind by language. I then recalled to you what was meant by Abstraction. Abstraction is no positive act; it is merely the negation of attention. We can fully attend only to a single thing at a time; and attention, therefore, concentrated on one object or one quality of an object, necessarily more or less abstracts our consciousness from others. Abstraction from, and attention to, are thus correlative terms, the one being merely the negation of the other. I noticed the improper use of the term *abstraction* by many philosophers, in applying it to that on which attention is converged.[α] This we may indeed be said to *prescind*,[β] but not to *abstract*. Thus let A, B, C, be three qualities of an object. We prescind A, in abstracting it from B and C; but we cannot, without impropriety, simply say that we abstract A. Thus by attending to one object to the abstraction from

α [Cf. Kant, *De Mundi Sensibilis Forma*, [§ 6; *Vermischte Schriften*, ii. 449: "Proprie dicendum esset ab *aliquibus abstrahere*, non *aliquid abstrahere*. . . . . Conceptus intellectualis *abstrahit* ab omni sensitivo, non *abstrahitur* a sensitivis, et forsitan rectius diceretur *abstrahens*, quam *abstractus*."—Ed.] Maine de Biran. [*Examen des Leçons de M. Laromiguière*, § 3, *Nouvelles Considérat.* p. 194.—Ed.] Bilfinger, *Dilucidationes*, § 262.]

β [On *Precision*, and its various kinds, see Derodon, *Logica*, pars ii. c. vi. § 11.—*Opera*, p. 238, ed. 1668; and Chauvin, *Lexicon Philosophicum* v. *Præcisio* (*Præcisive*).]

all others, we, in a certain sort, decompose or analyse the complex materials presented to us by Perception and Self-consciousness. This analysis or decomposition is of two kinds. In the first place, by concentrating attention on one integrant part of an object, we, as it were, withdraw or abstract it from the others. For example, we can consider the head of an animal to the exclusion of the other members. This may be called Partial or Concrete Abstraction. The process here noticed has, however, been overlooked by philosophers, insomuch that they have opposed the terms *concrete* and *abstract* as exclusive contraries. In the second place, we can rivet our attention on some particular mode of a thing, as its smell, its colour, its figure, its motion, its size, &c., and abstract it from the others. This may be called Modal Abstraction.

The abstraction we have been now speaking of is performed on individual objects, and is consequently particular. There is nothing necessarily connected with Generalisation in Abstraction. Generalisation is indeed dependent on abstraction, which it supposes; but abstraction does not involve generalisation. I remark this, because you will frequently find the terms *abstract* and *general* applied to notions, used as convertible. Nothing, however, can be more incorrect. "A person," says Mr Stewart, "who had never seen but one rose, might yet have been able to consider its *colour* apart from its other qualities; and, therefore, there may be such a thing as an idea which is at once abstract and particular. After having perceived this quality as belonging to a variety of individuals, we can consider it without reference to any of them, and thus form the notion of redness or whiteness in general, which may be called a *general abstract idea*. The words *abstract* and *general*, therefore, when

LECT. XXXV. applied to ideas, are as completely distinct from each other as any two words to be found in the language."[a]

I showed that abstraction implied comparison and judgment; for attention supposes preference, preference is a judgment, and a judgment is the issue of comparison.

I then proceeded to the process of Generalisation, which is still more obtrusively comparison, and nothing but comparison. Generalisation is the process through which we obtain what are called *general* or *universal* notions. A general notion is nothing but the abstract notion of a circumstance in which a number of individual objects are found to agree, that is, to resemble each other. In so far as two objects resemble each other, the notion we have of them is identical, and, therefore, to us the objects may be considered as the same. Accordingly, having discovered the circumstance in which objects agree, we arrange them by this common circumstance into classes, to which we also usually give a common name.

I explained how, in the prosecution of this operation, commencing with individual objects, we generalised these into a lowest class. Having found a number of such lowest classes, we then compare these again together, as we had originally compared individuals; we abstract their points of resemblance, and by these points generalise them into a higher class. The same process we perform upon these higher classes; and thus proceed, generalising class from classes, until we are at last arrested in the one highest class, that of *being*. Thus we find Peter, Paul, Timothy, &c., all agree in certain common attributes, and which distinguish them from other animated beings. We accord-

a [*Elements*, vol. i. c. iv. § I. *Works*, vol. ii. p. 166.—Ed.] So Whately, [*Logic*, b. I. § 6, p. 49; b. II. c. v. § 1, p. 122 (8th edit.)—Ed.]

ingly collect them into a class, which we call *man*. In like manner, out of the other animated beings which we exclude from *man*, we form the classes, *horse*, *dog*, *ox*, &c. These and *man* form so many lowest classes or species. But these species, though differing in certain respects, all agree in others. Abstracting from their diversities, we attend only to their resemblances; and as all manifesting life, sense, feeling, &c.,—this resemblance gives us a class, on which we bestow the name *animal*. Animal, or living sentient existences, we then compare with lifeless existences, and thus going on abstracting from differences, and attending to resemblances, we arrive at naked or undifferenced existence. Having reached the pinnacle of generalisation, we may redescend the ladder; and this is done by reversing the process through which we ascended. Instead of attending to the similarities, and abstracting from the differences, we now attend to the differences, and abstract from the similarities. And as the ascending process is called Generalisation, this is called Division or Determination;—division, because the higher or wider classes are cut down into lower or narrower;—determination, because every quality added on to a class limits or determines its extent, that is, approximates it more to some individual, real, or determinate existence.

Generalisation. Can we form an adequate idea of what is denoted by an abstract general term?

Having given you this necessary information in regard to the nature of Generalisation, I proceed to consider one of the most simple, and, at the same time, one of the most perplexed, problems in philosophy,—in regard to the object of the mind,—the object of consciousness, when we employ a general term. In the explanation of the process of generalisation all philosophers are at one; the only differences that arise among them relate to the point,—whether we can

LECT. XXXV.

form an adequate idea of that which is denoted by an abstract, or abstract and general term. In the discussion of this question, I shall pursue the following order: first of all, I shall state to you the arguments of the Nominalists,—of those who hold, that we are unable to form an idea corresponding to the abstract and general term; in the second place, I shall state to you the arguments of the Conceptualists,—of those who maintain that we are so competent; and, in the last, I shall show you that the opposing parties are really at one, and that the whole controversy has originated in the imperfection and ambiguity of our philosophical nomenclature. In this discussion I avoid all mention of the ancient doctrine of Realism. This is curious only in an historical point of view; and is wholly irrelevant to the question at issue among modern philosophers.

Order of discussion.

This controversy principally agitated in Britain and France.

This controversy has been principally agitated in this country, and in France, for a reason that I shall hereafter explain; and, to limit ourselves to Great Britain, the doctrine of Nominalism has, among others, been embraced by Hobbes, Berkeley, Hume, Principal Campbell, and Mr Stewart; while Conceptualism has found favour with Locke, Reid, and Brown.[a]

Two opinions which still divide philosophers.

Throwing out of view the antiquities of the question, (and this question is perhaps more memorable than any other in the history of philosophy),—laying, I say, out of account opinions which have been long exploded, there are two which still divide philosophers. Some maintain that every act and every object of mind is necessarily singular, and that the name is that alone which can pretend to generality. Others again hold that the mind is capable of forming notions, representations, correspondent in universality to the classes contained under, or expressed by, the general term.

a See below, pp. 297, 301.—ED.

Nominalism.

The former of these opinions,—the doctrine as it is called of Nominalism,—maintains that every notion, considered in itself, is singular, but becomes, as it were, general, through the intention of the mind to make it represent every other resembling notion, or notion of the same class. Take, for example, the term *man*. Here we can call up no notion, no idea, corresponding to the universality of the class or term. This is manifestly impossible. For as *man* involves contradictory attributes, and as contradictions cannot coexist in one representation, an idea or notion adequate to *man* cannot be realised in thought. The class *man* includes individuals, male and female, white and black and copper-coloured, tall and short, fat and thin, straight and crooked, whole and mutilated, &c., &c.; and the notion of the class must, therefore, at once represent all and none of these. It is, therefore, evident, though the absurdity was maintained by Locke,[α] that we cannot accomplish this; and, this being impossible, we cannot represent to ourselves the class *man* by any equivalent notion or idea. All that we can do is to call up some individual image, and consider it as representing, though inadequately representing, the generality. This we easily do, for as we can call into imagination any individual, so we can make that individual image stand for any or for every other which it resembles, in those essential points which constitute the identity of the class. This opinion, which, after Hobbes, has been in this country maintained, among others, by Berkeley,[β] Hume,[γ] Adam Smith,[δ]

α *Essay on Human Understanding*, b. iv. c. vii. § 9.—ED.

β *Principles of Human Knowledge*, Introd. § 10.—ED.

γ *Treatise of Human Nature*, part i. sect. vii.; *Works*, i. p. 31. *Essay on the Academical Philosophy*; *Works*, iv. p. 184.—ED.

δ *Dissertation concerning the first Formation of Languages*.—ED.

LECT. XXXV.

Campbell,[α] and Stewart,[β] appears to me not only true but self-evident.

The doctrine of Nominalism as stated by Berkeley.

No one has stated the case of the nominalists more clearly than Bishop Berkeley; and as his whole argument is, as far as it goes, irrefragable, I beg your attention to the following extract from his Introduction to the *Principles of Human Knowledge*.[γ]

Berkeley quoted.

"It is agreed, on all hands, that the qualities or modes of things do never really exist each of them apart by itself, and separated from all others, but are mixed, as it were, and blended together, several in the same object. But we are told, the mind, being able to consider each quality singly, or abstracted from those other qualities with which it is united, does by that means frame to itself abstract ideas. For example there is perceived by sight an object extended, coloured, and moved: this mixed or compound idea the mind resolving into its simple, constituent parts, and viewing each by itself, exclusive of the rest, does frame the abstract ideas of extension, colour, and motion. Not that it is possible for colour or motion to exist without extension; but only that the mind can frame to itself by *abstraction* the idea of colour exclusive of extension, and of motion exclusive of both colour and extension.

"Again, the mind having observed that in the particular extensions perceived by sense, there is something common and alike in all, and some other things peculiar, as this or that figure or magnitude, which distinguish them one from another; it considers apart

α *Philosophy of Rhetoric*, book II. c 7.—Ed.

β *Elements*, part II. c. iv. *Works*, vol. II. p. 170.—Ed.

γ Sections vii. viii. x. *Works*, i. 5 *et seq.*, 4to edit. Cf. *Encyclopædia Britannica*, art. *Metaphysics* vol. xiv. p. 622, 7th edit.—Ed.

or singles out by itself that which is common, making thereof a most abstract idea of extension, which is neither line, surface, nor solid, nor has any figure or magnitude, but is an idea entirely prescinded from all these. So likewise the mind, by leaving out of the particular colours perceived by sense, that which distinguishes them one from another, and retaining that only which is common to all, makes an idea of colour in abstract which is neither red, nor blue, nor white, nor any other determinate colour. And in like manner, by considering motion abstractedly not only from the body moved, but likewise from the figure it describes, and all particular directions and velocities, the abstract idea of motion is framed; which equally corresponds to all particular motions whatsoever that may be perceived by sense.

"Whether others have this wonderful faculty of *abstracting their ideas*, they best can tell: for myself I find, indeed, I have a faculty of imagining, or representing to myself the ideas of those particular things I have perceived, and of variously compounding and dividing them. I can imagine a man with two heads, or the upper parts of a man joined to the body of a horse. I can consider the hand, the eye, the nose, each by itself abstracted or separated from the rest of the body. But then whatever hand or eye I imagine, it must have some particular shape and colour. Likewise the idea of man that I frame to myself, must be either of a white, or a black, or a tawny, a straight, or a crooked, a tall, or a low, or a middle-sized man. I cannot by any effort of thought conceive the abstract idea above described. And it is equally impossible for me to form the abstract idea of motion distinct from the body moving, and which is neither swift nor

LECT. XXXV.

slow, curvilinear nor rectilinear; and the like may be said of all other abstract general ideas whatsoever.[α] To be plain, I own myself able to abstract in one sense, as when I consider some particular parts or qualities separated from others, with which though they are united in some object, yet it is possible they may really exist without them. But I deny that I can abstract one from another, or conceive separately, those qualities which it is impossible should exist so separated; or that I can frame a general notion by abstracting from particulars in the manner aforesaid. Which two last are the proper acceptations of *abstraction*. And there are grounds to think most men will acknowledge themselves to be in my case. The generality of men, which are simple and illiterate, never pretend to *abstract notions*. It is said they are difficult, and not to be attained without pains and study. We may therefore reasonably conclude that, if such there be, they are confined only to the learned."

Such is the doctrine of Nominalism, as asserted by Berkeley, and as subsequently acquiesced in by the principal philosophers of this country. Reid himself is, indeed, hardly an exception, for his opinion on this point is, to say the least of it, extremely vague.[β]

Conceptualism.

The counter-opinion, that of Conceptualism, as it is called, has, however, been supported by several philosophers of distinguished ability.

Locke.

Locke maintains the doctrine in its most revolting absurdity, boldly admitting that the general notion must be realised, in spite of the principle of Contradiction. "Does it not require," he says, "some pains and skill

α This argumentation is employed by Derodon, *Logica*, [pars II. c. vi. § 16. *Opera*, p. 236.—ED.], and others.

β For Reid's opinion, see *Intellectual Powers*, essay v., chap. ii. and vi. —ED.

to form the *general idea* of a triangle? (which is yet none of the most abstract, comprehensive, and difficult); for it must be neither oblique nor rectangle, neither equilateral, equicrural, nor scalenon; but all and none of these at once. In effect, it is something imperfect, that cannot exist; an idea wherein some parts of several different and inconsistent ideas are put together."[a] ✗✗

This doctrine was, however, too palpably absurd to obtain any advocates; and conceptualism, could it not find a firmer basis, behoved to be abandoned. Passing over Dr Reid's speculations on the question, which are, as I have said, wavering and ambiguous, I solicit your attention to the principal statement and defence of conceptualism by Dr Brown, in whom the doctrine has obtained a strenuous advocate. "If, then, the generalising process be, first, the perception or conception of two or more objects; secondly, the relative feeling of their resemblance in certain respects; thirdly, the designation of these circumstances of resemblance, by an appropriate name,—the doctrine of the Nominalists, which includes only two of these stages,—the perception of particular objects, and the invention of general terms, must be false, as excluding that relative suggestion of resemblance in certain respects, which is the second and most important step of the process; since it is this intermediate feeling alone that leads to the use of the term, which, otherwise, it would be impossible to limit to any set of objects. Accordingly, we found that, in their impossibility of accounting, on their own principles, for this limitation, which it is yet absolutely necessary to explain in some manner or other,—the Nominalists, to

Brown quoted.

a See above, p. 297, note a.—Ed.

LECT. XXXV. explain it, uniformly take for granted the existence of those very general notions, which they at the same time profess to deny,—that, while they affirm that we have no notion of a kind, species, or sort, independently of the general terms which denote them, they speak of our application of such terms only to objects of the same kind, species, or sort; as if we truly had some notions of these general circumstances of agreement to direct us,—and that they are thus very far from being Nominalists in the spirit of their argument, at the very moment when they are Nominalists in assertion, — strenuous opposers of those very general feelings, of the truth of which they avail themselves in their very endeavour to disprove them.

"If, indeed, it were the name which formed the class, and not that previous relative feeling, or general notion of resemblance of some sort, which the name denotes, then might anything be classed with anything, and classed with equal propriety. All which would be necessary, would be merely to apply the same name uniformly to the same objects; and, if we were careful to do this, John and a triangle might as well be classed together, under the name man, as John and William. Why does the one of those arrangements appear to us more philosophic than the other? It is because something more is felt by us to be necessary in classification, than the mere giving of a name at random. There is, in the relative suggestion that arises on our very perception or conception of objects, when we consider them together, a reason for giving the generic name to one set of objects rather than to another,—the name of man, for instance, to John and William, rather than to John and a triangle.

This reason is the feeling of the resemblance of the objects which we class,—that general notion of the relation of similarity in certain respects, which is signified by the general term,—and without which relative suggestion, as a previous state of the mind, the general term would as little have been invented, as the names of John and William would have been invented, if there had been no perception of any individual being whatever to be denoted by them."*

LECT. XXXV.

This part of Dr Brown's philosophy has obtained the most unmeasured encomium; it has been lauded as the most important step ever made in the philosophy of mind; and, as far as I am aware, no one has as yet made any attempt at refutation. I regret that in this, as in many other principal points of his doctrine, I find it impossible not to dissent from Dr Brown. An adequate refutation of his views would, indeed, require a more elaborate criticism than I am at present able to afford them; but I trust that the following hasty observations will be sufficient to evince, that the doctrine of Nominalism is not yet overthrown.

Brown's doctrine criticised.

Dr Brown has taken especial care that his theory of generalisation should not be misunderstood; for the following is the seventh, out of nine recapitulations, he has given us of it in his forty-sixth and forty-seventh Lectures. "If, then, the generalising process be, first, the perception or conception of two or more objects; secondly, the relative feeling of their resemblance in certain respects; thirdly, the designation of these circumstances of resemblance by an appropriate name, the doctrine of the Nominalists,

* *Philosophy of the Human Mind*, lecture xlvii. p. 303.—Ed.

LECT. XXXV. which includes only two of these stages,—the perception of particular objects, and the invention of general terms,—must be false, as excluding that relative suggestion of resemblance in certain respects, which is the second and most important step of the process; since it is this intermediate feeling alone that leads to the use of the term, which, otherwise, it would be impossible to limit to any set of objects."

This contains, in fact, both the whole of his own doctrine, and the whole ground of his rejection of that of the Nominalists. Now, upon this, I would, first of all, say, in general, that what in it is true is not new. But I hold it idle to prove that his doctrine is old and common, and to trace it to authors with whom Brown has shown his acquaintance, by repeatedly quoting them in his Lectures; it is enough to show that it is erroneous.

His confutation of Nominalism.

The first point I shall consider is his confutation of the Nominalists. In the passage I have just adduced, and in ten others, he charges the Nominalists with excluding "the relative suggestion of resemblance in certain respects, which is the second and most important step in the process." This, I admit, is a weighty accusation, and I admit at once that if it do not prove that his own doctrine is right, it would at least demonstrate theirs to be sublimely wrong. But is the charge well founded? Dr Brown, in a passage which I once read to you,[a] and with which he concludes his supposed exposition of what he calls "the series of Reid's wonderful misconceptions," wisely warns his pupils against according credit to all second-hand statements. "I trust," he says, "it will impress you with one important lesson, which could not be

[a] See above, vol. ii. lect. xxiii. p. 61.—Ed.

LECT. XXXV.

taught more forcibly than by the errors of so great a mind, that it will always be necessary for you to consult the opinions of authors, when their opinions are of sufficient importance to deserve to be accurately studied, in their own works, and not in the works of those who profess to give a faithful account of them. From my own experience, I can most truly assure you, that there is scarcely an instance in which, on examining the works of those authors whom it is the custom more to cite than to read, I have found the view which I had received of them faithful." No advice assuredly can be more sound, and I shall accordingly follow it now, as I have heretofore done, in application to his own reports. Let us see whether the nominalists, as he assures us, do really exclude the apprehension of resemblance in certain respects, as one step in their doctrine of generalisation. I turn first to Hobbes as the real father of this opinion,—to him, as Leibnitz truly says, "*nominalibus ipsis nominaliorem.*" The classical place of this philosopher on the subject is the fourth chapter of the *Leviathan;* and there we have the following passage—"One universal name is imposed on many things for their *similitude in some quality or other accident;* and whereas a proper name bringeth to mind one thing only, universals recall *any one* of those many." There are other passages to the same effect in Hobbes, but I look no further.

1. That the Nominalists allow the apprehension of resemblance, proved against Brown by reference to Hobbes.

The second great nominalist is Berkeley; and to him the doctrine chiefly owes the acceptation it latterly obtained. His doctrine on the subject is chiefly contained in the Introduction to the *Principles of Human Knowledge*, sect. 7, &c., and in the seventh Dialogue of the *Minute Philosopher*, sect. 5, &c. Out

Berkeley.

LECT. XXXV. of many similar passages, I select the two following. In both he is stating his own doctrine of nominalism. In the Introduction, sect. 22:—"To discern *the agreements or disagreements* that are between my ideas, to see what ideas are included in any compound idea, &c." In the *Minute Philosopher*, sect. 7:—"But may not words become general by being made to stand indiscriminately for all particular ideas, which, from a *mutual resemblance*, belong to the same kind, without the intervention of any abstract general idea?"

Hume. I next take down Hume. His doctrine on the point at issue is found in book i. part i. sect. 7 of the *Treatise of Human Nature*, entitled, *On Abstract Ideas*. This section opens with the following sentence:—"A great philosopher has disputed the received opinion in this particular, and has asserted that all general ideas are nothing but particular ones annexed to a certain term, which gives them a more extensive signification, and makes them recall upon occasion other individuals which are similar to them. As I look upon this to be one of the greatest and most valuable discoveries that has been made of late years in the republic of letters, I shall here endeavour to confirm it by some arguments, which I hope will put it beyond all doubt and controversy." In glancing over the subsequent exposition of the doctrine, I see the following:—"When we have found a *resemblance* among several objects, we apply the same name to all of them," &c. Again:—"As individuals are collected together and placed under a general term, with a view to that *resemblance* which they bear to each other," &c. In the last page and a half of the section, it is stated, no less than four times, that *perceived resemblance* is the foundation of classification.

LECT. XXXV.

Adam Smith.

Adam Smith's doctrine is to the same effect as his predecessor's. It is contained in his *Dissertation concerning the First Formation of Languages*, (appended to his *Theory of Moral Sentiments*), which literally is full of statements to the purport of the following, which alone I adduce :—"It is this application of the name of an individual to a great number of objects whose *resemblance* naturally recalls the idea of that individual, and of the name which expresses it, that seems originally to have given occasion to the formation of these classes and assortments, which in the schools are called *genera* and *species*, and of which the ingenious and eloquent Rousseau finds himself so much at a loss to account for the origin. What constitutes a species is merely a number of objects, bearing *a certain degree of resemblance* to one another, and on that account denominated by a single appellation, which may be applied to express any one of them."

Campbell. Stewart.

The assertion, that perceived resemblance is the principle of classification, is repeated *ad nauseam* by Principal Campbell and Mr Stewart. I shall quote only from the latter, and I take the first passage that strikes my eye :—"According to this view of the process of the mind, in carrying on general speculations, that idea which the ancient philosophers considered as the essence of an individual, is nothing more than the particular quality or qualities in which it *resembles* other individuals of the same class; and in consequence of which a generic name is applied to it."[a]

From the evidence I have already quoted, you will see how marvellously wrong is Brown's assertion, that the nominalists not only took no account of, but absolutely excluded from their statement of the pro-

[a] *Elements*, vol. I. c. iv. sect. ii. *Works*, vol. ii. p. 175.

LECT. XXXV.

cess of generalisation, the apprehension of the mutual similarity of objects. You will, therefore, not be surprised when I assure you, that not only no nominalist ever overlooked, ever excluded, the manifested resemblance of objects to each other, but that every nominalist explicitly founded his doctrine of classification on this resemblance, and on this resemblance alone.[a] No nominalist ever dreamt of disallowing the notion of relativity,—the conception of similarity between things,—this they maintain not less strenuously than the conceptualist; they only deny that this could ever constitute a general notion.

II. That Brown wrong in holding that the feeling (notion) of similitude is general, and constitutes the general notion,—proved by the following axioms.

But perhaps it may be admitted, that Brown is wrong in asserting that the nominalist excludes resemblance as an element of generalisation, and yet maintained, that he is right in holding, against the nominalists, that the notion, or, as he has it, the feeling of the similitude of objects in certain respects, is general, and constitutes what is called the general notion. I am afraid, however, that the misconception in regard to this point will be found not inferior to that in regard to the other.

1. Notion of similarity supposes notion of certain similar objects.

In the first place, then, resemblance is a relation; and a relation necessarily supposes certain objects as related terms. There can thus be no relation of resemblance conceived apart from certain resembling objects. This is so manifest, that a formal enunciation of the principle seems almost puerile. Let it, however, be laid down as a first axiom, that the notion

a [See Tellez, *Summa Phil. Universæ*, [pars. i. disp. iv. sect. i. subs. 8-16, vol. i. p. 49 *et seq.* (edit. 1644). Cf. sect. ii. subs. 1 *et seq.*, p. 65.—Ed.] Derodon, *Logica*, [pars. ii. c. v. art. 2, § 6, p. 211. Cf. art. 4, p. 221 *et seq.*—Ed.] Arriaga, *Logica*, [disp. vi. sect. i. subs. 1 *et seq.*, *Cursus Philosophicus*, p. 110 (edit. 1632).—Ed.] Mendoza, *Disp. Log.*, [disp. iii. § 1, *Disp. a Summulis ad Metaphysicam*, vol. i. p. 246.—Ed.] Fran. Bonæ Spei, *Logica*, [*De Porphyrianis Universalibus*, disp. i., *Commentarii in Arist. Phil.*, p. 53, (edit. 1652).—Ed.]

of similarity supposes the notion of certain similar objects. LECT. XXXV.

In the second place, objects cannot be similar without being similar in some particular mode or accident,—say in colour, in figure, in size, in weight, in smell, in fluidity, in life, &c. &c. This is equally evident, and this I lay down as a second axiom. 2. Similar objects are similar in some particular mode.

In the third place, I assume, as a third axiom, that a resemblance is not necessarily and of itself universal. On the contrary, a resemblance between two individual objects in a determinate quality, is as individual and determinate as the objects and their resembling qualities themselves. Who, for example, will maintain that my actual notion of the likeness of a particular snowball and a particular egg, is more general than the representations of the several objects and their resembling accidents of colour? 3. A resemblance not necessarily universal.

Now, let us try Dr Brown's theory on these grounds. In reference to the first, he does not pretend, that what he calls the general feeling of resemblance, can exist except between individual objects and individual representations. The universality, which he arrogates to this feeling, cannot accrue to it from any universality in the relative or resembling ideas. This neither he nor any other philosopher ever did or could pretend. They are supposed, *ex hypothesi*, to be individual,—singular. Brown's theory tested by these axioms.

Neither, in reference to the second axiom, does he pretend to derive the universality which he asserts to his feeling of resemblance, from the universality of the notion of the common quality, in which this resemblance is realised. He does not, with Locke and others, maintain this; on the contrary, it is on the admitted absurdity of such a foundation that he attempts to establish the doctrine of conceptualism on another ground.

LECT. XXXV.

But if the universality, assumed by Dr Brown for his "feeling of resemblance," be found neither in the resembling objects, nor in the qualities through which they are similar, we must look for it in the feeling of resemblance itself, apart from its actual realisation; and this in opposition to the third axiom we laid down as self-evident. In these circumstances, we have certainly a right to expect that Dr Brown should have brought us cogent proof for an assertion so contrary to all apparent evidence, that although this be the question which perhaps has been more ably, keenly, and universally agitated than any other, still no philosopher before himself was found even to imagine such a possibility. But in proof of this new paradox, Dr Brown has not only brought no evidence; he does not even attempt to bring any. He assumes and he asserts, but he hazards no argument. In this state of matters, it is perhaps superfluous to do more than to rebut assertion by assertion; and as Dr Brown is not *in possessorio*, and as his opinion is even opposed to the universal consent of philosophers, the counter assertion, if not overturned by reasoning, must prevail.

Possible grounds of Brown's supposition that the feeling of resemblance is universal.

But let us endeavour to conceive on what grounds it could possibly be supposed by Dr Brown, that the feeling of resemblance between certain objects, through certain resembling qualities, has in it anything of universal, or can, as he says, constitute the general notion. This to me is indeed not easy; and every hypothesis I can make is so absurd, that it appears almost a libel to attribute it, even by conjecture, to so ingenious and acute a thinker.

First.

In the first place, can it be supposed that Dr Brown believed that a feeling of resemblance between objects in a certain quality or respect was general because it was a relation? Then must every notion of a relation be a general notion; which neither he nor any other philosopher ever asserts. LECT. XXXV.

In the second place, does he suppose that there is anything in the feeling or notion of the particular relation called *similarity*, which is more general than the feeling or notion of any other relation? This can hardly be conceived. What is a feeling or notion of resemblance? Merely this; two objects affect us in a certain manner, and we are conscious that they affect us in the same way as a single object does, when presented at different times to our perception. In either case, we judge that the affections of which we are conscious are similar or the same. There is nothing general in this consciousness, or in this judgment. At all events, the relation recognised between the consciousness of similarity produced on us by two different eggs, is not more general than the feeling of similarity produced on us by the successive presentation of the same egg. If the one is to be called general, so is the other. Again, if the feeling or notion of resemblance be made general, so must the feeling or notion of difference. They are absolutely the same notion, only in different applications. You know the logical axiom,—the science of contraries is one. We know the like only as we know the unlike. Every affirmation of similarity is virtually an affirmation that difference does not exist; every affirmation of difference is virtually an affirmation that similarity is not to be found. But neither Brown nor any other philosopher has pretended, that the apprehension of difference is either general, or a ground of generalisation. On the contrary, the apprehension of difference is the negation of generalisation, and a descent from the universal to the particular. But if

LECT. XXXV. the notion or feeling of the dissimilarity is not general, neither is the feeling or notion of the similarity.

Third. In the third place, can it be that Dr Brown supposes the particular feeling or consciousness of similarity between certain objects in certain respects to be general, because we have, in general, a capacity of feeling or being conscious of similarity? This conjecture is equally improbable. On this ground every act of every power would be general; and we should not be obliged to leave Imagination, in order to seek for the universality which we cannot discover in the light and definitude of that faculty, in the obscurity and vagueness of another.

Fourth. In the fourth place, only one other supposition remains; and this may perhaps enable us to explain the possibility of Dr Brown's hallucination. A relation cannot be represented in Imagination. The two terms, the two relative objects, can be severally imaged in the sensible phantasy, but not the relation itself. This is the object of the Comparative Faculty, or of Intelligence Proper. To objects so different as the images of sense and the unpicturable notions of intelligence, different names ought to be given; and accordingly this has been done wherever a philosophical nomenclature of the slightest pretensions to perfection has been formed. In the German language, which is now the richest in metaphysical expressions of any living tongue, the two kinds of objects are carefully distinguished.[a] In our language, on the contrary, the terms *idea*, *conception*, *notion*, are used almost as convertible for either; and the vagueness and confusion which is thus produced, even within the narrow sphere of speculation to which the want of the dis-

[a] See *Reid's Works*, p. 107, note ‡, and 412, note.—Ed.

tinction also confines us, can be best appreciated by those who are conversant with the philosophy of the different countries.

Dr Brown seems to have had some faint perception of the difference between intellectual notions and sensible representations; and if he had endeavoured to signalise their contrast by a distinction of terms, he would have deserved well of English philosophy. But he mistook the nature of the intellectual notion, which connects two particular qualities by the bond of similarity, and imagined that there lurked under this intangible relation the universality which, he clearly saw, could not be found in a representation of the related objects, or of their resembling qualities. At least, if this do not assist us in accounting for his misconception, I do not know in what way we otherwise can.

Summary of the Author's doctrine of Generalisation.

What I have now said is, I think, sufficient in regard to the nature of Generalisation. It is notoriously a mere act of Comparison. We compare objects; we find them similar in certain respects, that is, in certain respects they affect us in the same manner; we consider the qualities in them, that thus affect us in the same manner, as the same; and to this common quality we give a name; and as we can predicate this name of all and each of the resembling objects, it constitutes them into a class. Aristotle has truly said that general names are only abbreviated definitions,[a] and definitions, you know, are judgments. For example, *animal* is only a compendious expression for *organised and animated body*; *man*, only a summary of *rational animal*, &c.

[a] *Rhet.*, iii. 6.—ED.

LECTURE XXXVI.

THE ELABORATIVE FACULTY.—GENERALISATION.—THE PRIMUM COGNITUM.

LECT. XXXVI.

Recapitulation.

We were principally employed, in our last Lecture, in considering Dr Brown's doctrine of Generalisation; and, in doing this, I first discussed his refutation of Nominalism, and, secondly, his own theory of Conceptualism. In reference to the former, I showed you that the ground on which he attempts to refute the Nominalists, is only an inconceivable mistake of his own. He rejects their doctrine as incomplete, because, he says, they take no account of the mutual resemblance of the classified objects. But so far are the nominalists from taking no account of the mutual resemblance of the classified objects, that their doctrine is notoriously founded on the apprehension of this similarity, and on the apprehension of this similarity alone. How Dr Brown could have run into this radical misrepresentation of so celebrated an opinion, is, I repeat, wholly inconceivable. Having proved to you by the authentic testimony of the British nominalists of principal celebrity, that Dr Brown had in his statement of their doctrine simply reversed it, I proceeded, in the second place, to test the accuracy of his own. Dr Brown repudiates the doctrine of Conceptualism as held by Locke and others. He admits that we can represent to ourselves no general notion of the common attribute or attributes which constitute a class; but he asserts that the generality, which cannot be realised in a notion of the resembling attribute, is realised in a notion of the resemblance itself. This theory, I endeavoured to make it evident, was altogether groundless. In the first place, the doctrine supposes that the notion, or, as he calls it, the feeling, of the mutual resemblance of particular objects in particular respects, is general. This, the very foundation of his theory, is not self-evidently true;—on the contrary, it stands obtrusively, self-evidently, false. It was primarily incumbent on Dr Brown to prove the reality of this basis. But he makes not even an attempt at this. He assumes all that is in question. To the noun-substantive, "feeling of resemblance," he prefixes the adjective, "general;" but he does not condescend to evince that the verbal collocations have any real connection.

But, in the second place, as it is not proved by Dr Brown, that our notion of the similarity of certain things in certain respects is general, so it can easily be shown against him that it is not.

The generality cannot be found in the relation of resemblance, apart from all resembling objects, and all circumstances of resemblance; for a resemblance only exists, and is only conceived, as between determinate objects, and in determinate attributes.[a] This is not denied by Dr Brown. On the contrary, he arrogates generality to what he calls the "feeling of similarity of certain objects in certain respects." These are the expressions he usually employs. So far, therefore, all

a If generality in relation of resemblance apart from particular objects and qualities, then only one general notion at all.—*Marginal Jotting.*

is manifest, all is admitted; a resemblance is only conceived, is only conceivable, as between particular objects, in particular qualities. Apart from these, resemblance is not asserted to be thinkable. This being understood, it is apparent, that the notion of the resemblance of certain objects in a certain attribute, is just the notion of that attribute itself; and if it be impossible, as Brown admits, to conceive that attribute generally, in other words, to have a general notion of it, it is impossible to have a general notion of the resemblance which it constitutes. For example, we have a perception or imagination of two figures resembling each other, in having three angles. Now here it is admitted, that if either the figures themselves be removed, or the attribute belonging to each, (of three angles), be thrown out of account, the notion of any resemblance is also annihilated. It is also admitted, that the notion of resemblance is realised through the notion of triangularity. In this all philosophers are at one. All likewise agree that the notion of similarity, and the notion of generality, are the same; though Brown, as we have seen, has misrepresented the doctrine of Nominalism on this point. But though all maintain that things are conceived similar only as conceived similar in some quality, and that their similarity in this quality alone constitutes them into a class, they differ in regard to their ulterior explanation. Let us suppose that, of our two figures, the one is a rectangled, and the other an equilateral, triangle; and let us hear, on this simple example, how the different theorists explain themselves. The nominalists say,—you can imagine a rectangular triangle alone, and an equilateral triangle alone, or you can imagine both at once; and, in this case, in the consciousness of their similarity, you may view either as the inadequate representative of both. But you cannot imagine a figure which shall adequately represent both *qua* triangle; that is, you cannot imagine a triangle which is neither an equilateral nor a rectangled triangle, and yet both at once. And as on our (the nominalist) doctrine, the similarity is only embodied in an individual notion, having relation to another, there is no general notion properly speaking at all.

The older Conceptualists, on the other hand, assert that it is possible to conceive a triangle neither equilateral nor rectangular,—but both at once. Dr Brown differs from nominalists and older conceptualists; he coincides with the nominalists in rejecting as absurd the hypothesis of the conceptualist, but he coincides with the conceptualist in holding, that there is a general notion adequate to the term triangle. This general notion he does not, however, place, with the conceptualist, in any general representation of the attribute triangle, but in the notion or feeling of resemblance between the individual representations of an equilateral and of a rectangled triangle. This opinion is, however, untenable. In the first place, there is here no generalisation; for what is called the common notion can only be realised in thought through notions of all the several objects which are to be classified. Thus, in our example, the notion of the similarity of the two figures, in being each triangular, supposes the actual perception or imagination of both together. Take out of actual perception, or actual representation, one or both of the triangles, and no similarity, that is, no general notion, remains. Thus, upon Dr Brown's doctrine, the general notion only exists in so far as the individual notions, from which it is general-

LECT. XXXVI.

ised, are present, that is, in so far as there is no generalisation at all. This is because resemblance is a relation; but a relation supposes two particular objects; and a relation between particular objects is just as particular as the objects themselves.

Brown's doctrine of general notions,—further considered.

But let us consider his doctrine in another point of view. In the example we have taken of the equilateral and rectangular triangles, triangularity is an attribute of each, and in each the conceived triangularity is a particular, not a general, notion. Now the resemblance between these figures lies in their triangularity, and the notion or feeling of resemblance in which Dr Brown places the generality,—must be a notion or feeling of triangularity,—triangularity must constitute their resemblance. This is manifest. For if it be not a notion of triangularity, it must be a notion of something else, and if a notion of something else, it cannot be a general notion of two figures as triangles. The notion of resemblance between the figures in question must, therefore, be a notion of triangularity. Now the triangularity thus conceived must be one notion,—one triangularity; for otherwise it could not be, (what is supposed), one common or general notion, but a plurality of notions. Again, this one triangularity must not be the triangularity, either of the equilateral triangle, or of the rectangular triangle alone; for, in that case, it would not be a general notion,—a notion common to both. But if it cannot be the triangularity of either, it must be the triangularity of both. Of such a triangularity, however, it is impossible to form a notion, as Dr Brown admits; for triangularity must be either rectangular or not rectangular; but as these are contradictory or exclusive attributes, we cannot conceive them together in the same notion, nor can we form a notion of triangularity except as the one or the other. LECT. XXXVI.

This being the case, the notion or feeling of similarity between the two triangles cannot be a notion or feeling of triangularity at all. But if it be not this, what can it otherwise possibly be? There is only one conceivable alternative. As a general notion, containing under it particular notions, it must be given up; but it may be regarded as a particular relation between the particular figures, and which supposes them to be represented, as the condition of being itself not represented, but conceived. And thus, by a different route, we arrive again at the same conclusion,—that Dr Brown has mistaken a particular, an individual, relation for a general notion. He clearly saw that all that is picturable in imagination is determinate and individual; he, therefore, avoided the absurdity involved in the doctrine of the old conceptualists; but he was not warranted, (if this were, indeed, the ground of his assumption), in assuming, that because a notion cannot be pictured in imagination, it is, therefore, general.

Instead of recapitulating what I stated in opposition to Dr Brown's views in my last Lecture, I have been led into a new line of argument; for, in fact, his doctrine is open to so many objections that, on what side soever we regard it, argument will not be wanting for its refutation. So far, therefore, from Nominalism being confuted by Brown, it is plain that, apart from the misconception he has committed, he is himself a nominalist.

I proceed now to a very curious question which has likewise divided philosophers. It is this,—Does Language originate in General Appellatives, or by Proper

The question,—Does Language originate in

General Appellatives or by Proper Names,—considered.

Names? Did mankind in the formation of language, and do children in their first applications of it, commence with the one kind of words, or with the other? The determination of this question,—the question of the *Primum Cognitum*, as it was called in the schools,—is not involved in the doctrine of Nominalism. Many illustrious philosophers have maintained, that all terms, as at first employed, are expressive of individual objects, and that these only subsequently obtain a general acceptation.

1. That all terms, as at first employed, expressive of individual objects,—maintained by Vives and others.

This opinion I find maintained by Vives,[α] Locke,[β] Rousseau,[γ] Condillac,[δ] Adam Smith,[ε] Steinbart,[ζ] Tittel,[η] Brown,[θ] and others.[ι] "The order of learning," (I translate from Vives), "is from the senses to the imagination, and from this to the intellect,—such is the order of life and of nature. We thus proceed from the simple to the complex, from the singular to the universal. This is to be observed in children, who first of all express the several parts of different things, and then conjoin them. Things general they call by a singular name; for instance, they call all smiths by the name of that individual *smith* whom they have first known, and all meats, *beef* or *pork*, as they have happened to have heard the one or the other first, when they begin to speak. Thereafter the mind collects universals from particulars, and then again reverts to particulars from universals." The same doctrine, without probably

α *De Anima*, lib. II., *De Discendi Ratione*,—*Opera*, vol. II. p. 530, Basileæ, 1555.—ED.

β See below, p. 321.—ED.

γ [See Toussaint, *De la Pensée*, c. x. p. 278-79.] *Discours sur l'Origine de l'Inégalité parmi les Hommes*, *Œuvres*, t. i. p. 268, ed. 1826.—ED.

δ See below, p. 321.—ED.

ε See below, p. 321.—ED.

ζ [*Anleitung des Verstandes*, § 46. Cf. § 83-89.]

η [*Erläuterungen der Philosophie*.] [*Logik*, p. 214 *et seq.* (edit. 1793).—ED.]

θ See below, p. 321.—ED.

ι Cf. Toletus, *In Phys. Arist.*, lib. I. c. I. t. 5, qu. 5, f. 10[b]. Conimbricenses *Ibid.*, lib. I. c. I. qu. 3, art. 2, p. 79; and qu. 4, art. 2, p. 89.—ED.

any knowledge of Vives, is maintained by Locke.[α] "There is nothing more evident than that the ideas of the persons children converse with, (to instance in them alone), are like the persons themselves, only particular. The ideas of the nurse and the mother are well framed in their minds; and, like pictures of them there, represent only those individuals. The names they first gave to them are confined to these individuals; and the names of *nurse* and *mamma*, the child uses, determine themselves to those persons. Afterwards, when time and a larger acquaintance have made them observe, that there are a great many other things in the world, that in some common agreements of shape, and several other qualities, resemble their father and mother, and those persons they have been used to, they frame an idea which they find those many particulars do partake in; and to that they give, with others, the name *man*, for example. And thus they come to have a general name, and a general idea."

Locke.

The same doctrine is advanced in many places of his works by Condillac.[β] Adam Smith has, however, the merit of having applied this theory to the formation of language; and his doctrine, which Dr Brown,[γ] absolutely, and Mr Stewart,[δ] with some qualification, adopts, is too important not to be fully stated, and in his own powerful language:—"The assignation," says Smith,[ε] "of particular names, to denote particular objects,—that is, the institution of nouns substantive,—

Condillac. Adam Smith. Brown. Stewart. Smith quoted.

α *Essay*, III. 3, 7.—ED.

β See *Essai sur l'Origine des Connoissances Humaines*, partie i. sect. iv. c. i., sect. v.; partie ii. sect. i. c. ix.; *Logique*, ch. iv. p. 36 *et seq.* (edit. Nieuport).—ED.

γ Lecture xlvii. p. 306 (edit. 1830).

δ *Elements*, vol. i. part ii. c. iv. *Works*, vol. ii. p. 159. Cf. *Elements*, vol. ii. part ii. c. ii. § 4. *Works*, p. 173.—ED.

ε *Considerations concerning the first Formation of Languages*, appended to *Theory of Moral Sentiments*.—ED.

LECT. XXXVI. would probably be one of the first steps towards the formation of language. Two savages, who had never been taught to speak, but had been bred up remote from the societies of men, would naturally begin to form that language by which they would endeavour to make their mutual wants intelligible to each other, by uttering certain sounds whenever they meant to denote certain objects. Those objects only which were most familiar to them, and which they had most frequent occasion to mention, would have particular names assigned to them. The particular cave whose covering sheltered them from the weather, the particular tree whose fruit relieved their hunger, the particular fountain whose water allayed their thirst, would first be denominated by the words *cave*, *tree*, *fountain*, or by whatever other appellations they might think proper, in that primitive jargon, to mark them. Afterwards, when the more enlarged experience of these savages had led them to observe, and their necessary occasions obliged them to make mention of other caves, and other trees, and other fountains, they would naturally bestow upon each of those new objects the same name by which they had been accustomed to express the similar object they were first acquainted with. The new objects had none of them any name of its own, but each of them exactly resembled another object, which had such an appellation. It was impossible that those savages could behold the new objects, without recollecting the old ones; and the name of the old ones, to which the new bore so close a resemblance. When they had occasion, therefore, to mention or to point out to each other any of the new objects, they would naturally utter the name of the correspondent old one, of which the idea could not fail, at that instant, to present itself to their memory in the strongest and liveliest manner. And thus those words, which were originally the proper names of individuals, would each of them insensibly become the common name of a multitude. A child that is just learning to speak, calls every person who comes to the house its papa, or its mamma; and thus bestows upon the whole species those names which it had been taught to apply to two individuals. I have known a clown who did not know the proper name of the river which ran by his own door. It was *the river*, he said, and he never heard any other name for it. His experience, it seems, had not led him to observe any other river. The general word *river*, therefore, was, it is evident, in his acceptance of it, a proper name signifying an individual object. If this person had been carried to another river, would he not readily have called it a river? Could we suppose any person living on the banks of the Thames so ignorant as not to know the general word *river*, but to be acquainted only with the particular word *Thames*, if he was brought to any other river, would he not readily call it *a Thames?* This, in reality, is no more than what they, who are well acquainted with the general word, are very apt to do. An Englishman, describing any great river which he may have seen in some foreign country, naturally says, that it is another Thames. The Spaniards, when they first arrived upon the coast of Mexico, and observed the wealth, populousness, and habitations of that fine country, so much superior to the savage nations which they had been visiting for some time before, cried out that it was another Spain. Hence, it was called New Spain; and this name has stuck to that unfortunate country ever since. We say, in the same manner, of a hero, that he is an Alexander; of an orator, that he is a Cicero; of a philosopher, that he is a Newton. This way of speaking, which the grammarians call an Antonomasia, and which is still extremely common, though now not at all necessary, demonstrates how much all mankind are naturally disposed to give to one object the name of any other which nearly resembles it; and thus, to denominate a multitude, by what originally was intended to express an individual.

"It is this application of the name of an individual to a great multitude of objects, whose resemblance naturally recalls the idea of that individual, and of the name which expresses it, that seems originally to have given occasion to the formation of those classes and assortments which, in the schools, are called *genera* and *species*."

2. An opposite doctrine maintained by many of the Schoolmen. Campanella. Leibnitz.

On the other hand, an opposite doctrine is maintained by many profound philosophers. A large section of the schoolmen[α] embraced it, and, among more modern thinkers, it is adopted by Campanella.[β] Campanella was an author profoundly studied by Leibnitz, who even places him on a line with, if not above, Bacon; and from him it is not improbable that Leibnitz may have taken a hint of his own doctrine on the subject. In his great work, the *Nouveaux Essais*, of which Stewart was not till very latterly aware, he

Leibnitz quoted.

says,[γ] that "general terms serve not only for the perfection of languages, but are even necessary for their essential constitution. For if by *particulars* be understood things individual, it would be impossible to

α Cf. Conimbricenses, *In Phys. Arist.*, lib. i. c. i. qu. 3, art. 1, p. 78; and qu. 4, art. 1, p. 87. Toletus, *Ibid.*, lib. i. c. i. text 3 et seq. f. 10ᵃ.—Ed.

β [See Tennemann, *Geschichte der Philosophie*, vol. ix. p. 334.]

γ Liv. iii. c. i. p. 297 (edit. Erdmann).—Ed.

speak, if there were only proper names, and no appellatives, that is to say, if there were only names for things individual, since, at every moment, we are met by new ones, when we treat of persons, of accidents, and especially of actions, which are those that we describe the most; but if by particulars be meant the lowest species (*species infimus*), besides that it is frequently very difficult to determine them, it is manifest that these are already universals, founded on similarity. Now, as the only difference of *species* and *genera* lies in a similarity of greater or less extent, it is natural to note every kind of similarity or agreement, and, consequently, to employ general terms of every degree; nay, the most general being less complex with regard to the essences which they comprehend, although more extensive in relation to the things individual to which they apply, are frequently the easiest to form, and are the most useful. It is likewise seen that children, and those who know but little of the language which they attempt to speak, or little of the subject on which they would employ it, make use of general terms, as *thing, plant, animal*, instead of using proper names, of which they are destitute. And it is certain that all *proper* or individual names have been originally *appellative* or general." In illustration of this latter most important doctrine, he, in a subsequent part of the work, says[a]:—"I would add, in conformity to what I have previously observed, that proper names have been originally appellative, that is to say, general in their origin, as Brutus, Cæsar, Augustus, Capito, Lentulus, Piso, Cicero, Elbe, Rhine, Rhur, Leine, Ocker, Bucephalus, Alps, Pyrenees, &c.," and, after illustrating this in detail, he concludes:

[a] Liv. iii. c. iii. p. 303 (edit. Erdmann).—Ed.

LECT. XXXVI. —"Thus I would make bold to affirm that almost all words have been originally general terms, because it would happen very rarely that men would invent a name, expressly and without a reason, to denote this or that individual. We may, therefore, assert that the names of individual things were names of species, which were given *par excellence*, or otherwise, to some individual, as the name *Great Head* to him of the whole town who had the largest, or who was the man of most consideration, of the Great Heads known. It is thus likewise that men give the names of genera to species, that is to say, that they content themselves with a term more general or vague to denote more particular classes, when they do not care about the differences. As, for example, we content ourselves with the general name *absinthium* (wormwood), although there are so many species of the plant that one of the Bauhins has filled a whole book with them."

Turgot. That this was likewise the opinion of the great Turgot, we learn from his biographer. "M. Turgot," says Condorcet,[a] "believed that the opinion was wrong, which held that in general the mind only acquired general or abstract ideas by the comparison of more particular ideas. On the contrary, our first ideas are very general, for seeing at first only a small number of qualities, our idea includes all the existences to which these qualities are common. As we acquire knowledge, our ideas become more particular, without ever reaching the last limit; and, what might have deceived the metaphysicians, it is precisely by this process that we learn that these ideas are more general than we had at first supposed."

Here are two opposite opinions, each having nearly

a [*Vie de M. Turgot*, Londres, 1786, p. 214.]

equal authority in its favour, maintained on both sides with equal ability and apparent evidence. Either doctrine would be held established were we unacquainted with the arguments in favour of the other.

LECT. XXXVI.

3. A third or intermediate opinion maintained,—that language at first expresses only the vague and confused.

But I have now to state to you a third opinion, intermediate between these, which conciliates both, and seems, moreover, to carry a superior probability in its statement. This opinion maintains, that as our knowledge proceeds from the confused to the distinct,—from the vague to the determinate,—so, in the mouths of children, language at first expresses neither the precisely general nor the determinately individual, but the vague and confused; and that, out of this the universal is elaborated by generification, the particular and singular by specification and individualisation.

That Perception commences with masses, already shown.

I formerly explained why I view the doctrine held by Mr Stewart and others in regard to perception in general, and vision in particular, as erroneous; inasmuch as they conceive that our sensible cognitions are formed by the addition of an almost infinite number of separate and consecutive acts of attentive perception, each act being cognisant of a certain *minimum sensibile*.[a] On the contrary, I showed that, instead of commencing with minima, perception commences with masses; that, though our capacity of attention be very limited in regard to the number of objects on which a faculty can be simultaneously directed, yet that these objects may be large or small. We may make, for example, a single object of attention either of a whole man, or of his face, or of his eye, or of the pupil of his eye, or of a speck upon the pupil. To each of these objects there can only be a certain amount

a See above, vol. i. lect. xiii. p. 243.—Ed.

LECT. XXXVI. of attentive perception applied, and we can concentrate it all on any one. In proportion as the object is larger and more complex, our attention can of course be less applied to any part of it, and, consequently, our knowledge of it in detail will be vaguer and more imperfect. But having first acquired a comprehensive knowledge of it as a whole, we can descend to its several parts, consider these both in themselves, and in relation to each other, and to the whole of which they are constituents, and thus attain to a complete and articulate knowledge of the object. We decompose and then we recompose.

The mind, in elaborating its knowledge, proceeds by analysis, from the whole to the parts.

But in this we always proceed first by decomposition or analysis. All analysis indeed supposes a foregone composition or synthesis, because we cannot decompose what is not already composite. But in our acquisition of knowledge, the objects are presented to us compounded; and they obtain a unity only in the unity of our consciousness. The unity of consciousness is, as it were, the frame in which objects are seen. I say, then, that the first procedure of mind in the elaboration of its knowledge is always analytical. It descends from the whole to the parts,—from the vague to the definite. Definitude, that is, a knowledge of minute differences, is not, as the opposite theory supposes, the first, but the last, term of our cognitions.

Illustrated.

Between two sheep an ordinary spectator can probably apprehend no difference, and if they were twice presented to him, he would be unable to discriminate the one from the other. But a shepherd can distinguish every individual sheep; and why? Because he has descended from the vague knowledge which we all have of sheep,—from the vague knowledge which makes every sheep, as it were, only a repetition of the same undifferenced unit,—to a definite knowledge of qualities by which each is contrasted from its neighbour. Now, in this example, we apprehend the sheep by marks not less individual than those by which the shepherd discriminates them; but the whole of each sheep being made an object, the marks by which we know it are the same in each and all, and cannot, therefore, afford the principle by which we can discriminate them from each other. Now this is what appears to me to take place with children. They first know,—they first cognise, the things and persons presented to them as wholes. But wholes of the same kind, if we do not descend to their parts, afford us no difference,—no mark by which we can discriminate the one from the other. Children, thus, originally perceiving similar objects,—persons, for example, —only as wholes, do at first hardly distinguish them. They apprehend first the more obtrusive marks that separate species from species, and, in consequence of the notorious contrast of dress, men from women; but they do not as yet recognise the finer traits that discriminate individual from individual. But, though thus apprehending individuals only by what we now call their specific or their generic qualities, it is not to be supposed that children know them by any abstract general attributes, that is, by attributes formed by comparison and attention. On the other hand, because their knowledge is not general, it is not to be supposed to be particular or individual, if by particular be meant a separation of species from species, and by individual the separation of individual from individual; for children are at first apt to confound individuals together, not only in name but in reality. "A child who has been taught to say *papa*, in pointing

LECT. XXXVI.

to his father, will give at first, as Locke, [and Aristotle before him], had remarked, the name of *papa* to all the men whom he sees.[a] As he only at first seizes on the more striking appearances of objects, they would appear to him all similar, and he denotes them by the same names. But when it has been pointed out to him that he is mistaken, or when he has discovered this by the consequences of his language, he studies to discriminate the objects which he had confounded, and he takes hold of their differences. The child commences, like the savage, by employing only isolated words in place of phrases; he commences by taking verbs and nouns only in their absolute state. But as these imperfect attempts at speech express at once many and very different things, and produce, in consequence, manifold ambiguities, he soon discovers the necessity of determining them with greater exactitude; he endeavours to make it understood in what respects the thing which he wishes to denote, is distinguished from those with which it is confounded; and, to succeed in this endeavour, he tries first to distinguish them himself. Thus when, at this age, the child seems to us as yet unoccupied, he is in reality very busy; he is devoted to a study which differs not in its nature from that to which the philosopher applies himself; the child, like the philosopher, observes, compares, and analyses."[β]

This doctrine maintained by Aristotle.

In support of this doctrine I can appeal to high authority; it is that maintained by Aristotle. Speaking of the order of procedure in physical science, he says, "We ought to proceed from the better known to

a Aristotle, *Phys. Ausc.*, i. 1. Cf. Locke, *Essay on the Human Understanding*, iii. 3, 7, who adduces the same instance, but not quite for the same purpose.—ED.

β Degerando, *Des Signes*, i. 156.

the less known, and from what is clearer to us to that which is clearer, in nature. But those things are first known and clearer, which are more complex and confused; for it is only by subsequent analysis that we attain to a knowledge of the parts and elements of which they are composed. We ought, therefore, to proceed from universals to singulars; for the whole is better known to sense than its parts; and the universal is a kind of whole, as the universal comprehends many things as its parts. Thus it is that names are at first better known to us than definitions; for the name denotes a whole, and that indeterminately; whereas the definition divides and explicates its parts. Children, likewise, at first call all men fathers and all women mothers; but thereafter they learn to discriminate each individual from another."[a]

J. C. Scaliger.

The subtle Scaliger teaches the same doctrine; and he states it better perhaps than any other philosopher:—

"Universalia magis, ac prius esse nota nobis. Sic enim patres a pueris omnes homines appellari. Quia æquivocationibus nomina communicantur ab ignaris etiam rebus differentibus definitione. Sic enim chirothecam meam, puerulus quidam manum appellabat. An ei pro chirothecæ specie manus species sese repræsentabat? Nequaquam. Sed judicium aberat, quod distingueret differentias. An vero summa genera nobis notiora? Non. Composita enim notiora nobis. Genera vero partes sunt specierum: quas in partes ipsæ species multa resolvuntur arte. Itaque eandem ob rationem ipsa genera, sub notione comprehensionis et prædicabilitatis, sunt notiora quam ipsæ species. Cognoscitur

a *Phys. Ausc.*, l. 1.—Ed. [Cf. *In loc. cit.* Philoponus, Themistius, Averroes, Simplicius, Pacius, Conimbricenses, Tolet.]

LECT. XXXVI.

animal. Animalium species quot ignorantur? Sunt enim species partes prædicabiles. Sic totum integrum nobis notius, quam partes e quibus constat. Omne igitur quodcunque sub totius notione sese offert, prius cognoscitur, quam ejus partes. Sic species constituta, prius quam constituentia: ut equus, prius quam animal domabile ad trahendum, et vehendum. Hoc enim postea scimus per resolutionem. Sic genus prædicabile, prius quam suæ species. Sic totum integrum, prius quam partes. Contrarius huic ordo Naturæ est."[a]

[a] *De Subtilitate*, Ex. cccvii. § 21. [Cf. Zabarella, *De Ordine Intelligendi*, c. i. (*De Rebus Naturalibus*, p. 1012), and *In Phys. Arist.*, lib. i. c. 1, text 5. Andreas Cæsalpinus, *Peripateticæ Quæstiones*, lib. i. qu. 1, p. 1 *et seq.* (edit. 1571). Philip Mocenicus, *Contemplationes*, cont. ii. pars ii. c. 16, p. 34 (ed. 1588). Piccolomineus, *Physica*, p. 1313 *et seq.* (ed. 1597). Biel, *In Sent.*, lib. i. dist. iii. qu. 5. Zimara, *De Primo Cognito*, in calce t. iv. *Aristotelis Operum Averrois* (Venet. 1560). Fonseca, *In Metaph. Arist.*, lib. i. c. ii. qu. 2, t. i. p. 157-172. Berigardus, *Circulus Pisanus*, pp. 5, 6 (edit. 1661). Fracastorius, *De Intellectione*, lib. i. sub fine, *Opera* (ed. 1584), f. 130a. Herbart, *Lehrbuch zur Psychologie*, § 141. Crousaz, *Logique*, t. iii. part i. sect. iii. c. 4, p. 141.]

LECTURE XXXVII.

THE ELABORATIVE FACULTY.—JUDGMENT AND REASONING.

LECT. XXXVII.

Judgment and Reasoning.

In our last Lecture, I terminated the consideration of the faculty of Comparison in its process of Generalisation. I am to-day to consider it in those of its operations, which have obtained the special names of Judgment and Reasoning.

Acts of Comparison.

In these processes the act of Comparison is a judgment of something more than a mere affirmation of the existence of a phænomenon, — something more than a mere discrimination of one phænomenon from another; and, accordingly, while it has happened, that the intervention of judgment in every, even the simplest, act of primary cognition, as monotonous and rapid, has been overlooked, the name has been exclusively limited to the more varied and elaborate comparison of one notion with another, and the enouncement of their agreement or disagreement. It is in the discharge of this, its more obtrusive, function, that we are now about to consider the Elaborative Faculty.

Judgment and Reasoning, necessary from the limitation of the human mind.

Considering the Elaborative Faculty as a mean of discovering truth, by a comparison of the notions we have obtained from the Acquisitive Powers, it is evident that, though this faculty be the attribute by which man is distinguished as a creation higher than the animals, it is equally the quality which marks his

LECT. XXXVII. inferiority to superior intelligences. Judgment and Reasoning are rendered necessary by the imperfection of our nature. Were we capable of a knowledge of things and their relations at a single view, by an intuitive glance, discursive thought would be a superfluous act. It is by such an intuition that we must suppose that the Supreme Intelligence knows all things at once.

Our knowledge commences with the vague and confused.

I have already noticed that our knowledge does not commence with the individual, and the most particular, objects of knowledge,—that we do not rise in any regular progress from the less to the more general, first considering the qualities which characterise individuals, then those which belong to species and genera, in regular ascent. On the contrary, our knowledge commences with the vague and confused, in the way which Aristotle has so well illustrated in the passage alleged to you.[a] This I may further explain by another analogy. We perceive an object approaching from a distance. At first we do not know whether it be a living or an inanimate thing. By degrees we become aware that it is an animal, but of what kind,—whether man or beast,—we are not as yet able to determine. It continues to advance, we discover it to be a quadruped, but of what species we cannot yet say. At length, we perceive that it is a horse, and again, after a season, we find that it is Bucephalus. Thus, as I formerly observed, children, first of all, take note of the generic differences, and they can distinguish species long before they are able to discriminate individuals. In all this, however, I must again remark, that our knowledge does not properly commence with the general, but with the vague and confused. Out of

Illustrated.

a See above, p. 330.—Ed.

this the general and the individual are both equally evolved. LECT. XXXVII.

"In consequence of this genealogy of our knowledge we usually commence by bestowing a name upon a whole object, or congeries of objects, of which, however, we possess only a partial and indefinite conception. In the sequel, this vague notion becomes somewhat more determinate; the partial idea which we had becomes enlarged by new accessions; by degrees, our conception waxes fuller, and represents a greater number of attributes. With this conception, thus amplified and improved, we compare the last notion which has been acquired, that is to say, we compare a part with its whole, or with the other parts of this whole, and finding that it is harmonious,—that it dovetails and naturally assorts with other parts, we acquiesce in this union; and this we denominate an act of Judgment. Act of judgment,—what.

"In learning Arithmetic, I form the notion of the number *six*, as surpassing *five* by a single unit, and as surpassed in the same proportion by *seven*. Then I find that it can be divided into two equal halves, of which each contains three units. By this procedure, the notion of the number six becomes more complex; the notion of an even number is one of its parts. Comparing this new notion with that of the number, six becomes fuller by this addition. I recognise that the two notions suit,—in other words, I judge that six is an even number. Illustrated.

"I have the conception of a triangle, and this conception is composed in my mind of several others. Among these partial notions, I select that of two sides greater than the third, and this notion, which I had at first, as it were, taken apart, I reunite with the others from

LECT. XXXVII. which it had been separated, saying the triangle contains always two sides, which together are greater than the third.

"When I say, body is divisible; among the notions which concur in forming my conception of body, I particularly attend to that of divisible, and finding that it really agrees with the others, I judge accordingly that body is divisible.

Subject. Predicate. Copula.

"Every time we judge, we compare a total conception with a partial, and we recognise that the latter really constitutes a part of the former. One of these conceptions has received the name of *subject*, the other that of *attribute* or *predicate*."[a] The verb which connects these two parts is called the *copula*. *The quadrangle is a double triangle; nine is an odd number; body is divisible.* Here *quadrangle, nine, body*, are subjects; *a double triangle, an odd number, divisible*, are predicates. The whole mental judgment, formed by the subject, predicate, and copula, is called, when enounced in words, *proposition*.

Proposition.

How the parts of a proposition are to be discriminated.

"In discourse, the parts of a proposition are not always found placed in logical order; but to discover and discriminate them, it is only requisite to ask,—What is the thing of which something else is affirmed or denied? The answer to this question will point out the subject; and we shall find the predicate if we inquire,—What is affirmed or denied of the matter of which we speak?

"A proposition is sometimes so enounced that each of its terms may be considered as subject and as predicate. Thus, when we say,—*Death is the wages of sin;* we may regard *sin* as the subject of which we predicate *death*, as one of its consequences, and we

a Crousaz, [*Logique*, tom. iii. part ii. c. 1. pp. 178, 181.—ED.]

may likewise view *death* as the subject of which we predicate *sin*, as the origin. In these cases, we must consider the general tenor of the discourse, and determine from the context what is the matter of which it principally treats."

What Judgment involves.

"In fine, when we judge we must have, in the first place, at least two notions; in the second place, we compare these; in the third, we recognise that the one contains or excludes the other; and, in the fourth, we acquiesce in this recognition."[a]

Reasoning,—what.

Simple Comparison or Judgment is conversant with two notions, the one of which is contained in the other. But it often happens that one notion is contained in another not immediately, but mediately, and we may be able to recognise the relation of these to each other only through a third, which, as it immediately contains the one, is immediately contained in the other. Take the notions A, B, C.—A contains B; B contains C;—A, therefore, also contains C. But as, *ex hypothesi*, we do not at once and directly know C as contained in A, we cannot immediately compare them together, and judge of their relation. We, therefore, perform a double or complex process of comparison; we compare B with A, and C with B, and then C with A, through B. We say B is a part of A; C is a part of B; therefore, C is a part of A. This double act of comparison has obtained the name of *Reasoning;* the term *Judgment* being left to express the simple act of comparison, or rather its result.

Illustrated.

If this distinction between Judgment and Reasoning were merely a verbal difference to discriminate the simpler and more complex act of comparison, no objection could be raised to it on the score of pro-

a Crousaz, [*Logique*, t. iii. part ii. c. i. pp. 181, 186.—Ed.]

LECT. XXXVII.

priety, and its convenience would fully warrant its establishment. But this distinction has not always been meant to express nothing more. It has, in fact, been generally supposed to mark out two distinct faculties.

Reasoning, Deductive and Inductive.

Reasoning is either from the whole to its parts; or from all the parts, discretively, to the whole they constitute, collectively. The former of these is Deductive; the latter is Inductive Reasoning. The statement you will find, in all logical books, of reasonings from certain parts to the whole, or from certain parts to certain parts, is erroneous. I shall first speak of the reasoning from the whole to its parts,—or of the Deductive Inference.

Deductive Reasoning,—its axiom. Two phases of Deductive Reasoning, determined by two kinds of whole and parts.

1°, It is self-evident, that whatever is the part of a part, is a part of the whole. This one axiom is the foundation of all reasoning from the whole to the parts. There are, however, two kinds of whole and parts; and these constitute two varieties, or rather two phases of deductive reasoning. This distinction, which is of the most important kind, has nevertheless been wholly overlooked by logicians, in consequence of which the utmost perplexity and confusion have been introduced into the science.

Subject or predicate may be considered severally as whole and as part.

I have formerly stated that a proposition consists of two terms,—the one called subject, the other predicate; the subject being that of which some attribute is said, the predicate being the attribute so said. Now, in different relations, we may regard the subject as the whole, and the predicate as its part, or the predicate as the whole and the subject as its part.

Illustrated.

Let us take the proposition,—*milk is white*. Now, here we may either consider the predicate *white* as one of a number of attributes, the whole complement of which constitutes the subject *milk*. In this point of view, the predicate is a part of the subject. Or, again, we may consider the predicate *white* as the name of a class of objects, of which the subject is one. In this point of view, the subject is a part of the predicate.

Comprehension and Extension of notions, as applied to Reasoning.

You will remember the distinction, which I formerly stated, of the twofold quantity of notions or terms. The Breadth or Extension of a notion or term corresponds to the greater number of subjects contained under a predicate; the Depth, Intension, or Comprehension of a notion or term, to the greater number of predicates contained in a subject. These quantities or wholes are always in the inverse ratio of each other. Now, it is singular, that logicians should have taken this distinction between notions, and yet not have thought of applying it to reasoning. But so it is, and this is not the only oversight they have committed in the application of the very primary principles of their science. The great distinction we have established between the subject and predicate considered severally, as, in different relations, whole and as part, constitutes the primary and principal division of Syllogisms, both Deductive and Inductive; and its introduction wipes off a complex mass of rules and qualifications, which the want of it rendered necessary. I can of course, at present, only explain in general the nature of this distinction; its details belong to the science of the Laws of Thought, or Logic, of which we are not here to treat.

1. Deductive Reasoning in the whole of Comprehension,—in which the

I shall first consider the process of that Deductive Inference in which the subject is viewed as the whole, the predicate as the part. In this reasoning, the whole is determined by the Comprehension,

subject is viewed as the whole, the predicate as the part. This whole either Physical or Mathematical.

and is, again, either a Physical or Essential whole, or an Integral or Mathematical whole.[a] A Physical or Essential whole is that which consists of not really separable parts, of or pertaining to its substance. Thus, man is made up of two substantial parts,—a mind and a body; and each of these has again various qualities, which, though separable only by mental abstraction, are considered as so many parts of an essential whole. Thus the attributes of respiration, of digestion, of locomotion, of colour, are so many parts of the whole notion we have of the human body; cognition, feeling, desire, virtue, vice, &c., so many parts of the whole notion we have of the human mind; and all these together, so many parts of the whole notion we have of man. A Mathematical, or Integral, or Quantitative whole, is that which has part out of part, and which, therefore, can be really partitioned. The Integral or, as it ought to be called, Integrate whole (*totum integratum*), is composed of integrant parts (*partes integrantes*), which are either homogeneous, or heterogeneous. An example of the former is given in the division of a square into two triangles; of the latter, of the animal body into head, trunk, extremities, &c.

These wholes, (and there are others of less importance which I omit), are varieties of that whole which we may call a Comprehensive, or Metaphysical; it might be called a Natural whole.

Canon of Deductive reasoning in the whole of comprehension.

This being understood, let us consider how we proceed when we reason from the relation beween a comprehensive whole and its parts. Here, as I have said, the subject is the whole, the predicate its part; in

a See Eugenios, [Λογική, c. iv. pp. 196, 203 (1766).—Ed.] [Cf. Burgersdyck, *Institut. Logicæ*, lib. I. c. xiv. p. 52 *et seq.*, edit. 1660.]

other words, the predicate belongs to the subject. Now, here it is evident, that all the parts of the predicate must also be parts of the subject; in other terms, all that belongs to the predicate must also belong to the subject. In the words of the scholastic adage,—*Nota notæ est nota rei ipsius; Predicatum predicati est predicatum subjecti.* An example of this reasoning:—

Europe contains England;

England contains Middlesex;

Therefore, Europe contains Middlesex.

In other words;—England is an integrant part of Europe; Middlesex is an integrant part of England; therefore, Middlesex is an integrant part of Europe. This is an example from a mathematical whole and parts. Again:—

Socrates is just, (that is, Socrates contains justice as a quality);

Justice is a virtue, (that is, justice contains virtue as a constituent part);

Therefore, Socrates is virtuous.

In other words;—Justice is an attribute or essential part of Socrates; virtue is an attribute or essential part of justice; therefore, virtue is an attribute or essential part of Socrates. This is an example from a physical or essential whole and parts.

What I have now said will be enough to show, in general, what I mean by a deductive reasoning, in which the subject is the whole, the predicate the part.

2. Deductive Reasoning in the whole of Extension, in which the subject is viewed

I proceed, in the second place, to the other kind of Deductive Reasoning,—that in which the subject is the part, the predicate is the whole. This reasoning proceeds under that species of whole which has been called the Logical or Potential or Universal. This

LECT. XXXVII. whole is determined by the Extension of a notion; the genera having species, and the species individuals, as their parts. Thus *animal* is a universal whole, of which *bird* and *beast* are immediate, *eagle* and *sparrow*, *dog* and *horse*, mediate, parts; while *man*, which, in relation to animal, is a part, is a whole in relation to Peter, Paul, Socrates, &c. The parts of a logical or universal whole, I should notice, are called the *subject parts*.

as the part, the predicate as the whole.

From what you now know of the nature of generalisation, you are aware that general terms are terms expressive of attributes which may be predicated of many different objects; and inasmuch as these objects resemble each other in the common attribute, they are considered by us as constituting a class. Thus, when I say, that a horse is a quadruped; Bucephalus is a horse; therefore, Bucephalus is a quadruped;—I virtually say,—*horse* the subject is a part of the predicate *quadruped*, *Bucephalus* the subject is part of the predicate *horse*; therefore, *Bucephalus* the subject is part of the predicate *quadruped*. In the reasoning under this whole, you will observe that the same word, as it is whole or part, changes from predicate to subject; *horse*, when viewed as a part of *quadruped*, being the subject of the proposition; whereas when viewed as a whole, containing *Bucephalus*, it becomes the predicate.

Inductive Reasoning,—its axiom.

Such is a general view of the process of Deductive Reasoning, under the two great varieties determined by the two different kinds of whole and parts. I now proceed to the counter-process,—that of Inductive Reasoning. The deductive is founded on the axiom, that what is part of the part, is also part of the containing whole; the inductive on the principle, that what is true of every constituent part belongs, or does not belong, to the constituted whole.

Of two kinds, as it proceeds in the whole of Comprehension or of Extension.

Induction, like deduction, may be divided into two kinds, according as the whole and parts about which it is conversant, are a Comprehensive or Physical or Natural, or an Extensive or Logical, whole. Thus, in the former ;—

Gold is a metal, yellow, ductile, fusible in *aqua regia*, of a certain specific gravity, and so on ;

These qualities constitute this body, (are all its parts) ;

Therefore, this body is gold.

In the latter ;—Ox, horse, dog, &c., are animals,—that is, are contained under the class animal ;

Ox, horse, dog, &c., constitute, (are all the constituents of), the class quadruped.

Therefore, quadruped is contained under animal.

Deductive and Inductive illation must be of an absolute necessity.

Both in the deductive and inductive processes the inference must be of an absolute necessity, in so far as the mental illation is concerned ; that is, every consequent proposition must be evolved out of every antecedent proposition with intuitive evidence. I do not mean by this, that the antecedent should be necessarily true, or that the consequent be really contained in it ; it is sufficient that the antecedent be assumed as true, and that the consequent be, in conformity to the laws of thought, evolved out of it as its part or its equation. This last is called Logical or Formal or Subjective truth ; and an inference may be subjectively or formally true, which is objectively or really false.

Account of Induction by Logicians, erroneous.

The account given of Induction in all works of Logic is utterly erroneous. Sometimes we find this inference described as a precarious, not a necessary,

LECT. XXXVII. reasoning. It is called an illation from some to all. But here *the some*, as it neither contains nor constitutes *the all*, determines no necessary movement, and a conclusion drawn under these circumstances is logically vicious. Others again describe the inductive process thus:—

What belongs to some objects of a class belongs to the whole class;

This property belongs to some objects of the class;

Therefore, it belongs to the whole class.

This account of induction, which is the one you will find in all the English works on Logic, is not an inductive reasoning at all. It is, logically considered, a deductive syllogism; and, logically considered, a syllogism radically vicious. It is logically vicious to say, that, because some individuals of a class have certain common qualities apart from that property which constitutes the class itself, therefore the whole individuals of the class should partake in these qualities. For this there is no logical reason,—no necessity of thought. The probability of this inference, and it is only probable, is founded on the observation of the analogy of nature, and, therefore, not upon the laws of thought, by which alone reasoning, considered as a logical process, is exclusively governed. To become a formally legitimate induction, the objective probability must be clothed with a subjective necessity, and *the some* must be translated into *the all* which it is supposed to represent.

In Extension and Comprehension, the analysis of the one corresponds to the synthesis of the other.

In the deductive syllogism we proceed by analysis,—that is, by decomposing a whole into its parts; but as the two wholes with which reasoning is conversant are in the inverse ratio of each other, so our analysis in the one will correspond to our synthesis in the other. For example, when I divide a whole of extension into its parts,—when I divide a genus into the species, a species into the individuals, it contains,—I do so by adding new differences, and thus go on accumulating in the parts a complement of qualities which did not belong to the wholes. This, therefore, which, in point of extension, is an analysis, is, in point of comprehension, a synthesis. In like manner, when I decompose a whole of comprehension, that is, decompose a complex predicate into its constituent attributes, I obtain by this process a simpler and more general quality, and thus this, which, in relation to a comprehensive whole, is an analysis, is, in relation to an extensive whole, a synthesis.

As the deductive inference is Analytic, the inductive is Synthetic. But as induction, equally as deduction, is conversant with both wholes, so the Synthesis of induction on the comprehensive whole is a reversed process to its synthesis on the extensive whole.

Confusion among philosophers from not having observed this.

From what I have now stated, you will, therefore, be aware, that the terms *analysis* and *synthesis*, when used without qualification, may be employed, at cross purposes, to denote operations precisely the converse of each other. And so it has happened. Analysis, in the mouth of one set of philosophers, means precisely what synthesis denotes in the mouth of another; nay, what is even still more frequent, these words are perpetually converted with each other by the same philosopher. I may notice, what has rarely, if ever, been remarked, that *synthesis* in the writings of the Greek logicians is equivalent to the *analysis* of modern philosophers: the former, regarding the extensive whole as the principal, applied analysis, κατ' ἐξοχήν,

LECT. XXXVII.

to its division;[a] the latter, viewing the comprehensive whole as the principal, in general limit analysis to its decomposition. This, however, has been overlooked, and a confusion the most inextricable prevails in regard to the use of these words, if the thread to the labyrinth is not obtained.

[a] Thus the Platonic method of Division is called Analytical. See Laertius, iii. 24. Compare *Discussions*, p. 173.—ED. [Cf. Zabarella, *In Post. Analyt.*, lib. ii. c. xii. texts 70, 81. *Opera Logica*, pp. 1190, 1212.]

LECTURE XXXVIII.

THE REGULATIVE FACULTY.

LECT. XXXVIII.

The Regulative Faculty. Peculiarity of sense in which the term Faculty is here employed.

I NOW enter upon the last of the Cognitive Faculties,—the faculty which I denominated the Regulative. Here the term *faculty*, you will observe, is employed in a somewhat peculiar signification, for it is employed not to denote the proximate cause of any definite energy, but the power the mind has of being the native source of certain necessary or *a priori* cognitions; which cognitions, as they are the conditions, the forms, under which our knowledge in general is possible, constitute so many fundamental laws of intellectual nature. It is in this sense that I call the power which the mind possesses of modifying the knowledge it receives, in conformity to its proper nature, its Regulative Faculty. The Regulative Faculty is, however, in fact, nothing more than the complement of such laws,—it is the *locus principiorum*. It thus corresponds to what was known in the Greek philosophy under the name of νοῦς, when that term was rigorously used. To this faculty has been latterly applied the name *Reason*; but this term is so vague and ambiguous, that it is almost unfitted to convey any definite meaning. The term *Common Sense* has likewise been applied to designate the place of principles. This word is also ambiguous. In the first

Designations of the Regulative Faculty.—Νοῦς, Reason.

Common Sense,—its various meanings.

LECT. XXXVIII.

place, it was the expression used in the Aristotelic philosophy to denote the Central or Common Sensory, in which the different external senses met and were united.[α] In the second place, it was employed to signify a sound understanding applied to vulgar objects, in contrast to a scientific or speculative intelligence, and it is in this signification that it has been taken by those who have derided the principle on which the philosophy, which has been distinctively denominated the Scottish, professes to be established. This is not, however, the meaning which has always or even principally been attached to it; and an incomparably stronger case might be made out in defence of this expression than has been done by Reid, or even by Mr Stewart. It is in fact a term of high antiquity, and very general acceptation. We find it in Cicero,[β] in several passages not hitherto observed. It is found in the meaning in question in Phædrus,[γ] and not in the signification of community of sentiment, which it expresses in Horace[δ] and Juvenal.[ε] "Natura," says Tertullian,[ζ] speaking of the universal consent of mankind to the immortality of the soul,—"Natura pleraque suggeruntur quasi de *publico sensu*, quo animam Deus dotare dignatus est." And in the same meaning the term *Sensus Communis* is employed by St Augustin.[η] In modern times it is to be found in the philosophical writings of every country of Europe. In Latin it is used by the German Melanchthon, Victorinus,[ι] Keckermannus,[κ] Christian Thomasius,[λ]

Authorities for the use of the term Common Sense as equivalent to Νοῦς.

α See *De Anima*, iii. 2, 7. Cf. *In loc. cit.*, Conimbricenses, pp. 373, 407. —Ed.

β See *Reid's Works*, p. 774.—Ed.

γ L. i. f. 7.—Ed.

δ *Sat.*, i. 3, 66. But see *Reid's Works*, p. 774.—Ed.

ε *Sat.*, viii. 73.—Ed.

ζ See *Reid's Works*, p. 776.—Ed.

η *Ibid.*, p. 776.—Ed.

θ *Ibid.*, p. 778.—Ed.

ι [Victorinus Strigelius, *Hypomnemata in Dialect. Melanchthonis*, pp. 798, 1010, ed. 1566.]

κ See *Reid's Works*, p. 780.—Ed.

λ *Ibid.*, p. 785.—Ed.

Leibnitz,[α] Wolf,[β] and the Dutch De Raei,[γ] by the Gallo-Portuguese Antonius Goveanus,[δ] the Spanish Nunnesius,[ε] the Italian Genovesi,[ζ] and Vico,[η] and by the Scottish Abercromby;[θ] in French by Balzac,[ι] Chanet,[κ] Pascal,[λ] Malebranche,[μ] Bouhours, Barbeyrac;[ν] in English by Sir Thomas Browne,[ξ] Toland,[ο] Charleton.[π] These are only a few of the testimonies I could adduce in support of the term Common Sense for the faculty in question; in fact, so far as use and wont may be allowed to weigh, there is perhaps no philosophical expression in support of which a more numerous array of authorities may be alleged. The expression, however, is certainly exceptionable, and it can only claim toleration in the absence of a better.

I may notice that Pascal and Hemsterhuis[ρ] have applied *Intuition* and *Sentiment* in this sense; and Jacobi[σ] originally employed *Glaube*, (*Belief* or *Faith*), in the same way, though he latterly superseded this expression by that of *Vernunft*, (*Reason*.)

Noetic and Dianoetic,—how to be employed.

Were it allowed in metaphysical philosophy, as in physical, to discriminate scientific differences by scientific terms, I would employ the word *noetic*, as derived

α See *Reid's Works*, p. 785.—Ed.

β *Ibid.*, p. 790.—Ed.

γ See *Clavis Philosophiæ Naturalis Aristotelico-Cartesianæ*, Dissert. I. *De Cognitione Vulgari et Philosophica*, p. 7. "Communis facultas omnium hominum;" Dissert. II. *De Prærogativis in Genere*, §§ iv. v. pp. 34, 35. "Communes Notiones;" § x. p. 41. "Communis Sensus."—Ed.

δ See *Reid's Works*, p. 779.—Ed.

ε *Ibid.*—Ed.

ζ *Ibid.*, p. 790.—Ed.

η *Ibid.*—Ed.

θ *Ibid.*, p. 785.—Ed.

ι *Ibid.*, p. 782.—Ed.

κ *Ibid.*—Ed.

λ *Ibid.*, p. 783.—Ed.

μ *Ibid.*, p. 784.—Ed.

ν *Des Droits de la Puissance Souveraine*, *Recueil de Discours*, t. I. pp. 36, 37. A translation from the Latin of Noodt, in which *mens sana* and *sensus communis* are both rendered by *le sens commun*.—Ed.

ξ See *Reid's Works*, p. 782.—Ed.

ο *Ibid.*, p. 785.—Ed.

π Charleton uses the term in its Aristotelian signification, as denoting the central or common sensory and its function. See his *Immortality of the Human Soul demonstrated by the Light of Nature* (1657), pp. 92, 98, 158.—Ed.

ρ See *Reid's Works*, p. 792.—Ed.

σ *Ibid.*, p. 793.—Ed.

LECT. XXXVIII.

Nomenclature of the cognitions due to the Regulative Faculty.

from νοῦς, to express all those cognitions that originate in the mind itself, *dianoetic* to denote the operations of the Discursive, Elaborative, or Comparative Faculty. So much for the nomenclature of the faculty itself.

On the other hand, the cognitions themselves, of which it is the source, have obtained various appellations. They have been denominated *κοιναὶ προλήψεις, κοιναὶ ἔννοιαι, φυσικαὶ ἔννοιαι, πρῶται ἔννοιαι, πρῶτα νοήματα; naturæ judicia, judicia communibus hominum sensibus infixa, notiones* or *notitiæ connatæ* or *innatæ, semina scientiæ, semina omnium cognitionum, semina æternitatis, zopyra, (living sparks), præcognita necessaria, anticipationes; first principles, common anticipations, principles of common sense, self-evident* or *intuitive truths, primitive notions, native notions, innate cognitions, natural knowledges (cognitions), fundamental reasons, metaphysical* or *transcendental truths, ultimate* or *elemental laws of thought, primary* or *fundamental laws of human belief,* or *primary laws of human reason, pure* or *transcendental* or *a priori cognitions, categories of thought, natural beliefs, rational instincts,* &c. &c.[a]

Importance of the distinction of native and adventitious knowledge.

The history of opinions touching the acceptation, or rejection, of such native notions, is, in a manner, the history of philosophy; for as the one alternative, or the other, is adopted in this question, the character of a system is determined. At present I content myself with stating that, though from the earliest period of philosophy, the doctrine was always common, if not always predominant, that our knowledge originated, in part at least, in the mind, yet it was only at a very recent date that the criterion was explicitly enounced, by which the native may be

[a] See *Reid's Works*, Note A, § v. p. 755 *et seq.*—Ed.

discriminated from the adventitious elements of knowledge. Without touching on some ambiguous expressions in more ancient philosophers, it is sufficient to say that the character of universality and necessity, as the quality by which the two classes of knowledge are distinguished, was first explicitly proclaimed by Leibnitz. It is true, indeed, that, previously to him, Descartes all but enounced it. In the notes of Descartes on the *Programma* of 1647, (which you will find under Letter XCIX. of the First Part of his *Epistolæ*), in arguing against the author who would derive all our knowledge from observation or tradition, he has the following sentence:—"I wish that our author would inform me what is that corporeal motion which is able to form in our intellect any common notion,—for example, things that are equal to the same thing are equal to each other, or any other of the same kind; for all those motions are particular, but these notions are universal, having no affinity with motions, and holding no relation to them." Now, had he only added the term *necessary* to universal, he would have completely anticipated Leibnitz. I have already frequently had occasion incidentally to notice, that we should carefully distinguish between those notions or cognitions which are primitive facts, and those notions or cognitions which are generalised or derivative facts. The former are given us; they are not, indeed, obtrusive,—they are not even cognisable of themselves. They lie hid in the profundities of the mind, until drawn from their obscurity by the mental activity itself employed upon the materials of experience. Hence it is, that our knowledge has its commencement in sense, external or internal, but its origin in intellect. "Cognitio omnia a sensibus exordium, a

Criterion of necessity first enounced by Leibnitz.

Partially anticipated by Descartes.

LECT. XXXVIII. mente originem habet primum."[α] The latter, the derivative cognitions, are of our own fabrication; we form them after certain rules; they are the tardy result of Perception and Memory, of Attention, Reflection, Abstraction. The primitive cognitions, on the contrary, seem to leap ready armed from the womb of reason, like Pallas from the head of Jupiter; sometimes the mind places them at the commencement of its operations, in order to have a point of support and a fixed basis, without which the operations would be impossible; sometimes they form, in a certain sort, the crowning,—the consummation, of all the intellectual operations. The derivative or generalised notions are an artifice of intellect,—an ingenious mean of giving order and compactness to the materials of our knowledge. The primitive and general notions are the root of all principles,—the foundation of the whole edifice of human science. But how different soever be the two classes of our cognitions, and however distinctly separated they may be by the circumstance,—that we cannot but think the one, and can easily annihilate the other in thought,—this discriminative quality was not explicitly signalised till done by Leibnitz. The older philosophers are at best undeveloped. Descartes made the first step towards a more perspicuous and definite discrimination. He frequently enounces that our primitive notions, (besides being clear and distinct), are universal. But this universality is only a derived circumstance;—a notion is universal, (meaning thereby that a notion is common to all mankind), because it is necessary to the thinking mind,—because the mind cannot but think it. Spinoza, in one passage of his treatise *De Emendatione Intellectus*,[β] says:—"The ideas which we form clear and distinct, appear

And by Spinoza.

α See above, vol. ii. lect. xxi. p. 27.—Ed. β *Opera Posthuma*, p. 391.

so to follow from the sole necessity of our nature, that they seem absolutely to depend from our sole power [of thought]; the confused ideas on the contrary," &c. This is anything but explicit; and, as I said, Leibnitz is the first by whom the criterion of necessity,—of the impossibility not to think so and so,—was established as a discriminative type of our native notions, in contrast to those which we educe from experience, and build up through generalisation.

The enouncement of this criterion, a great step in the science of mind.

The enouncement of this criterion was, in fact, a great discovery in the science of mind; and the fact that a truth so manifest, when once proclaimed, could have lain so long unnoticed by philosophers, may warrant us in hoping that other discoveries of equal importance may still be awaiting the advent of another Leibnitz. Leibnitz has, in several parts of his works, laid down the distinction in question; and, what is curious, almost always in relation to Locke. In the fifth volume of his works by Dutens,[a] in an Epistle to Bierling of 1710, he says, (I translate from the Latin):—

Leibnitz quoted.

"In Locke there are some particulars not ill expounded, but upon the whole he has wandered far from the gate,[β] nor has he understood the nature of the intellect, (natura mentis). Had he sufficiently considered the difference between necessary truths or those apprehended by demonstration, and those which become known to us by induction alone,—he would have seen that those which are necessary, could only be approved to us by principles native to the mind, (menti insitis); seeing that the senses indeed inform us what may take place, but not what necessarily takes place. Locke has not observed, that the notions of being, of substance, of one and the

[a] P. 358. [β] This refers to Aristotle's *Metaphysics* [A Minor, c. I.—ED.]

LECT. XXXVIII.

same, of the true, of the good, and many others, are innate to our mind, because our mind is innate to itself, and finds all these in its own furniture. It is true, indeed, that there is nothing in the intellect which was not previously in the sense,—except the intellect itself." He makes a similar observation in reference to Locke, in Letter XI., to his friend Mr Burnet of Kemnay.[a] And in his *Nouveaux Essais*, (a detailed refutation of Locke's Essay, and not contained in the collected edition of his works by Dutens), he repeatedly enforces the same doctrine. In one place he says,[β]—"Hence there arises another question, viz.:—Are all truths dependent on experience, that is to say, on induction and examples? Or are there some which have another foundation? For if some events can be foreseen before all trial has been made, it is manifest that we contribute something on our part. The senses, although necessary for all our actual cognitions, are not, however, competent to afford us all that cognitions involve; for the senses never give us more than examples, that is to say, particular or individual truths. Now all the examples which confirm a general truth, how numerous soever they may be, are insufficient to establish the universal necessity of this same truth; for it does not follow that what has happened will happen always in like manner. For example; the Greeks and Romans and other nations have always observed that during the course of twenty-four hours, day is changed into night, and night into day. But we should be wrong, were we to believe that the same rule holds everywhere, as the contrary has been observed during a residence in

Leibnitz further quoted.

a *Opera*, vol. vi. p. 274 (edit. Dutens).

β Avant Propos, p. 5 (edit. Raspe).

Nova Zembla. And he again would deceive himself, who should believe that, in our latitudes at least, this was a truth necessary and eternal; for we ought to consider, that the earth and the sun themselves have no necessary existence, and that there will perhaps a time arrive when this fair star will, with its whole system, have no longer a place in creation,—at least under its present form. Hence it appears, that the necessary truths, such as we find them in Pure Mathematics, and particularly in Arithmetic and Geometry, behove to have principles the proof of which does not depend upon examples, and, consequently, not on the evidence of sense; howbeit, that without the senses, we should never have found occasion to call them into consciousness. This is what it is necessary to distinguish accurately, and it is what Euclid has so well understood, in demonstrating by reason what is sufficiently apparent by experience and sensible images. Logic, likewise, with Metaphysics and Morals, the one of which constitutes Natural Theology, the other Natural Jurisprudence, are full of such truths; and, consequently, their proof can only be derived from internal principles, which we call innate. It is true, that we ought not to imagine that we can read in the soul, these eternal laws of reason, *ad aperturam libri*, as we can read the edict of the Prætor without trouble or research; but it is enough, that we can discover them in ourselves by dint of attention, when the occasions are presented to us by the senses. The success of the observation serves to confirm reason, in the same way as proofs serve in Arithmetic to obviate erroneous calculations, when the computation is long. It is hereby, also, that the cognitions of men differ from those of beasts. The beasts are purely empirical, and only regulate themselves by examples; for as far as we can judge, they never attain to the formation of necessary judgments, whereas, men are capable of demonstrative sciences, and herein the faculty which brutes possess of drawing inferences is inferior to the reason which is in men." And, after some other observations, he proceeds:—"Perhaps our able author," (he refers to Locke), "will not be wholly alien from my opinion. For after having employed the whole of his first book to refute innate cognitions, taken in a certain sense, he, however, avows, at the commencement of the second and afterwards, that ideas which have not their origin in Sensation, come from Reflection. Now reflection is nothing else than an attention to what is in us, and the senses do not inform us of what we already carry with us. This being the case, can it be denied that there is much that is innate in our mind, seeing that we are as it were innate to ourselves, and that there are in us existence, unity, substance, duration, change, action, perception, pleasure, and a thousand other objects of our intellectual notions? These same objects being immediate, and always present to our understanding, (although they are not always perceived by reason of our distractions and our wants), why should it be a matter of wonder, if we say that these ideas are innate in us, with all that is dependent on them? In illustration of this, let me make use likewise of the simile of a block of marble which has veins, rather than of a block of marble wholly uniform, or of blank tablets, that is to say, what is called a *tabula rasa* by philosophers; for if the mind resembled these blank tablets, truths would be in us, as the figure of Hercules is in a piece of marble, when the marble is altogether indifferent to the reception of this figure or of any other. But if we suppose that there are veins in the stone, which would mark out the figure of Hercules by preference to other figures, this stone would be more determined thereunto, and Hercules would exist there, innately in a certain sort; although it would require labour to discover the veins, and to clear them by polishing and the removal of all that prevents their manifestation. It is thus that ideas and truths are innate in us; like our inclinations, dispositions, natural habitudes or virtualities, and not as actions; although these virtualities be always accompanied by some corresponding actions, frequently, however, unperceived.

"It seems that our able author [Locke] maintains, that there is nothing virtual in us, and even nothing of which we are [not] always actually conscious. But this cannot be strictly intended, for in that case his opinion would be paradoxical, since even our acquired habits and the stores of our memory are not always in actual consciousness, nay, do not always come to our aid when wanted; while again, we often call them to mind on any trifling occasion which suggests them to our remembrance, like as it only requires us to be given the commencement of a song to help us to the recollection of the rest. He, therefore, limits his thesis in other places, saying that there is at least nothing in us which we have not, at some time or other, acquired by experience and perception." And in another remarkable passage,[a] Leibnitz says, "The mind is not only capable of knowing pure and necessary truths, but likewise of discovering them in itself; and if it possessed only the simple capacity of receiving cognitions, or the passive power of knowledge, as indetermined as that of the

[a] *Nouveaux Essais*, p. 36 (edit. Raspe). [Liv. I. § 5.—Ed.]

LECT. XXXVIII. wax to receive figures, or a blank tablet to receive letters, it would not be the source of necessary truths, as I am about to demonstrate that it is: for it is incontestable, that the senses could not suffice to make their necessity apparent, and that the intellect has, therefore, a disposition, as well active as passive, to draw them from its own bosom, although the senses be requisite to furnish the occasion, and the attention to determine it upon some in preference to others. You see, therefore, these very able philosophers, who are of a different opinion, have not sufficiently reflected on the consequences of the difference that subsists between necessary or eternal truths and the truths of experience, as I have already observed, and as all our contestation shows. The original proof of necessary truths comes from the intellect alone, while other truths are derived from experience or the observations of sense. Our mind is competent to both kinds of knowledge, but it is itself the source of the former; and how great soever may be the number of particular experiences in support of a universal truth, we should never be able to assure ourselves for ever of its universality by induction, unless we knew its necessity by reason. . . . . . . The senses may register, justify, and confirm these truths, but not demonstrate their infallibility and eternal certainty."

And in speaking of the faculty of such truths, he says: "It is not a naked faculty, which consists in the mere possibility of understanding them; it is a disposition, an aptitude, a preformation, which determines our mind to elicit, and which causes that they can be elicited; precisely as there is a difference between the figures which are bestowed indifferently on stone or marble, and those which veins mark out or are disposed to mark out, if the sculptor avail himself of the indications."[a] I have quoted these passages from Leibnitz, not only for their own great importance, as the first full and explicit enouncement, and certainly not the least able illustrations, of one of the most momentous principles in philosophy; but, likewise, because the *Nouveaux Essais*, from which they are principally extracted, though of all others the most important psychological work of Leibnitz, was wholly unknown, not only to the other philosophers of this country, but even to Mr Stewart, prior to the last years of his life.[β]

Reid discriminated native from adventitious knowledge by the same difference, independently of Leibnitz.

We have thus seen that Leibnitz was the first philosopher who explicitly established the quality of necessity as the criterion of distinction between empirical and *a priori* cognitions. I may, however, remark, what is creditable to Dr Reid's sagacity, that he founded the same discrimination on the same difference: and I am disposed to think, that he did this without being aware of his coincidence with Leibnitz; for he does not seem to have studied the system of that philosopher in his own works; and it was not till Kant had shown the importance of the criterion, by its application in his hands, that the attention of the learned was called to the scattered notices of it in the writings of Leibnitz. In speaking of the principle of causality, Dr Reid says:—"We are next to consider whether we may not learn this truth from experience,—That

a *Nouv. Essais*, liv. i. § 11. See above, vol. ii. lect. xxix. p. 195.—Ed.

β The reason of this was, that it was not published till long after the death of its author, and it is not included in the collected edition of the works of Leibnitz by Dutens. In consequence of its republication in *Leibnitzii Opera Philosophica*, by Erdmann, it is now easily procured.

LECT. XXXVIII.

effects which have all the marks and tokens of design, must proceed from a designing cause."

Reid quoted.

"I apprehend that we cannot learn this truth from experience, for two reasons.

"*First*, Because it is a necessary truth, not a contingent one. It agrees with the experience of mankind since the beginning of the world, that the area of a triangle is equal to half the rectangle under its base and perpendicular. It agrees no less with experience, that the sun rises in the east and sets in the west. So far as experience goes, these truths are upon an equal footing. But every man perceives this distinction between them,—that the first is a necessary truth, and that it is impossible it should not be true; but the last is not necessary, but contingent, depending upon the will of Him who made the world. As we cannot learn from experience that twice three must necessarily make six, so neither can we learn from experience that certain effects must proceed from a designing and intelligent cause. Experience informs us only of what has been, but never of what must be."*

And in speaking of our belief in the principle that an effect manifesting design must have had an intelligent cause, he says:—"It has been thought, that, although this principle does not admit of proof from abstract reasoning, it may be proved from experience, and may be justly drawn by induction, from instances that fall within our observation.

"I conceive this method of proof will leave us in great uncertainty, for these three reasons:

"1*st*, Because the proposition to be proved is not

* *Int. Powers*, Essay vi. chap. vi. *Coll. Works*, p. 459.

a contingent but a *necessary* proposition. It is not that things which begin to exist commonly have a cause, or even that they always in fact have a cause; but that they must have a cause, and cannot begin to exist without a cause.

"Propositions of this kind, from their nature, are incapable of proof by induction. Experience informs us only of what *is* or *has been*, not of what *must be*; and the conclusion must be of the same nature with the premises.

"For this reason, no mathematical proposition can be proved by induction. Though it should be found by experience in a thousand cases, that the area of a plane triangle is equal to the rectangle under the altitude and half the base, this would not prove that it must be so in all cases, and cannot be otherwise; which is what the mathematician affirms.

"In like manner, though we had the most ample experimental proof, that things which have begun to exist had a cause, this would not prove that they must have a cause. Experience may show us what is the established course of nature, but can never show what connections of things are in their nature necessary.

"*2dly*, General maxims, grounded on experience, have only a degree of probability proportioned to the extent of our experience, and ought always to be understood so as to leave room for exceptions, if future experience shall discover any such.

"The law of gravitation has as full a proof from experience and induction as any principle can be supposed to have. Yet, if any philosopher should, by clear experiment, show that there is a kind of matter

LECT. XXXVIII.

in some bodies which does not gravitate, the law of gravitation ought to be limited by that exception.

"Now, it is evident that men have never considered the principle of the necessity of causes, as a truth of this kind which may admit of limitation or exception; and therefore it has not been received upon this kind of evidence.

"3*dly*, I do not see that experience could satisfy us that every change in nature actually has a cause.

"In the far greatest part of the changes in nature that fall within our observation, the causes are unknown; and, therefore, from experience, we cannot know whether they have causes or not.

"Causation is not an object of sense. The only experience we can have of it, is in the consciousness we have of exerting some power in ordering our thoughts and actions. But this experience is surely too narrow a foundation for a general conclusion, that all things that have had or shall have a beginning, must have a cause.

"For these reasons, this principle cannot be drawn from experience, any more than from abstract reasoning."[a]

Hume arrived at the same conclusion.

It ought, however, to be noticed that Mr Hume's acuteness had arrived at the same conclusion. "As to past experience," he observes, "it can be allowed to give direct and certain information of those precise objects only, and that precise period of time, which fell under its cognisance; but why this experience should be extended to future times and to other objects,—this is the main question on which I would insist."[β]

a *Intellectual Powers*, Essay vi. chap. vi. *Coll. Works*, pp. 455, 456. Reid has several other passages to the same effect in the same chapter of this Essay.

β *Inquiry concerning the Human Understanding*, § iv. *Philosophical Works*, vol. iv. p. 42.—Ed.

The philosopher, however, who has best known how to turn the criterion to account, is Kant; and the general success with which he has applied it, must be admitted even by those who demur to many of the particular conclusions which his philosophy would establish.

Philosophers divided in regard to what cognitions ought to be classed as ultimate; and what as modifications of the ultimate.

But though it be now generally acknowledged, by the profoundest thinkers, that it is impossible to analyse all our knowledge into the produce of experience, external or internal, and that a certain complement of cognitions must be allowed as having their origin in the nature of the thinking principle itself; they are not at one in regard to those which ought to be recognised as ultimate and elemental, and those which ought to be regarded as modifications or combinations of these. Reid and Stewart, (the former in particular), have been considered as too easy in their admission of primary laws; and it must be allowed that the censure, in some instances, is not altogether unmerited. But it ought to be recollected, that those who thus agree in reprehension are not in unison in regard to the grounds of censure; and they wholly forget that our Scottish philosophers made no pretension to a final analysis of the primary laws of human reason,—that they thought it enough to classify a certain number of cognitions as native to the mind, leaving it to their successors to resolve these into simpler elements.

Reid and Stewart have been censured for their too easy admission of first principles.

Reid quoted in self-vindication.

"The most general phænomena," says Dr Reid,* "we can reach, are what we call Laws of Nature. So that the laws of nature are nothing else but the most general facts relating to the operations of nature, which include a great many particular facts under them. And if, in any case, we should give the name

* *Inquiry*, chap. vi. § 13, *Works*, p. 163.—Ed.

LECT. XXXVIII.

of a law of nature to a general phænomenon, which human industry shall afterwards trace to one more general, there is no great harm done. The most general assumes the name of a law of nature when it is discovered; and the less general is contained and comprehended in it." In another part of his work, he has introduced the same remark. "The labyrinth may be too intricate, and the thread too fine, to be traced through all its windings; but, if we stop where we can trace it no farther, and secure the ground we have gained, there is no harm done; a quicker eye may in time trace it farther."[α] The same view has been likewise well stated by Mr Stewart.[β] "In all the other sciences, the progress of discovery has been gradual, from the less general to the more general laws of nature; and it would be singular indeed, if, in this science, which but a few years ago was confessedly in its infancy, and which certainly labours under many disadvantages peculiar to itself, a step should all at once be made to a single principle, comprehending all the particular phænomena which we know. As the order established in the intellectual world seems to be regulated by laws analogous to those which we trace among the phænomena of the material system; and as in all our philosophical inquiries, (to whatever subject they may relate), the progress of the mind is liable to be affected by the same tendency to a premature generalisation, the following extract from an eminent chemical writer may contribute to illustrate the scope and to confirm the justness of some of the foregoing reflections. 'Within

Stewart quoted to the same effect.

α *Inquiry into the Human Mind*, c. i. § 2. *Works*, p. 99.—Ed.

β *Philosophical Essays*, Prel. Diss. c. i. *Works*, vol. v. p. 13. Cf. *Elements*, vol. i. c. v. p. 2, § 4. *Works*, vol. ii. pp. 342, 343.—Ed.

the last fifteen or twenty years, several new metals and new earths have been made known to the world. The names that support these discoveries are respectable, and the experiments decisive. If we do not give our assent to them, no single proposition in chemistry can for a moment stand. But whether all these are really simple substances, or compounds not yet resolved into their elements, is what the authors themselves cannot possibly assert; nor would it, in the least, diminish the merit of their observations, if future experiments should prove them to have been mistaken, as to the simplicity of these substances. This remark should not be confined to later discoveries; it may as justly be applied to those earths and metals with which we have been long acquainted.' 'In the dark ages of chemistry, the object was to rival nature; and the substance which the adepts of those days were busied to create, was universally allowed to be simple. In a more enlightened period, we have extended our inquiries and multiplied the number of the elements. The last task will be to simplify; and by a closer observation of nature, to learn from what a small store of primitive materials, all that we behold and wonder at was created.'"

That the list of the primary elements of human reason, which our two philosophers have given, has no pretence to order; and that the principles which it contains are not systematically deduced by any ambitious process of metaphysical ingenuity, is no valid ground of disparagement. In fact, which of the vaunted classifications of these primitive truths can stand the test of criticism? The most celebrated, and by far the most ingenious, of these,—the scheme of Kant,—though the truth of its details may be admitted,

That Reid and Stewart offer no systematic deduction of the primary elements of human reason, is no valid ground for disparaging their labours.

LECT. XXXVIII.

is no longer regarded as affording either a necessary deduction or a natural arrangement of our native cognitions; and the reduction of these to system still remains a problem to be resolved.

Philosophers have not yet established the principle on which our ultimate cognitions are to be classified, and reduced to system.

In point of fact, philosophers have not yet purified the antecedent conditions of the problem,—have not yet established the principles on which its solution ought to be undertaken. And here I would solicit your attention to a circumstance, which shows how far philosophers are still removed from the prospect of an ultimate decision. It is agreed, that the quality of necessity is that which discriminates a native from an adventitious element of knowledge. When we find, therefore, a cognition which contains this discriminative quality, we are entitled to lay it down as one which could not have been obtained as a generalisation from experience. This I admit. But when philosophers lay it down not only as native to the mind, but as a positive and immediate datum of an intellectual power, I demur. It is evident that the quality of necessity in a cognition may depend on two different and opposite principles, inasmuch as it may either be the result of a power, or of a powerlessness, of the thinking principle. In the one case, it will be a Positive, in the other a Negative, necessity. Let us take examples of these opposite cases. In an act of perceptive consciousness, I think, and cannot but think, that I and that something different from me exist,—in other words, that my perception, as a modification of the ego, exists, and that the object of my perception, as a modification of the non-ego, exists. In these circumstances, I pronounce Existence to be a native cognition, because I find that I cannot think except under the condition of thinking all that I am con-

Necessity,—either Positive, or Negative, as it results from a power, or from a powerlessness of mind.

The first order of Necessity,—the Positive,—illustrated, by the act of Perception.

scious of to exist. Existence is thus a form, a category, of thought. But here, though I cannot but think existence, I am conscious of this thought as an act of power,—an act of intellectual force. It is the result of strength, and not of weakness.

By an arithmetical example.

In like manner, when I think $2 \times 2 = 4$, the thought, though inevitable, is not felt as an imbecility; we know it as true, and, in the perception of the truth, though the act be necessary, the mind is conscious that the necessity does not arise from impotence. On the contrary, we attribute the same necessity to God. Here, therefore, there is a class of natural cognitions, which we may properly view as so many positive exertions of the mental vigour, and the cognitions of this class we consider as Positive. To this class will belong the notion of Existence and its modifications, the principles of Identity, and Contradiction, and Excluded Middle, the intuitions of Space and Time, &c.

The second order of necessity,—the Negative. This not recognised by philosophers.

But besides these, there are other necessary forms of thought, which, by all philosophers, have been regarded as standing precisely on the same footing, which to me seem to be of a totally different kind. In place of being the result of a power, the necessity which belongs to them is merely a consequence of the impotence of our faculties. But if this be the case, nothing could be more unphilosophical than to arrogate to these negative inabilities the dignity of positive energies. Every rule of philosophising would be violated. The law of Parcimony prescribes, that principles are not to be multiplied without necessity, and that an hypothetical force be not postulated to explain a phænomenon which can be better accounted for by an admitted impotence. The phænomenon of a heavy body rising from the earth, may warrant us in the assumption of a special power; but it would surely be absurd to devise a special power, (that is, a power besides gravitation), to explain the phænomenon of its descent.

Illustrated.

Now, that the imbecility of the human mind constitutes a great negative principle, to which sundry of the most important phænomena of intelligence may be referred, appears to me incontestable; and though the discussion is one somewhat abstract, I shall endeavour to give you an insight into the nature and application of this principle.

Principles referred to in the discussion.

I begin by the statement of certain principles, to which it is necessary in the sequel to refer.

1. The Law of Non-Contradiction.

The highest of all logical laws, in other words, the supreme law of thought, is what is called the principle of Contradiction, or more correctly the principle of Non-Contradiction.[a] It is this:—A thing cannot be and not be at the same time,—*Alpha est, Alpha non est*, are propositions which cannot both be true at once.

2. The Law of Excluded Middle.

A second fundamental law of thought, or rather the principle of Contradiction viewed in a certain aspect, is called the principle of Excluded Middle, or, more fully, the principle of Excluded Middle between two Contradictories. A thing either is or it is not,—*Aut est Alpha aut non est;* there is no medium; one must be true, both cannot. These principles require, indeed admit of, no proof. They prove everything, but are proved by nothing. When I, therefore, have occasion to speak of these laws by name, you will know to what principle I refer.

Grand law of thought,—That the conceivable lies between two contradictory extremes.

Now, then, I lay it down as a law which, though not generalised by philosophers, can be easily proved to be true by its application to the phænomena;—That all that is conceivable in thought, lies between two

[a] See Appendix, II. Ed.

extremes, which, as contradictory of each other, cannot both be true, but of which, as mutual contradictories, one must. For example, we conceive space,—we cannot but conceive space. I admit, therefore, that Space, indefinitely, is a positive and necessary form of thought. But when philosophers convert the fact, that we cannot but think space, or, to express it differently, that we are unable to imagine anything out of space,—when philosophers, I say, convert this fact with the assertion, that we have a notion,—a positive notion, of absolute or of infinite space, they assume, not only what is not contained in the phænomenon, nay, they assume what is the very reverse of what the phænomenon manifests. It is plain, that space must either be bounded or not bounded. These are contradictory alternatives; on the principle of Contradiction, they cannot both be true, and, on the principle of Excluded Middle, one must be true. This cannot be denied, without denying the primary laws of intelligence. But though space must be admitted to be necessarily either finite or infinite, we are able to conceive the possibility, neither of its finitude, nor of its infinity.

Established and illustrated, by reference to Space,—1°, as a Maximum.

Space either bounded or not bounded.

We are altogether unable to conceive space as bounded,—as finite; that is, as a whole beyond which there is no further space. Every one is conscious that this is impossible. It contradicts also the supposition of space as a necessary notion; for if we could imagine space as a terminated sphere, and that sphere not itself enclosed in a surrounding space, we should not be obliged to think everything in space; and, on the contrary, if we did imagine this terminated sphere as itself in space, in that case we should not have actually conceived all space as a bounded whole. The one

Space as absolutely bounded inconceivable.

contradictory is thus found inconceivable; we cannot conceive space as positively limited.

Space, as infinitely unbounded, inconceivable.

On the other hand, we are equally powerless to realise in thought the possibility of the opposite contradictory; we cannot conceive space as infinite, as without limits. You may launch out in thought beyond the solar walk, you may transcend in fancy even the universe of matter, and rise from sphere to sphere in the region of empty space, until imagination sinks exhausted;—with all this what have you done? You have never gone beyond the finite, you have attained at best only to the indefinite, and the indefinite, however expanded, is still always the finite. As Pascal energetically says, "Inflate our conceptions as we may, with all the finite possible we cannot make one atom of the infinite."[α] "The infinite is infinitely incomprehensible."[β] Now then, both contradictories are equally inconceivable, and could we limit our attention to one alone, we should deem it at once impossible and absurd, and suppose its unknown opposite as necessarily true. But as we not only can, but are constrained to consider both, we find that both are equally incomprehensible; and yet though unable to view either as possible, we are forced by a higher law to admit that one, but one only, is necessary.

Though both these contradictory alternatives are inconceivable, one or other is yet necessary.

Space, 2°, as a Minimum.

That the conceivable lies always between two inconceivable extremes, is illustrated by every other relation of thought. We have found the maximum of space incomprehensible, can we comprehend its minimum? This is equally impossible. Here, likewise, we recoil from one inconceivable contradictory

α *Pensées*, Première Partie, art. iv. § 1, (vol. ii. p. 64, edit. Faugère.) Pascal's words are:—"Nous avons beau enfler nos conceptions au delà des espaces imaginables; nous n'enfantons que des atomes, au prix de la réalité des choses."—Ed.

β *Ibid.*, Sec. Part., art. iii. § 1.—Ed.

only to infringe upon another. Let us take a portion of space however small, we can never conceive it as the smallest. It is necessarily extended, and may, consequently, be divided into a half or quarters, and each of these halves or quarters may again be divided into other halves or quarters, and this *ad infinitum*. But if we are unable to construe to our mind the possibility of an absolute minimum of space, we can as little represent to ourselves the possibility of an infinite divisibility of any extended entity.

LECT. XXXVIII.

An absolute minimum of space, and its infinite divisibility, alike inconceivable.

In like manner Time;—this is a notion even more universal than space, for while we exempt from occupying space the energies of mind, we are unable to conceive these as not occupying time. Thus, we think everything, mental and material, as in time, and out of time we can think nothing. But, if we attempt to comprehend time, either in whole or in part, we find that thought is hedged in between two incomprehensibles. Let us try the whole. And here let us look back,—let us consider time *a parte ante*. And here we may surely flatter ourselves that we shall be able to conceive time as a whole, for here we have the past period bounded by the present; the past cannot, therefore, be infinite or eternal, for a bounded infinite is a contradiction. But we shall deceive ourselves. We are altogether unable to conceive time as commencing; we can easily represent to ourselves time under any relative limitation of commencement and termination, but we are conscious to ourselves of nothing more clearly, than that it would be equally possible to think without thought, as to construe to the mind an absolute commencement, or an absolute termination, of time, that is, a beginning and an end, beyond which time is conceived as

Further illustration by reference to Time;—1°, as a Maximum.

1. Time, *a parte ante*, as an absolute whole, inconceivable.

non-existent. Goad imagination to the utmost, it still sinks paralysed within the bounds of time, and time survives as the condition of the thought itself in which we annihilate the universe. On the other hand, the concept of past time as without limit, — without commencement, is equally impossible. We cannot conceive the infinite regress of time; for such a notion could only be realised by the infinite addition in thought of finite times, and such an addition would itself require an eternity for its accomplishment. If we dream of effecting this, we only deceive ourselves by substituting the indefinite for the infinite, than which no two notions can be more opposed. The negation of a commencement of time involves, likewise, the affirmation, that an infinite time has, at every moment, already run; that is, it implies the contradiction, that an infinite has been completed. For the same reasons, we are unable to conceive an infinite progress of time; while the infinite regress and the infinite progress taken together, involve the triple contradiction of an infinite concluded, of an infinite commencing, and of two infinities, not exclusive of each other.

2. Time, as an infinite regress, inconceivable.

3. Time, as an infinite progress, inconceivable.

Time, T, as a Minimum. The moment of time either divisible to infinity, or composed of certain absolutely smallest parts. Both alternatives inconceivable.

Now take the parts of time,—a moment, for instance; this we must conceive, as either divisible to infinity, or that it is made up of certain absolutely smallest parts. One or other of these contradictories must be the case. But each is, to us, equally inconceivable. Time is a protensive quantity, and, consequently, any part of it, however small, cannot, without a contradiction, be imagined as not divisible into parts, and these parts into others *ad infinitum*. But the opposite alternative is equally impossible; we cannot think this infinite division. One is necessarily true;

but neither can be conceived possible. It is on the inability of the mind to conceive either the ultimate indivisibility, or the endless divisibility of space and time, that the arguments of the Eleatic Zeno against the possibility of motion are founded, — arguments which at least show, that motion, however certain as a fact, cannot be conceived possible, as it involves a contradiction.

LECT. XXXVIII.

This grand principle called the Law of the Conditioned.

The same principle could be shown in various other relations, but what I have now said is, I presume, sufficient to make you understand its import. Now the law of mind, that the conceivable is in every relation bounded by the inconceivable, I call the Law of the Conditioned. You will find many philosophers who hold an opinion the reverse of this,—maintaining that the absolute is a native or necessary notion of intelligence. This, I conceive, is an opinion founded on vagueness and confusion. They tell us we have a notion of absolute or infinite space, of absolute or infinite time. But they do not tell us in which of the opposite contradictories this notion is realised. Though these are exclusive of each other, and though both are only negations of the conceivable on its opposite poles, they confound together these exclusive inconceivables into a single notion; suppose it positive; and baptise it with the name of absolute. The sum, therefore, of what I have now stated is, that the Conditioned is that which is alone conceivable or cogitable; the Unconditioned, that which is inconceivable or incogitable. The conditioned or the thinkable lies between two extremes or poles; and these extremes or poles are each of them unconditioned, each of them inconceivable, each of them exclusive or contradictory of the other. Of these two repugnant opposites, the one is that of

The contrary opinion founded on vagueness and confusion.

Sum of the author's doctrine.

Unconditional or Absolute Limitation; the other that of Unconditional or Infinite Illimitation. The one we may, therefore, in general call the Absolutely Unconditioned, the other, the Infinitely Unconditioned; or, more simply, the Absolute and the Infinite; the term *absolute* expressing that which is finished or complete, the term *infinite* that which cannot be terminated or concluded. These terms, which, like the Absolute and Infinite themselves, philosophers have confounded, ought not only to be distinguished, but opposed as contradictory. The notion of either unconditioned is negative:—the absolute and the infinite can each only be conceived as a negation of the thinkable. In other words, of the absolute and infinite we have no conception at all. On the subject of the unconditioned,—the absolute and infinite, it is not necessary for me at present further to dilate.

The author's doctrine both the one true and the only orthodox inference.

I shall only add in conclusion, that, as this is the one true, it is the only orthodox, inference. We must believe in the infinity of God; but the infinite God cannot by us, in the present limitation of our faculties, be comprehended or conceived. A Deity understood, would be no Deity at all; and it is blasphemy to say that God only is as we are able to think Him to be. We know God, according to the finitude of our faculties; but we believe much that we are incompetent properly to know. The Infinite, the infinite God, is what, to use the words of Pascal, is infinitely inconceivable. Faith,—Belief,—is the organ by which we apprehend what is beyond our knowledge. In this, all Divines and Philosophers, worthy of the name, are found to coincide; and the few who assert to man a knowledge of the infinite, do this on the daring, the extravagant, the paradoxical supposition, either that

Human Reason is identical with the Divine, or that Man and the Absolute are one.

To assert that the infinite can be thought, but only inadequately thought, is contradictory.

The assertion has, however, sometimes been hazarded, through a mere mistake of the object of knowledge or conception; as if that could be an object of knowledge, which was not known; as if that could be an object of conception, which was not conceived.

It has been held, that the infinite is known or conceived, though only a part of it, (and every part, be it observed, is *ipso facto* finite), can be apprehended; and Aristotle's definition of the infinite has been adopted by those who disregard his declaration, that the infinite, *qua* infinite, is beyond the reach of human understanding.[a] To say that the infinite can be thought, but only inadequately thought, is a contradiction *in adjecto*; it is the same as saying that the infinite can be known, but only known as finite.

The Scriptures explicitly declare that the infinite is for us now incognisable;—they declare that the finite, and the finite alone, is within our reach. It is said, (to cite one text out of many), that "*now* I know *in part*," (*i.e.* the finite); "but *then*" (*i.e.* in the life to come), "shall I know even as I am known,"[β] (*i.e.* without limitation).[γ]

a *Phys.*, I. 4, 6 (Bekker): Τὸ μὲν ἄπειρον ᾗ ἄπειρον ἄγνωστον. The definition occurs, *Phys.*, iii. 6, 11: Ἄπειρον μὲν οὖν ἐστὶν οὗ κατὰ ποσὸν λαμβάνουσιν αἰεί τι λαβεῖν ἔστιν ἔξω. To the ἄπειρον is opposed the ὅλον and τέλειον; for it is added;—Οὗ δὲ μηδὲν ἔξω, τοῦτ' ἐστὶ τέλειον καὶ ὅλον. See *Discussions*, p. 27.—ED.

β 1 *Corinthians*, xiii. 12.

γ See Appendix, III.—ED.

LECTURE XXXIX.

THE REGULATIVE FACULTY.—LAW OF THE CONDITIONED, IN ITS APPLICATIONS.—CAUSALITY.

LECT. XXXIX.

Law of the Conditioned in its applications.

I HAVE been desirous to explain to you the principle of the Conditioned, as out of it we are able not only to explain the hallucination of the Absolute, but to solve some of the most momentous, and hitherto most puzzling, problems of mind. In particular, this principle affords us, I think, a solution of the two great intellectual principles of Cause and Effect, and of Substance and Phænomenon or Accident. Both are only applications of the principle of the Conditioned, in different relations.

Causality—the problem, and attempts at solution.

Of all questions in the history of philosophy, that concerning the nature and genealogy of the notion of Causality, is, perhaps, the most famous; and I shall endeavour to give you a comprehensive, though necessarily a very summary, view of the problem, and of the attempts which have been made at its solution. This, however imperfect in detail, may not be without advantage; for there is not, as far as I am aware, in any work a generalised survey of the various actual and possible opinions on the subject.

The phænomenon of Causality,—what.

But before proceeding to consider the different attempts to explain the phænomenon, it is proper to state and to determine what the phænomenon to be explained really is. Nor is this superfluous, for we shall find that some philosophers, instead of accommodating their solutions to the problem, have accommodated the problem to their solutions.

What appears to us to begin to be, is necessarily thought by us as having previously existed under another form.

"When we are aware of something which begins to be, we are, by the necessity of our intelligence, constrained to believe that it has a Cause. But what does the expression, *that it has a cause*, signify? If we analyse our thought, we shall find that it simply means, that as we cannot conceive any new existence to commence, therefore, all that now is seen to arise under a new appearance, had previously an existence under a prior form. We are utterly unable to realise in thought the possibility of the complement of existence being either increased or diminished. We are unable, on the one hand, to conceive nothing becoming something,—or, on the other, something becoming nothing. When God is said to create out of nothing, we construe this to thought by supposing that He evolves existence out of Himself; we view the Creator as the cause of the universe. "Ex nihilo nihil, in nihilum nil posse reverti,"[β] expresses, in its purest form, the whole intellectual phænomenon of causality.

Hence an absolute tautology between the effect and its causes. This illustrated.

There is thus conceived an absolute tautology between the effect and its causes. We think the causes to contain all that is contained in the effect; the effect to contain nothing which was not contained in the causes. Take an example. A neutral salt is an effect of the conjunction of an acid and alkali. Here we do not, and here we cannot, conceive that, in effect, any new existence has been added, nor can we conceive that any has been taken away. But another example:—Gunpowder is the effect of a mixture of sulphur, charcoal, and nitre, and these three substances are again the effect,—result, of simpler constituents,

α Cf. *Discussions*, p. 609.—Ed.

β Persius, iii. 84. [Cf. Rixner, *Geschichte der Philosophie*, I. p. 83, § 62.]

and these constituents again of simpler elements, either known or conceived to exist. Now, in all this series of compositions, we cannot conceive that aught begins to exist. The gunpowder, the last compound, we are compelled to think, contains precisely the same quantum of existence that its ultimate elements contained prior to their combination. Well, we explode the powder. Can we conceive that existence has been diminished by the annihilation of a single element previously in being, or increased by the addition of a single element which was not heretofore in nature? "Omnia mutantur; nihil interit,"[α]—is what we think, what we must think. This then is the mental phænomenon of causality,—that we necessarily deny in thought that the object which appears to begin to be, really so begins; and that we necessarily identify its present with its past existence.

Not necessary to the notion of Causality, that we should know the particular causes of the particular effect.

Here it is not requisite that we should know under what form, under what combinations, this existence was previously realised, in other words, it is not requisite that we should know what are the particular causes of the particular effect. The discovery of the connection of determinate causes and determinate effects is merely contingent and individual,—merely the datum of experience; but the principle that every event should have its causes, is necessary and universal, and is imposed on us as a condition of our human intelligence itself. This last is the only phænomenon to be explained. Nor are philosophers, in general, really at variance in their statement of the problem. However divergent in their mode of explanation, they are at one in regard to the matter to be explained.[β] But there is one exception. Dr Brown has given a very different account

α Ovid, *Met.*, xv. 165.—Ed.

β On the nature and origin of the notion Causality, see Platner. *Phil. Aph.*, i. § 815 *et seq.*—Ed.

of the phænomenon in question. To this statement of it, I beg to solicit your attention; for as his theory is solely accommodated to his view of the phænomenon, so his theory is refuted by showing that his view of the phænomenon is erroneous. To prevent misconception, I shall exhibit to you his doctrine in his own words:*—

LECT. XXXIX.

Brown's account of the phænomenon of Causality.

"Why is it, then, we believe that continual similarity of the future to the past, which constitutes, or at least is implied in, our notion of power? A stone tends to the earth,—a stone will always tend to the earth,—are not the same proposition; nor can the first be said to involve the second. It is not to experience, then, alone that we must have recourse for the origin of the belief, but to some other principle which converts the simple facts of experience into a general expectation or confidence, that is afterwards to be physically the guide of all our plans and actions.

Brown quoted.

"This principle, since it cannot be derived from experience itself, which relates only to the past, must be an original principle of our nature. There is a tendency in the very constitution of the mind from which the experience arises,—a tendency, that, in everything which it adds to the mere facts of experience, may truly be termed instinctive; for though that term is commonly supposed to imply something peculiarly mysterious, there is no more real mystery in it than in any of the simplest successions of thought, which are all, in like manner, the results of a natural tendency of the mind to exist in certain states, after existing in certain other states. The belief is, a state or feeling of the mind as easily conceivable as any other state of it,—a new feeling, aris-

* *Phil. of the Human Mind*, Lect. vi. p. 34, edit. 1830.

LECT. XXXIX. ing in certain circumstances, as uniformly as, in certain other circumstances, there arise other states or feelings of the mind, which we never consider as mysterious; those, for example, which we term the sensations of sweetness or of sound. To have our nerves of taste or hearing affected in a certain manner, is not, indeed, to taste or hear, but it is immediately afterwards to have those particular sensations; and this merely because the mind was originally so constituted, as to exist directly in the one state after existing in the other. To observe, in like manner, a series of antecedents and consequents, is not, in the very feeling of the moment, to believe in the future similarity, but, in consequence of a similar original tendency, it is immediately afterwards to believe that the same antecedents will invariably be followed by the same consequents. That this belief of the future is a state of mind very different from the mere perception or memory of the past, from which it flows, is indeed true; but what resemblance has sweetness, as a sensation of the mind, to the solution of a few particles of sugar on the tongue; or the harmonics of music, to the vibration of particles of air? All which we know, in both cases, is, that these successions regularly take place; and in the regular successions of nature, which could not, in one instance more than in another, have been predicted without experience, nothing is mysterious, or everything is mysterious. . . . .

"It is more immediately our present purpose to consider, What it truly is which is the object of inquiry, when we examine the physical successions of events, in whatever manner the belief of their similarity of sequence may have arisen? Is it the mere series of regular antecedents and consequents themselves? or, Is it anything more mysterious, which must be supposed to intervene and connect them by some invisible bondage?

"We see in nature one event followed by another; the fall of a spark on gunpowder, for example, followed by the deflagration of the gunpowder: and, by a peculiar tendency of our constitution, which we must take for granted, whatever be our theory of power, we believe, that, as long as all the circumstances continue the same, the sequence of events will continue the same; that the deflagration of gunpowder, for example, will be the invariable consequence of the fall of a spark on it; in other words, we believe the gunpowder to be susceptible of deflagration on the application of a spark, and a spark to have the power of deflagrating gunpowder.

"There is nothing more, then, understood in the train of events, however regular, than the regular order of antecedents and consequents which compose the train; and between which if anything else existed, it would itself be a part of the train. All that we mean, when we ascribe to one substance a susceptibility of being affected by another substance, is that a certain change will uniformly take place in it when that other is present;—all that we mean, in like manner, when we ascribe to one substance a power of affecting another substance, is, that, where it is present, a certain change will uniformly take place in that other substance. Power, in short, is significant not of anything different from the invariable antecedent itself, but of the mere invariableness of the order of its appearance in reference to some invariable consequent,—the invariable antecedent being denominated a *cause*, the invariable consequent an

LECT. XXXIX.

effect. To say, that water has the power of dissolving salt, and to say that salt will always melt when water is poured upon it, are to say precisely the same thing;—there is nothing in the one proposition, which is not exactly and to the same extent enunciated in the other."

Now, in explaining to you the doctrine of Dr Brown, I am happy to avail myself of the assistance of my late lamented friend, Dr Brown's successor, whose metaphysical acuteness was not the least remarkable of his many brilliant qualities.

Wilson quoted on Brown's doctrine of Causality.

"Now, the distinct and full purport of Dr Brown's doctrine, it will be observed, is this,—that when we apply in this way the words *cause* and *power*, we attach no other meaning to the terms than what he has explained. By the word *cause*, we mean no more than that in this instance the spark falling is the event immediately prior to the explosion: including the belief that in all cases hitherto, when a spark has fallen on gunpowder, (of course, supposing other circumstances the same), the gunpowder has kindled; and that whenever a spark shall again so fall, the grains will again take fire. The present immediate priority, and the past and future invariable sequence of the one event upon the other, are all the ideas that the mind can have in view in speaking of the event in that instance as a cause;—and in speaking of the power in the spark to produce this effect, we mean merely to express the invariableness with which this has happened and will happen.

"This is the doctrine; and the author submits it to this test:—'Let any one,' he says, 'ask himself what it is which he means by the term 'power,' and without contenting himself with a few phrases that signify nothing, reflect before he give his answer,— and he will find that he means nothing more than that, in all similar circumstances, the explosion of gunpowder will be the immediate and uniform consequence of the application of a spark.'

"This test, indeed, is the only one to which the question can be brought. For the question does not regard causes themselves, but solely the ideas of cause, in the human mind. If, therefore, every one to whom this analysis of the idea that is in his mind when he speaks of a cause, is proposed, finds, on comparing it with what passed in his mind, that this is a complete and full account of his conception, there is nothing more to be said, and the point is made good. By that sole possible test the analysis is, in such a case, established. If, on the contrary, when this analysis is proposed, as containing all the ideas which we annex to the words cause and power, the minds of most men cannot satisfy themselves that it is complete, but are still possessed with a strong suspicion that there is something more, which is not here accounted for,— then the analysis is not yet established, and it becomes necessary to inquire, by additional examination of the subject, what that more may be.

"Let us then apply the test by which Dr Brown proposes that the truth of his views shall be tried. Let us ask ourselves, what we mean when we say, that the spark has power to kindle the gunpowder,— that the powder is susceptible of being kindled by the spark. Do we mean only that whenever they come together this will happen? Do we merely predict this simple and certain futurity?

"We do not fear to say, that when we speak of a power in one substance to produce a change in another,

LECT. XXXIX.

and of a susceptibility of such change in that other, we express more than our belief that the change has taken and will take place. There is more in our mind than a conviction of the past and a foresight of the future. There is, besides this, the conception included of a fixed constitution of their nature, which determines the event,—a constitution, which, while it lasts, makes the event a necessary consequence of the situation in which the objects are placed. We should say then, that there are included in these terms, 'power,' and 'susceptibility of change,' two ideas which are not expressed in Dr Brown's analysis,—one of necessity, and the other of a constitution of things, in which that necessity is established. That these two ideas are not expressed in the terms of Dr Brown's analysis, is seen by quoting again his words:—'He will find that he means nothing more than that, in all similar circumstances, the explosion of gunpowder will be the immediate and uniform consequence of the application of a spark.'

"It is certain, from the whole tenor of his work, that Dr Brown has designed to exclude the idea of necessity from his analysis."[a]

Fundamental defect in Brown's theory.

Now this admirably expresses what I have always felt is the grand and fundamental defect in Dr Brown's theory,—a defect which renders that theory *ab initio* worthless. Brown professes to explain the phænomenon of causality, but, previously to explanation, he evacuates the phænomenon of all that desiderates explanation. What remains in the phænomenon, after the quality of necessity is thrown, or rather silently allowed to drop out, is only accidental,—only a consequence of the essential circumstance.

[a] Professor Wilson, in *Blackwood's Magazine*, vol. xl. p. 122 *et seq.*

Classification of opinions on the nature and origin of the Principle of Causality.

The opinions in regard to the nature and origin of the principle of Causality, in so far as that principle is viewed as a subjective phænomenon,—as a judgment of the human mind,—fall into two great categories. The first category (A) comprehends those theories which consider this principle as Empirical or *a posteriori*, that is, as derived from experience; the other (B) comprehends those which view it as Pure or *a priori*, that is, as a condition of intelligence itself. These two primary genera are, however, severally subdivided into various subordinate classes.

The former category (A), under which this principle is regarded as the result of experience, contains two classes, inasmuch as the causal judgment may be supposed founded either (a) on an Original, or (b) on a Derivative, cognition. Each of these again is divided into two, according as the principle is supposed to have an objective, or a subjective, origin. In the former case, that is, where the cognition is supposed to be original and underived, it is Objective, or rather Objectivo-Objective, when held to consist in an immediate perception of the power or efficacy of causes in the external and internal worlds (1); and Subjective, or rather Objectivo-Subjective, when viewed as given in a self-consciousness alone of the power or efficacy of our own volitions (2). In the latter case, that is, where the cognition is supposed to be derivative, if objective, it is viewed as a product of Induction and Generalisation (3); if subjective, of Association and Custom (4).

In like manner, the latter category (B), under which the causal principle is considered not as a result, but as a condition, of experience, is variously divided and subdivided. In the first place, the opinions under

LECT. XXXIX. this category fall into two classes, inasmuch as some regard the causal judgment (c) as an Ultimate or Primary law of mind, while others regard it (d) as a Secondary or Derived. Those who hold the former doctrine, in viewing it as a simple original principle, hold likewise that it is a positive act,—an affirmative datum, of intelligence. This class is finally subdivided into two opinions. For some hold that the causal judgment, as necessary, is given in what they call "the principle of Causality," that is, the principle which declares that everything which begins to be, must have its cause (5); whilst at least one philosopher, without explicitly denying that the causal judgment is necessary, would identify it with the principle of our "Expectation of the Constancy of nature" (6).

Those who hold that it can be analysed into a higher principle, also hold that it is not of a positive but of a negative character. These, however, are divided into two classes. By some it has been maintained, that the principle of Causality can be resolved into the principle of Contradiction (7), which, as I formerly stated to you, ought in propriety to be called the principle of Non-Contradiction. On the other hand, it may be, (though it never has been), argued, that the judgment of Causality can be analysed into what I called the principle of the Conditioned,—the principle of Relativity (8). To one or other of these eight heads, all the doctrines that have been actually maintained in regard to the origin of the principle in question, may be referred; and the classification is the better worthy of your attention, as in no work will you find any attempt at even an enumeration of the various theories, actual and possible, on this subject.

The following is a tabular view of the theories in regard to the principle of Causality:—

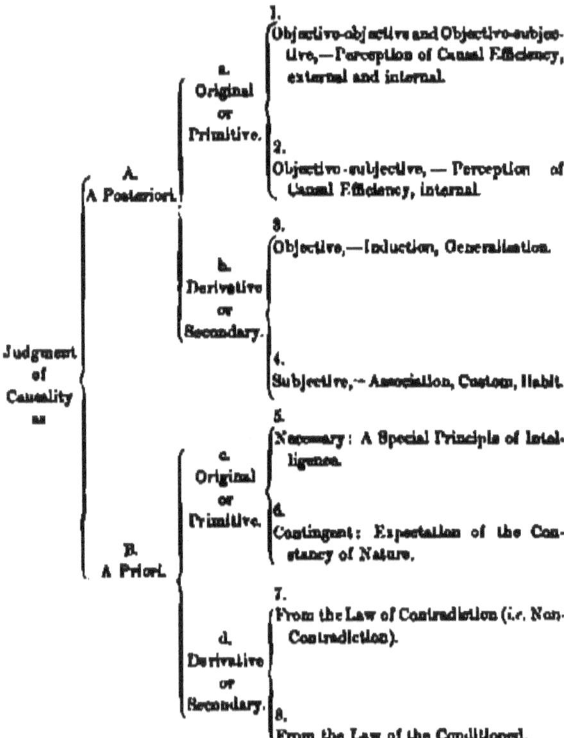

These eight doctrines considered in general.

An adequate discussion of these several heads, and a special consideration of the differences of the individual opinions which they comprehend, would far exceed our limits. I shall, therefore, confine myself to a few observations on the value of these eight doctrines in general, without descending to the particular modifications under which they have been maintained by particular philosophers.

LECT. XXXIX.

I. Objective-Objective and Objective-Subjective. Perception of causal efficiency, external and internal. Refuted on two grounds.

Of these, the first,—that which asserts that we have a perception of the causal agency, as we have a perception of the existence of external objects,—this opinion has been always held in combination with the second,—that which maintains that we are self-conscious of efficiency; though the second has been frequently held by philosophers who have abandoned the first as untenable.

Considering them together, that is, as forming the opinion that we directly and immediately apprehend the efficiency of causes, both external and internal,—this opinion is refuted by two objections. The first is, that we have no such apprehension,—no such knowledge; the second, that if we had, this being merely empirical,—merely conversant with individual instances, could never account for the quality of necessity and universality, which accompanies the judgment of causality. In regard to the first of these objections, it is now universally admitted that we have no perception of the connection of cause and effect in the external world. For example, when one billiard-ball is seen to strike another, we perceive only that the impulse of the one is followed by the motion of the other, but have no perception of any force or efficiency in the first, by which it is connected with the second, in the relation of causality. Hume was the philosopher who decided the opinion of the world on this point. He was not, however, the first who stated the fact, or even the reasoner who stated it most clearly. He, however, believed himself, or would induce us to believe, that in this he was original. Speaking of this point, "I am sensible," he says, "that of all the paradoxes, which I have had, or shall hereafter have, occasion to advance, in the course of this treatise, the present one is the most violent, and that it is merely by dint of solid proof and reasoning I can ever hope

That we have no perception of the connection of cause and effect in the external world,—maintained by Hume.

it will have admission, and overcome the inveterate prejudices of mankind. Before we are reconciled to this doctrine, how often must we repeat to ourselves, that the simple view of any two objects or actions, however related, can never give us any idea of power, or of a connection betwixt them; that this idea arises from the repetition of their union: that the repetition neither discovers nor causes anything in the objects, but has an influence only on the mind, by that customary transition it produces: that this customary transition is, therefore, the same with the power and necessity; which are consequently qualities of perceptions, not of objects, and are internally felt by the soul, and not perceived externally in bodies?"[α]

And, before him, by many philosophers.

I could adduce to you a whole army of philosophers previous to Hume, who had announced and illustrated the fact.[β] As far as I have been able to trace it, this doctrine was first promulgated towards the commencement of the twelfth century, at Bagdad, by Algazel, (El Gazeli), a pious Mahommedan philosopher, who not undeservedly obtained the title of Imaun of the World. Algazel did not deny the reality of causation, but he maintained that God was the only efficient cause in nature;[γ] and that second causes were not properly causes, but only occasions, of the effect. That we have no perception of any real agency of one body on an-

Algazel,—probably the first.

α *Treatise of Human Nature*, b. i. part iii. § 14, vol. i. p. 291, orig. edit.

β Cf. Sturm, *Physica Electiva*, c. iv. p. 163 (edit. 1697). Stewart, *Elements*, i., *Works*, ii. Note C, p. 476. *Elements*, ii., *Works*, iii. Note O, p. 319.—Ed. [See Le Clerc, *Ontologia*, c. 2. § 8-4. *Opera Phil.*, i. p. 318. Chev. Ramsay, *Philos. Prin. of Natural and Revealed Religion*, p. 109; Glasgow, 1748. That Aristotle did not acknowledge that sense had any perception of the causal connection, is shown by his denying sense as principle of science, i. e. διότι, (see *Post. An.*, i. c. 31; and *ibi*, Zabarella), and by his denying that sense is principle of wisdom, as ignorant of cause, (see *Met.*, i. c. 1, and *ibi*, Fonseca. See also Conimbricenses, *In Org.*, ii. p. 436.)]

γ See Averroes, *Destructio Destructionis*, *Aristotelis Opera*, Venet. 1550, vol ix. p. 56. Quoted by Tennemann, *Gesch. der Phil.*, vol. viii. p. 405.—Ed.

other, is a truth which has not more clearly been stated or illustrated by any subsequent philosopher than by him who first proclaimed it. The doctrine of Algazel was adopted by that great sect among the Mussulman doctors, who were styled *those speaking in the law, (loquentes in lege)*, that is, the law of Mahommed. From the Eastern Schools the opinion passed to those of the West; and we find it a problem which divided the scholastic philosophers, whether God were the only efficient, or whether causation could be attributed to created existences.[a] After the revival of letters, the opinion of Algazel was maintained by many individual thinkers, though it no longer retained the same prominence in the schools. It was held, for example, by Malebranche,[β] and his illustration from the collision of two billiard-balls is likewise that of Hume, who probably borrowed from Malebranche both the opinion and the example.

Mussulman doctors.

The Schoolmen.

Malebranche.

But there are many philosophers who surrender the external perception, and maintain our internal consciousness, of causation or power. This opinion was, in one chapter of his *Essay*,[γ] advanced by Locke, and, at a very recent date, it has been amplified and enforced with distinguished ability by the late M. Maine de Biran,[δ]—one of the acutest metaphysicians of France. On this doctrine, the notion of cause is not given to us by the observation of external phænomena,

II. Objective-Subjective. Perception of causal efficiency, internal.

Locke.

M. de Biran.

a [See Biel, *In Sent.*, lib. iv. dist. 1, q. 1. D'Ailly, *Ibid.*, dist. 2 q. 23; referred to by Scheibler, *Opera Metaphysica*, lib. ii. c. iii. tit. 19, p. 121 (edit. 1665). See also Sturm, *Phys. Elect.*, c. iv. p. 128 *et seq.* Poiret, *Œconomia Divina*, l. vi. § 6, p. 66 *et seq.* (edit. 1705).]

β [*Recherche de la Vérité*, liv. vi. part ii. c. iii.]

γ Book II. c. xxi. § 5.—Ed.

δ See *Examen des Leçons de Philosophie*, § viii., *Nouvelles Considérations*, p. 241; and *Réponses aux Arguments contre l'Apperception Immédiate d'une Liaison Causale entre le Vouloir et la Motion*, &c., *Nouv. Con.* p. 363 (edit. 1834). Cf. Préface, by M. Cousin, p. 34; and *Cours de l'Histoire de la Philosophie* (xviiie Siècle), t. ii. leç. xix. p. 231 (edit. 1829).—Ed.

which, as considered only by the senses, manifest no causal efficiency, and appear to us only as successive; it is given to us within, in reflection, in the consciousness of our operations and of the power which exerts them,—viz., the will. I make an effort to move my arm, and I move it. When we analyse attentively the phænomenon of effort, which M. de Biran considers as the type of the phænomena of volition, the following are the results:—1°, The consciousness of an act of will; 2°, The consciousness of a motion produced; 3°, A relation of the motion to the volition. And what is this relation? Not a simple relation of succession. The will is not for us a pure act without efficiency,—it is a productive energy; so that in a volition there is given to us the notion of cause, and this notion we subsequently transport,—project out from our internal activities, into the changes of the external world.

Shown to be untenable.

1. No consciousness of causal connection between volition and motion.

[a] This reasoning, in so far as regards the mere empirical fact of our consciousness of causality, in the relation of our will as moving and of our limbs as moved, is refuted by the consideration, that between the overt fact of corporeal movement of which we are cognisant, and the internal act of mental determination of which we are also cognisant, there intervenes a numerous series of intermediate agencies of which we have no knowledge; and, consequently, that we can have no consciousness of any causal connection between the extreme links of this chain,—the volition to move and the limb moving, as this hypothesis asserts. No one is immediately conscious, for example, of moving his arm through his volition. Previously to this ultimate movement, muscles, nerves, a multitude of solid and

a See *Reid's Works*, p. 866; *Discuss.*, p. 612.—Ed.

LECT. XXXIX. fluid parts, must be set in motion by the will, but of this motion we know, from consciousness, absolutely nothing. A person struck with paralysis is conscious of no inability in his limb to fulfil the determinations of his will; and it is only after having willed and finding that his limbs do not obey his volition that he learns by this experience, that the external movement does not follow the internal act. But as the paralytic learns after the volition that his limbs do not obey his mind; so it is only after volition that the man in health learns, that his limbs do obey the mandates of his will.

2. And even if this admitted, fails to account for the judgment of Causality. But, independently of all this, the second objection above mentioned is fatal to the theory which would found the judgment of causality on any empirical cognition, whether of the phænomena of mind or of the phænomena of matter. Admitting that causation were cognisable, and that perception and self-consciousness were competent to its apprehension, still as these faculties could only take note of individual causations, we should be wholly unable, out of such empirical acts, to evolve the quality of necessity and universality, by which this notion is distinguished. Admitting that we had really observed the agency of any number of causes, still this would not explain to us, how we are unable to think a manifestation of existence without thinking it as an effect. Our internal experience, especially in the relation of our volitions to their effects, may be useful in giving us a clearer notion of causality; but it is altogether incompetent to account for what in it there is of the quality of necessity. So much for the two theories at the head of the Table.

As the first and second opinions have been usually associated, so also have the third and fourth,—that is, the doctrine that our notion of causality is the offspring of the objective principle of Induction or Generalisation, and the doctrine, that it is the offspring of the subjective principle of Association or Custom.

LECT. XXXIX.

III. Objective—Induction, Generalisation.

In regard to the former,—the third, it is plain that the observation, that certain phænomena are found to succeed certain other phænomena, and the generalisation consequent thereon, that these are reciprocally causes and effects, could never of itself have engendered not only the strong but the irresistible belief, that every event must have its cause. Each of these observations is contingent; and any number of observed contingencies will never impose upon us the feeling of necessity,—of our inability to think the opposite. Nay more; this theory evolves the absolute notion of causality out of the observation of a certain number of uniform consecutions among phænomena. But we find no difficulty whatever in conceiving the reverse of all or any of the consecutions we have observed; and yet the general notion of causality, which, *ex hypothesi*, is their result, we cannot possibly think as possibly unreal. We have always seen a stone fall to the ground, when thrown into the air, but we find no difficulty in representing to ourselves the possibility of one or all stones gravitating from the earth; only we cannot conceive the possibility of this, or any other event, happening without a cause.

IV. Subjective—Association.

Nor does the latter,—the fourth theory,—that of Custom or Association,—afford a better solution. The attribute of necessity cannot be derived from custom. Allow the force of custom to be great as may be, still it is always limited to the customary, and the customary has nothing whatever in it of the necessary. But we have here to account not for a strong, but for an

LECT. XXXIX.

absolutely irresistible, belief. On this theory, also, the causal judgment, when association is recent, should be weak, and should only gradually acquire its full force in proportion as custom becomes inveterate. But do we find that the causal judgment is weaker in the young, stronger in the old? There is no difference. In either case there is no less and more; the necessity in both is absolute. Mr Hume patronised the opinion, that the notion of causality is the offspring of experience engendered upon custom.[a] But those have a sorry insight into the philosophy of that great thinker, who suppose that this was a dogmatic theory of his own. On the contrary, in his hands, it was a mere reduction of dogmatism to absurdity by showing the inconsistency of its results. To the Lockian sensualism, Hume proposed the problem,—to account for the phænomenon of necessity in our notion of the causal nexus. That philosophy afforded no other principle through which even the attempt at a solution could be made;—and the principle of custom, Hume shows, could not furnish a real necessity. The alternative was plain. Either the doctrine of sensualism is false, or our nature is a delusion. Shallow thinkers adopted the latter alternative, and were lost; profound thinkers, on the contrary, were determined to lay a deeper foundation of philosophy than that of the superficial edifice of Locke; and thus it is that Hume became the cause or the occasion of all that is of principal value in our more recent metaphysics. Hume is the parent of the philosophy of Kant, and, through Kant, of the whole philosophy of Germany; he is the parent of the philosophy of Reid and Stewart in Scotland,

a [On Hume's theory, see Platner, *Phil. Aph.*, i. § 850, p. 485-6; edit. 1793.]

and of all that is of pre-eminent note in the metaphysics of France and Italy.—But to return.

I now come to the second category (B), and to the first of the four particular heads which it likewise contains,—the opinion, namely, that the judgment, that everything that begins to be must have a cause, is a simple primary datum, a positive revelation of intelligence. To this head are to be referred the theories on causality of Descartes, Leibnitz, Reid, Stewart, Kant, Fichte, Cousin, and the majority of recent philosophers. This is the fifth theory in order.

V. A special principle of intelligence.

Dr Brown has promulgated a doctrine of Causality, which may be numbered as the sixth; though perhaps it is hardly deserving of distinct enumeration. He actually identifies the causal judgment, which to us is necessary, with the principle by which we are merely inclined to believe in the uniformity of nature's operations.

VI. Expectation of the constancy of nature.

Superseding any articulate consideration of this opinion, and reverting to the fifth, much might be said in relation to the several modifications of this opinion, as held by different philosophers; but I must content myself with a brief criticism of the doctrine in reference to its most general features.

Now it is manifest, that, against the assumption of a special principle, which this doctrine makes, there exists a primary presumption of philosophy. This is the law of Parcimony, which forbids, without necessity, the multiplication of entities, powers, principles, or causes; above all, the postulation of an unknown force, where a known impotence can account for the effect. We are, therefore, entitled to apply Occam's razor to this theory of causality, unless it be proved impossible to explain the causal judgment at a cheaper

Fifth opinion criticised. Primary presumption of philosophy against assumption of special principle of causality.

LECT. XXXIX.

rate, by deriving it from a higher and that a negative origin. On a doctrine like the present is thrown the onus of vindicating its necessity, by showing that, unless a special and positive principle be assumed, there exists no competent mode to save the phænomena. It can only, therefore, be admitted provisorily; and it falls of course, if the phænomenon it would explain can be explained on less onerous conditions.

VII. The principle of Non-Contradiction.

Leaving, therefore, this theory to stand or fall according as the two remaining opinions are or are not found insufficient, I proceed to the consideration of these. The first,—the seventh, is a doctrine that has long been exploded. It attempts to establish the principle of Causality upon the principle of Contradiction. Leibnitz was too acute a metaphysician to attempt to prove the principle of Sufficient Reason or Causality, which is an ampliative or synthetic principle, by the principle of Contradiction, which is merely explicative or analytic. But his followers were not so wise. Wolf,[α] Baumgarten,[β] and many other Leibnitzians, paraded demonstrations of the law of the Sufficient Reason on the ground of the law of Contradiction; but the reasoning always proceeds on a covert assumption of the very point in question. The same argument is, however, at an earlier date, to be found in Locke,[γ] and modifications of it in Hobbes[δ] and Clarke.[ε] Hume,[ζ] who was only aware of the

α [*Ontologia*, § 70.]

β [*Metaphysik*, § 18.] [Cf. Walch, *Lexikon*, v. *Zureichender Grund*. Zedler, *Lexikon*, v. *Causalität*.]

γ [*Essay*, book iv. c. 10, § 3. *Works*, i. p. 291.] [This is doubtless the passage of Locke which is criticised by Hume (*Treat. of Hum. Nat.*, b. i. part iii. § 3); but it will hardly bear the interpretation put upon it by Hume and Sir W. Hamilton.—Ed.]

δ *Of Liberty and Necessity*, *Works*, edit. Molesworth, vol. iv. p. 276.—Ed.

ε [*Demonstration*, p. 9, *alibi*. See also 'S Gravesande, *Introd. ad Phil.*, § 80.]

ζ *Treat. of Hum. Nature*, book i. part iii. § 3. Cf. Reid, *Works*, p. 455. Stewart, *Works*, i. p. 441.—Ed.

argument as in the hands of the English metaphysicians, has given it a refutation, which has earned the approbation of Reid; and by foreign philosophers its emptiness, in the hands of the Wolfian metaphysicians, has frequently been exposed.* Listen to the pretended demonstration: — Whatever is produced without a cause, is produced by nothing; in other words, has nothing for its cause. But nothing can no more be a cause than it can be something. The same intuition that makes us aware, that nothing is not something, shows us that everything must have a real cause of its existence.—To this it is sufficient to say, that the existence of causes being the point in question, the existence of causes must not be taken for granted, in the very reasoning which attempts to prove their reality. In excluding causes we exclude all causes; and consequently exclude nothing considered as a cause; it is not, therefore, allowable, contrary to that exclusion, to suppose nothing as a cause, and then from the absurdity of that supposition to infer the absurdity of the exclusion itself. If everything must have a cause, it follows that, upon the exclusion of other causes, we must accept of nothing as a cause. But it is the very point at issue, whether everything must have a cause or not; and, therefore, it violates the first principles of reasoning to take this quæsitum itself as granted. This opinion is now universally abandoned.

Fallacy of the supposed demonstration.

The eighth and last opinion is that which regards the judgment of causality as derived; and derives it not from a power, but from an impotence, of mind; in a word, from the principle of the Conditioned. I do not think it possible, without a detailed exposition

VIII. The Law of the Conditioned.

* [See Walch, *Lexikon*, v. *Zureichender Grund*. Biedermann, *Acta Scholastica*, t. vii. p. 120. Schwab, *Preisschriften über die Metaphysik*, p. 119. Lossius, *Lexikon*, v. *Causalität*, i. p. 669.]

LECT. XXXIX.

of the various laws or categories of thought, to make you fully understand the grounds and bearings of this opinion. In attempting to explain, you must, therefore, allow me to take for granted certain laws of thought, to which I have only been able incidentally to allude. Those, however, which I postulate, are such as are now generally admitted by all philosophers who allow the mind itself to be a source of cognitions; and the only one which has not been recognised by them, but which, as I endeavoured briefly to prove to you in my last Lecture, must likewise be taken into account, is the Law of the Conditioned,—the law that the conceivable has always two opposite extremes, and that these extremes are equally inconceivable. That the conditioned is to be viewed, not as a power, but as a powerlessness, of mind, is evinced by this,—that the two extremes are contradictories, and, as contradictories, though neither alternative can be conceived,—thought as possible, one or other must be admitted to be necessary.

Judgment of Causality, how deduced from this law.

Categories of thought. Existence.

Philosophers, who allow a native principle to the mind at all, allow that Existence is such a principle. I shall, therefore, take for granted Existence as the highest category or condition of thought. As I noticed to you in my last Lecture,[a] no thought is possible except under this category. All that we perceive or imagine as different from us, we perceive or imagine as objectively existent. All that we are conscious of as an act or modification of self, we are conscious of only as subjectively existent. All thought, therefore, implies the thought of existence; and this is the veritable exposition of the enthymeme of Descartes,—*Cogito ergo sum*. I cannot think that I think, without thinking that I exist,—I cannot be

[a] P. 366.—Ed.

conscious, without being conscious that I am. Let existence, then, be laid down as a necessary form of thought. As a second category or subjective condition of thought, I postulate that of Time. This, likewise, cannot be denied me. It is the necessary condition of every conscious act; thought is only realised to us as in succession, and succession is only conceived by us under the concept of time. Existence and existence in time is thus an elementary form of our intelligence.

Time.

But we do not conceive existence in time absolutely or infinitely,—we conceive it only as conditioned in time; and Existence Conditioned in Time expresses at once and in relation, the three categories of thought, which afford us in combination the principle of Causality. This requires some explanation.

The Conditioned.

When we perceive or imagine an object, we perceive or imagine it—1°, As existent, and, 2°, As in Time; Existence and Time being categories of all thought. But what is meant by saying, I perceive, or imagine, or, in general, think, an object only as I perceive, or imagine, or, in general, think it to exist? Simply this,—that, as thinking it, I cannot but think it to exist, in other words, that I cannot annihilate it in thought. I may think away from it, I may turn to other things; and I can thus exclude it from my consciousness; but, actually thinking it, I cannot think it as non-existent, for as it is thought, so is it thought existent.

Existence Conditioned in Time affords the principle of Causality.

But a thing is thought to exist, only as it is thought to exist in time. Time is present, past, and future. We cannot think an object of thought as non-existent *de presenti*,—as actually an object of thought. But can we think that quantum of existence of which an object, real or ideal, is the complement, as non-exist-

LECT. XXXIX. ent, either in time past, or in time future? Make the experiment. Try to think the object of your thought as non-existent in the moment before the present.—You cannot. Try it in the moment before that.—You cannot. Nor can you annihilate it by carrying it back to any moment, however distant in the past. You may conceive the parts of which this complement of existence is composed, as separated; if a material object, you can think it as shivered to atoms, sublimated into æther; but not one iota of existence can you conceive as annihilated, which subsequently you thought to exist. In like manner try the future,—try to conceive the prospective annihilation of any present object,—of any atom of any present object.—You cannot. All this may be possible, but of it we cannot think the possibility. But if you can thus conceive neither the absolute commencement nor the absolute termination of anything that is once thought to exist, try, on the other hand, if you can conceive the opposite alternative of infinite non-commencement, of infinite non-termination. To this you are equally impotent. This is the category of the Conditioned, as applied to the category of Existence under the category of Time.

But in this application is the principle of Causality not given? Why, what is the law of Causality? Simply this,—that when an object is presented phænomenally as commencing, we cannot but suppose that the complement of existence, which it now contains, has previously been;—in other words, that all that we at present come to know as an effect must previously have existed in its causes; though what these causes are we may perhaps be altogether unable even to surmise.

LECTURE XL.

THE REGULATIVE FACULTY.—LAW OF THE CONDITIONED, IN ITS APPLICATIONS.—CAUSALITY.

LECT. XL.

Recapitulation.

Our last Lecture was principally occupied in giving a systematic view and a summary criticism of the various opinions of philosophers, regarding the origin of that inevitable necessity of our nature, which compels us to refuse any real commencement of existence to the phænomena which arise in and around us; in other words, that necessity of our nature, under which we cannot but conceive everything that occurs, to be an effect, that is, to be something consequent, which, as wholly derived from, may be wholly refunded into, something antecedent. The opinions of philosophers with regard to the genealogy of this claim of thought, may be divided into two *summa genera* or categories; as all opinions on this point view the Causal Judgment either, 1°, As resting immediately or mediately on experience, or, 2°, As resting immediately or mediately on a native principle of the mind itself;—in short, all theories of causality either make it *a posteriori* or Empirical, or make it *a priori* or Pure.

I shall not again enumerate the various subordinate doctrines into which the former category is subdivided; and, in relation to all of these, it is enough to say that they are one and all wholly worthless, as wholly in-

LECT. XL. capable of accounting for the quality of necessity, by which we are conscious that the causal judgment is characterised.

The opinions which fall under the second category are not obnoxious to this sweeping objection, (except Brown's), as they are all equally competent to save the phænomenon of a subjective necessity. Of the three opinions, (I discount Brown's), under this head, one supposes that the law of Causality is a positive affirmation, and a primary fact of thought, incapable of all further analysis. The other two, on the contrary, view it as a negative principle, and as capable of resolution into a higher law.

Of these, the first opinion (the sixth) is opposed *in limine*, by the presumption of philosophy against the multiplication of special principles. By the law of Parcimony, the assumption of a special principle can only be legitimated by its necessity; and that necessity only emerges if the phænomenon to be explained can be explained by no known and ordinary causes. The possible validity of this theory, therefore, depends on the two others being actually found incompetent. As postulating no special, no new, no positive principle, and professing to account for the phænomenon upon a common and a negative ground, they possess a primary presumption in their favour; and if one or other be found to afford us a possible solution of the problem, we need not, nay, we are not entitled to, look beyond.

Of these two theories, the one (the seventh) attempts to analyse the principle of Causality into the principle of Contradiction; the other (the eighth), into the principle of the Conditioned. The former has been long exploded, and is now universally abandoned. The attempt to demonstrate that a negation of causes involves an affirmation of two contradictory propositions, has been shown to be delusive, as the demonstration only proceeds on a virtual assumption of the point in question. The field, therefore, is left open for the last (the eighth), which endeavours to analyse the mental law of Causality into the mental law of the Conditioned. This theory, which has not hitherto been proposed, is recommended by its extreme simplicity. It postulates no new, no special, no positive principle. It only supposes that the mind is limited; and the law of limitation, the law of the Conditioned, in one of its applications, constitutes the law of Causality. The mind is necessitated to think certain forms; and, under these forms, thought is only possible in the interval between two contradictory extremes, both of which are absolutely inconceivable, but one of which, on the principle of Excluded Middle, is necessarily true. In reference to the present subject, it is only requisite to specify two of these forms,—Existence and Time. I showed you that thought is only possible under the native conceptions,—the *a priori* forms,—of existence and time; in others words, the notions of existence and time are essential elements of every act of intelligence. But while the mind is thus astricted to certain necessary modes or forms of thought, in these forms it can only think under certain conditions. Thus, while obliged to think under the thought of time, it cannot conceive, on the one hand, the absolute commencement of time, and it cannot conceive, on the other, the infinite non-commencement of time; in like manner, on the one hand, it cannot conceive an absolute minimum of time, nor yet, on the other, can it conceive the infinite

The law of Causality constituted by the law of the Conditioned.

The law of the Conditioned.

LECT. XI.

divisibility of time. Yet these form two pairs of contradictories, that is, of counter-propositions, which, if our intelligence be not all a lie, cannot both be true, but of which, on the same authority, one necessarily must be true. This proves: 1°, That it is not competent to argue, that what cannot be comprehended as possible by us, is impossible in reality; and, 2°, That the necessities of thought are not always positive powers of cognition, but often negative inabilities to know. The law of mind, that all that is positively conceivable, lies in the interval between two inconceivable extremes, and which, however palpable when stated, has never been generalised, as far as I know, by any philosopher, I call the Law or Principle of the Conditioned.

This law in its application to a thing thought under Existence and Time, affords the phænomenon of Causality.

Thus, the whole phænomenon of causality seems to me to be nothing more than the law of the Conditioned, in its application to a thing thought under the form or mental category of Existence, and under the form or mental category of Time. We cannot know, we cannot think, a thing, except as existing, that is, under the category of existence; and we cannot know or think a thing as existing, except in time. Now the application of the law of the conditioned to any object, thought as existent, and thought as in time, will give us at once the phænomenon of causality. And thus:—An object is given us, either by sense or suggestion, — imagination. As known, we cannot but think it existent, and in time. But to say that we cannot but think it to exist, is to say, that we are unable to think it non-existent, that is, that we are unable to annihilate it in thought. And this we cannot do. We may turn aside from it; we may occupy our attention with other objects; and we may thus exclude it from our thoughts. This is certain: we need not think it; but it is equally certain, that thinking it, we cannot think it not to exist. This will be at once admitted of the present; but it may possibly be denied of the past and future. But if we make the experiment, we shall find the mental annihilation of an object equally impossible under time past, present, or future. To obviate misapprehension, however, I must make a very simple observation. When I say that it is impossible to annihilate an object in thought,—in other words, to conceive it as non-existent,—it is of course not meant that it is impossible to imagine the object wholly changed in form. We can figure to ourselves the elements of which it is composed, distributed and arranged and modified in ten thousand forms,—we can imagine anything of it, short of annihilation. But the complement, the quantum, of existence, which is realised in any object,—that we cannot represent to ourselves, either as increased, without abstraction from other bodies, or as diminished, without addition to them. In short, we are unable to construe it in thought, that there can be an atom absolutely added to, or an atom absolutely taken away from, existence in general. Make the experiment. Form to yourselves a notion of the universe; now, can you conceive that the quantity of existence, of which the universe is the sum, is either amplified or diminished? You can conceive the creation of a world as lightly as you can conceive the creation of an atom. But what is a creation? It is not the springing of nothing into something. Far from it:—it is conceived, and is by us conceivable, merely as the evolution of a new form of existence, by the fiat of the

Annihilation and Creation,—as conceived by us.

LECT. XL.

Deity. Let us suppose the very crisis of creation. Can we realise it to ourselves, in thought, that, the moment after the universe came into manifested being, there was a larger complement of existence in the universe and its Author together, than there was, the moment before, in the Deity himself alone? This we cannot imagine. What I have now said of our conceptions of creation, holds true of our conceptions of annihilation. We can conceive no real annihilation,—no absolute sinking of something into nothing. But, as creation is cogitable by us only as an exertion of divine power, so annihilation is only to be conceived by us as a withdrawal of the divine support. All that there is now actually of existence in the universe, we conceive as having virtually existed, prior to creation, in the Creator; and in imagining the universe to be annihilated by its Author, we can only imagine this, as the retractation of an outward energy into power. All this shows how impossible it is for the human mind to think aught that it thinks, as non-existent either in time past or in time future.

Our inability to think aught as extruded from Space gives the law of Ultimate Incompressibility.

[* Our inability to think, what we have once conceived existent in Time, as in time becoming non-existent, corresponds with our inability to think, what we have conceived existent in Space, as in space becoming non-existent. We cannot realise it to thought, that a thing should be extruded, either from the one quantity or the other. Hence, under extension, the law of Ultimate Incompressibility; under protension, the law of Cause and Effect.]

The infinite regress of Time as

We have been hitherto speaking only of one inconceivable extreme of the conditioned, in its application

* Supplied from *Discussions*, p. 620.—ED.

to the category of existence in the category of time,—the extreme of absolute commencement; the other is equally incomprehensible, that is, the extreme of infinite regress or non-commencement. With this latter we have, however, at present nothing to do. [a Indeed, as not obtrusive, the Infinite figures far less in the theatre of mind, and exerts a far inferior influence in the modification of thought than the Absolute. It is, in fact, both distant and delitescent; and in place of meeting us at every turn, it requires some exertion on our part to seek it out.] It is the former alone,—it is the inability we experience of annihilating in thought an existence in time past, in other words, our utter impotence of conceiving its absolute commencement, that constitutes and explains the whole phænomenon of causality. An object is presented to our observation which has phænomenally begun to be. Well, we cannot realise it in thought that the object, that is, this determinate complement of existence, had really no being at any past moment; because this supposes that, once thinking it as existent, we could again think it as non-existent, which is for us impossible. What, then, can we do? That the phænomenon presented to us began, as a phænomenon, to be,—this we know by experience; but that the elements of its existence only began, when the phænomenon they constitute came into being,—this we are wholly unable to represent in thought. In these circumstances, how do we proceed?—How must we proceed? There is only one possible mode. We are compelled to believe that the object, (that is, a certain *quale* and *quantum* of being), whose phænomenal rise into existence we have witnessed, did really exist, prior to this rise,

[Margin: less inconceivable than its absolute commencement.]

[Margin: Our inability to conceive existence as absolutely beginning in time, constitutes the phænomenon of causality.]

a Supplied from *Discussions*, p. 621.—Ed.

LECT. XL.

under other forms; [a and by *form*, be it observed, I mean any mode of existence, conceivable by us or not]. But to say that a thing previously existed under different forms, is only in other words to say, that a thing had causes. I have already noticed to you the error of philosophers in supposing that anything can have a single cause. Of course, I speak only of Second Causes. Of the causation of the Deity we can form no possible conception. Of second causes, I say, there must always be at least a concurrence of two to constitute an effect. Take the example of vapour. Here to say that heat is the cause of evaporation, is a very inaccurate,—at least a very inadequate, expression. Water is as much the cause of evaporation as heat. But heat and water together are the causes of the phænomenon. Nay, there is a third concause which we have forgot,—the atmosphere. Now, a cloud is the result of these three concurrent causes or constituents; and, knowing this, we find no difficulty in carrying back the complement of existence, which it contains prior to its appearance. But on the hypothesis, that we are not aware what are the real constituents or causes of the cloud, the human mind must still perforce suppose some unknown, some hypothetical, antecedents, into which it mentally refunds all the existence which the cloud is thought to contain.

Of Second Causes there must be at least a concurrence of two, to constitute an effect.

Nothing can be a greater error in itself, or a more fertile cause of delusion, than the common doctrine, that the causal judgment is elicited only when we apprehend objects in consecution, and uniform consecution. Of course, the observation of such succession prompts and enables us to assign particular causes to particular effects. But this consideration ought to

To suppose that the causal judgment is elicited only by objects in uniform consecution, is erroneous.

a Supplied from *Discussions*, p. 621.—Ed.

be carefully distinguished from the law of Causality, absolutely, which consists not in the empirical attribution of this phænomenon, as cause, to that phænomenon, as effect, but in the universal necessity of which we are conscious, to think causes for every event, whether that event stand isolated by itself, and be by us referable to no other, or whether it be one in a series of successive phænomena, which, as it were, spontaneously arrange themselves under the relation of effect and cause. [a Of no phænomenon, as observed, need we think *the* cause; but of every phænomenon, must we think *a* cause. The former we may learn through a process of induction and generalisation; the latter we must always and at once admit, constrained by the condition of Relativity. On this, not sunken rock, Dr Brown and others have been shipwrecked.]

LECT. XL.

This doctrine of Causality seems to me preferable to any other, for the following, among other, reasons:—

The author's doctrine of Causality, to be preferred.

In the first place, to explain the phænomenon of the Causal Judgment, it postulates no new, no extraordinary, no express principle. It does not even found upon a positive power; for, while it shows that the phænomenon in question is only one of a class, it assigns, as their common cause, only a negative impotence. In this, it stands advantageously contrasted with the one other theory which saves the phænomenon, but which saves it only by the hypothesis of a special principle, expressly devised to account for this phænomenon alone. Nature never works by more, and more complex, instruments than are necessary;—μηδὲν περιττῶς; and to assume a particular force, to perform what can be better explained by a

1°. From its simplicity.

a Supplied from *Discussions*, p. 622.—Ed.

LECT. XL.

general imbecility, is contrary to every rule of philosophising.

2°. Averting scepticism.

But, in the second place, if there be postulated an express and positive affirmation of intelligence to account for the fact, that existence cannot absolutely commence, we must equally postulate a counter affirmation of intelligence, positive and express, to explain the counter fact, that existence cannot infinitely not commence. The one necessity of mind is equally strong as the other; and if the one be a positive doctrine, an express testimony of intelligence, so also must be the other. But they are contradictories; and, as contradictories, they cannot both be true. On this theory, therefore, the root of our nature is a lie! By the doctrine, on the contrary, which I propose, these contradictory phænomena are carried up into the common principle of a limitation of our faculties. Intelligence is shown to be feeble but not false; our nature is, thus, not a lie, nor the Author of our nature a deceiver.

3°. Avoiding the alternative of fatalism or inconsistency.

In the third place, this simpler and easier doctrine avoids a serious inconvenience, which attaches to the more difficult and complex. It is this:—To suppose a positive and special principle of causality, is to suppose, that there is expressly revealed to us, through intelligence, the fact that there is no free causation, that is, that there is no cause which is not itself merely an effect; existence being only a series of determined antecedents and determined consequents. But this is an assertion of Fatalism. Such, however, most of the patrons of that doctrine will not admit. The assertion of absolute necessity, they are aware, is virtually the negation of a moral universe, consequently of the Moral Governor of a moral universe,—in a word, Atheism. Fatalism and Atheism are, indeed, convertible terms. The only valid arguments for the existence of a God, and for the immortality of the soul, rest on the ground of man's moral nature;[a] consequently, if that moral nature be annihilated, which in any scheme of necessity it is, every conclusion, established on such a nature, is annihilated also. Aware of this, some of those who make the judgment of causality a special principle,—a positive dictate of intelligence,—find themselves compelled, in order to escape from the consequences of their doctrine, to deny that this dictate, though universal in its deliverance, should be allowed to hold universally true; and, accordingly, they would exempt from it the facts of volition. Will, they hold to be a free cause, that is, a cause which is not an effect; in other words, they attribute to will the power of absolute origination. But here their own principle of causality is too strong for them. They say that it is unconditionally given, as a special and positive law of intelligence, that every origination is only an apparent, not a real, commencement. Now, to exempt certain phænomena from this law, for the sake of our moral consciousness, cannot validly be done. For, in the first place, this would be to admit that the mind is a complement of contradictory revelations. If mendacity be admitted of some of our mental dictates, we cannot vindicate veracity to any. "Falsus in uno, falsus in omnibus." Absolute scepticism is hence the legitimate conclusion. But, in the second place, waiving this conclusion, what right have we, on this doctrine, to subordinate the positive affirmation of causality to our consciousness of moral liberty, —what right have we, for the interest of the latter, to

a See above, vol. i. lect. ii. p. 25 et seq.—Ed.

LECT. XL.

derogate from the universality of the former? We have none. If both are equally positive, we have no right to sacrifice to the other the alternative, which our wishes prompt us to abandon.

Advantages of the Author's doctrine further shown.

But the doctrine which I propose is not exposed to these difficulties. It does not suppose that the judgment of Causality is founded on a power of the mind to recognise as necessary in thought what is necessary in the universe of existence; it, on the contrary, founds this judgment merely on the impotence of the mind to conceive either of two contradictories, and, as one or other of two contradictories must be true, though both cannot, it shows that there is no ground for inferring from the inability of the mind to conceive an alternative as possible, that such alternative is really impossible. At the same time, if the causal judgment be not an affirmation of mind, but merely an incapacity of positively thinking the contrary, it follows that such a negative judgment cannot stand in opposition to the positive consciousness,—the affirmative deliverance, that we are truly the authors,—the responsible originators, of our actions, and not merely links in the adamantine series of effects and causes. It appears to me that it is only on this doctrine that we can philosophically vindicate the liberty of the will,—that we can rationally assert to man a "fatis avolsa voluntas." How the will can possibly be free must remain to us, under the present limitation of our faculties, wholly incomprehensible. We cannot conceive absolute commencement; we cannot, therefore, conceive a free volition. But as little can we conceive the alternative on which liberty is denied, on which necessity is affirmed. And in favour of our moral nature, the fact that we are free, is given us in the consciousness of an uncompromising law of Duty, in the consciousness of our moral accountability; and this fact of liberty cannot be redargued on the ground that it is incomprehensible, for the doctrine of the Conditioned proves, against the necessitarian, that something may, nay must, be true, of which the mind is wholly unable to construe to itself the possibility; whilst it shows that the objection of incomprehensibility applies no less to the doctrine of fatalism than to the doctrine of moral freedom. If the deduction, therefore, of the Causal Judgment, which I have attempted, should speculatively prove correct, it will, I think, afford a securer and more satisfactory foundation for our practical interests, than any other which has ever yet been promulgated.[a]

[a] Here, in the manuscript, occurs the following sentence, with mark of deletion:—"But of this we shall have to speak, when we consider the question of the Liberty or Necessity of our Volitions, under the Third Great Class of the Mental Phænomena,—the Conative." The author does not, however, resume the consideration of this question in these Lectures. It will also be observed that Sir W. Hamilton does not pursue the application of the Law of the Conditioned to the principle of Substance and Phænomenon, as proposed at the outset of the discussion. See above, p. 376. On Causality, and on Liberty and Necessity, see further in *Discussions*, p. 625 *et seq.*, and Appendix, IV.—Ed.

LECTURE XLI.

SECOND GREAT CLASS OF MENTAL PHÆNOMENA,—THE FEELINGS: THEIR CHARACTER, AND RELATION TO THE COGNITIONS AND CONATIONS.

LECT. XLI.

Second Great Class of mental phænomena,—the Feelings.

HAVING concluded our consideration of the First Great Class of the Phænomena revealed to us by consciousness,—the phænomena of Knowledge,—we are now to enter on the Second of these Classes,—the class which comprehends the phænomena of Pleasure and Pain, or, in a single word, the phænomena of Feeling.[a] Before, however, proceeding to a discussion of this class of mental appearances, considered in themselves, there are several questions of a preliminary character, which it is proper to dispose of. Of these, two naturally present themselves in the very threshold of our inquiry. The first is,—Do the phænomena of Pleasure and Pain constitute a distinct order of internal states, so that we are warranted in establishing the capacity of Feeling as one of the fundamental powers of the human mind?

Two preliminary questions regarding the Feelings.

The second is,—In what position do the Feelings stand by reference to the Cognitions and the Conations; and, in particular, whether ought the Feelings or the Conations to be considered first, in the order of science?

[a] See above, vol I. lect. xi. p. 182.—Ed.

Of these questions, the former is by no means one that can be either superseded or lightly dismissed. This is shown, both by the very modern date at which the analysis of the Feelings into a separate class of phænomena was proposed, and by the controversy to which this analysis has given birth.

I. Do the phænomena of Pleasure and Pain constitute a distinct order of internal states?

The Feelings were not recognised as the manifestations of any fundamental power, until a very recent period.

Until a very recent epoch, the feelings were not recognised by any philosopher as the manifestations of any fundamental power. The distinction taken in the Peripatetic School, by which the mental modifications were divided into Gnostic or Cognitive, and Orectic or Appetent, and the consequent reduction of all the faculties to the *Facultas cognoscendi* and the *Facultas appetendi*, was the distinction which was long most universally prevalent, though under various, but usually less appropriate, denominations. For example, the modern distribution of the mental powers into those of the Understanding and those of the Will, or into Powers Speculative and Powers Active, —these are only very inadequate, and very incorrect, versions of the Peripatetic analysis, which, as far as it went, was laudable for its conception, and still more laudable for its expression. But this Aristotelic division of the internal states, into the two categories of Cognitions and of Appetencies, is exclusive of the Feelings, as a class co-ordinate with the two other genera; nor was there, in antiquity, any other philosophy which accorded to the feelings the rank denied to them in the analysis of the Peripatetic school. An attempt has, indeed, been made to show that, by Plato, the capacity of Feeling was regarded as one of the three fundamental powers; but it is only by a total perversion of Plato's language, by a total reversion of the whole analogy of his psychology, that any

Peripatetic division of the mental phænomena.

LECT. XLI.

Recognition of the Feelings by modern philosophers.

colour can be given to this opinion. Kant, as I have formerly observed, was the philosopher to whom we owe this tri-logical classification. But it ought to be stated, that Kant only placed the keystone in the arch, which had been raised by previous philosophers among his countrymen. The phænomena of Feeling had, for thirty years prior to the reduction of Kant, attracted the attention of the German psychologists, and had by them been considered as a separate class of mental states.

Sulzer, Mendelssohn, Kæstner, Meiners, Eberhard, Platner.

This had been done by Sulzer[α] in 1751, by Mendelssohn[β] in 1763, by Kæstner[γ] in 1763 (?), by Meiners[δ] in 1773, by Eberhard[ε] in 1776, and by Platner[ζ] in 1780 (?).

Kant,—the first to establish the trichotomy of the mental powers.

It remained, however, for Kant to establish, by his authority, the decisive trichotomy of the mental powers. In his *Critique of Judgment* (*Kritik der Urtheilskraft*), and, likewise, in his *Anthropology*, he treats of the capacities of Feeling, apart from, and along with, the faculties of Cognition and Conation.[η] At the same time, he called

α See *Untersuchung über den Ursprung der angenehmen und unangenehmen Empfindungen*; first published in the Memoirs of the Berlin Academy, in 1751 and 1762. See *Vermischte philosophische Schriften*, i. p. 1. Leipsic, 1800. Cf. his *Allgemeine Theorie der schönen Künste*, 1771.—Ed. [For a summary and criticism of the former work, see Reinhold, *Über die bisherigen Begriffe vom Vergnügen*. *Vermischte Schriften*, i. p. 206. Jena, 1796.]

β *Briefe über die Empfindungen*, 1755.—Ed.

γ See *Nouvelle Theorie des Plaisirs*, par M. Sulzer; *avec des Réflexions sur l'Origine du Plaisir*, par M. Kæstner, de l'Académie Royale de Berlin, 1767, first published in the Memoirs of the Academy in 1749. See below, p. 461.—Ed.

δ See *Abriss der Psychologie*, 1773,—Ed.

ε See *Allgemeine Theorie des Denkens und Empfindens*, read before the Royal Society of Berlin in 1776; new edit. 1786. Cf. *Theorie der schönen Wissenschaften*, 2d edit. Halle, 1786.—Ed.

ζ The threefold division of the mental phænomena forms the basis of the psychological part of Platner's *Neue Anthropologie*, 1790; see book ii. The first edition (*Anthropologie*) appeared in 1772-4. Cf. *Phil. Aphorismen*, vol. i. b. i. §§ 27-43. edit. 1793. Kant's *Kritik der Urtheilskraft* was first published in 1790; the *Anthropologie*, though written before it, was only first published in 1798.—Ed.

η See above, vol. i. lect. xi. p. 186.—Ed.

attention to their great importance in the philosophy of mind, and more precisely and more explicitly than any of his predecessors did he refer them to a particular power,—a power which constituted one of the three fundamental phænomena of mind.

Kant's doctrine controverted by some philosophers of note.

This important innovation necessarily gave rise to controversy. It is true that the Kantian reduction was admitted, not only by the great majority of those who followed the impulsion which Kant had given to philosophy, but, likewise, by the great majority of the psychologists of Germany, who ranged themselves in hostile opposition to the principles of the Critical School. A reaction was, however, inevitable; and while, on the one hand, the greater number were disposed to recognise the Feelings in their new rank, as one of the three grand classes of the mental phænomena; a smaller number,—but among them some philosophers of no mean account,—endeavoured, however violent the procedure, to reannex them, as secondary manifestations, to one or other of the two co-ordinate classes,—the Cognitions and the Conations.

Meaning of the term Feeling.

Before proceeding to consider the objections to the classification in question, it is proper to premise a word in reference to the meaning of the term by which the phænomena of Pleasure and Pain are designated,—the term *Feeling;* for this is an ambiguous expression, and on the accident of its ambiguity have been founded some of the reasons against the establishment of the class of phænomena, which it is employed to denote.

Easy to convey a clear knowledge of the meaning of words which denote phæno-

It is easy to convey a clear and distinct knowledge of what is meant by a word, when that word denotes some object which has an existence external to the mind. I have only to point out the object, and to

LECT. XLI.

say, that such or such a thing is signified by such or such a name; for example, this is called a *house*, that a *rainbow*, this a *horse*, that an *ox*, and so forth. In these cases, the exhibition of the reality is tantamount to a definition; or, as an old logician expresses it, "Cognitio omnis intuitiva est definitiva."[a] The same, however, does not hold in regard to an object which lies within the mind itself. What was easy in the one case becomes difficult in the other. For although he to whom I would explain the meaning of a term, by pointing out the object which it is intended to express, has, at least may have, that very object present in his mind, still I cannot lay my finger on it,—I cannot give it to examine by the eye,—to smell, to taste, to handle. Thus it is that misunderstandings frequently occur in reference to this class of objects, inasmuch as one attaches a different meaning to the word from that in which another uses it; and we ought not to be surprised that, in the nomenclature of our mental phænomena, it has come to pass, that, in all languages, one term has become the sign of a plurality of notions, while at the same time a single notion is designated by a plurality of terms. This vacillation in the application and employment of language, as it originates in the impossibility, anterior to its institution, of approximating different minds to a common cognition of the same internal object; so this ambiguity, when once established, reacts powerfully in perpetuating the same difficulty; insomuch that a principal, if not the very greatest, impediment in the progress of the philosopher of mind, is the vagueness and uncertainty of the instrument of thought itself.

(Margin: means external to the mind. — Not so with respect to words denoting objects that lie within the mind.)

a Cf. Melanchthon, *Erotemata Dialectices*, lib. i., Pr. *De Definitione*, who quotes it as an old saying: "Vetus enim dictum est, et dignum memoria: Omnis intuitiva notitia est definitio."—ED. [Cf. Keckermann, *Opera*, t. i. p. 198. Facciolati, *Institutiones Logicæ*, pars i. c. iii. note 5.]

LECT. XLI.

Feeling, Gefühl, Αἴσθησις, ambiguous.

A remarkable example of this, and one extending to all languages, is seen in the words most nearly correspondent to the very indeterminate expression *feeling*. In English, this, like all others of a psychological application, was primarily of a purely physical relation, being originally employed to denote the sensations we experience through the sense of Touch, and in this meaning it still continues to be employed. From this, its original relation to matter and the corporeal sensibility, it came, by a very natural analogy, to express our conscious states of mind in general, but particularly in relation to the qualities of pleasure and pain, by which they are characterised. Such is the fortune of the term in English; and precisely similar is that of the cognate term *Gefühl* in German. The same, at least a similar, history might be given of the Greek term αἴσθησις, and of the Latin *sensus, sensatio*, with their immediate and mediate derivatives in the different Romanic dialects of modern Europe,—the Italian, Spanish, French, and English dialects. In applying the term *feeling* to the mental states, strictly in so far as these manifest the phænomena of pleasure and pain, it is, therefore, hardly necessary to observe, that the word is used, not in all the meanings in which it can be employed, but in a certain definite relation, were it not that a very unfair advantage has been taken of this ambiguity of the expression. *Feeling*, in one meaning, is manifestly a cognition; but this affords no ground for the argument, that *feeling*, in every signification, is also a cognition. This reasoning has, however, been proposed, and that by a philosopher from whom so paltry a sophism was assuredly not to be expected.

It being, therefore, understood that the word is ambiguous, and that it is only used because no pre-

LECT. XLI.

Can we discriminate in consciousness certain states which cannot be reduced to those of Cognition or Conation?

ferable can be found, the question must be determined by the proof or disproof of the affirmation,—that I am able to discriminate in consciousness certain states, certain qualities of mind, which cannot be reduced to those either of Cognition or Conation; and that I can enable others, in like manner, to place themselves in a similar position, and observe for themselves these states or qualities, which I call *Feelings*. Let us take an example. In reading the story of Leonidas and his three hundred Spartans at Thermopylæ, what do we experience? Is there nothing in the state of mind, which the narrative occasions, other than such as can be referred either to the cognition or to will and desire? Our faculties of knowledge are called certainly into exercise; for this is, indeed, a condition of every other state. But is the exultation which we feel at this spectacle of human virtue, the joy which we experience at the temporary success, and the sorrow at the final destruction of this glorious band,—are these affections to be reduced to states either of cognition or of conation in either form? Are they not feelings,—feelings partly of pleasure, partly of pain?

This question decided in the affirmative by an appeal to experience.

Take another, and a very familiar, instance. You are all probably acquainted with the old ballad of *Chevy Chase*, and you probably recollect the fine verse of the original edition, so lamentably spoiled in the more modern versions:—

> "For Widdrington my soul is sad,
> That ever he slain should be,
> For when his legs were stricken off,
> He kneeled and fought on his knee."[a]

[a] For Wetharryngton my harte was wo,
That ever he slayne shulde be;
For when both his leggis wear hewyne in to,
He knyled and fought on hys kne."
—*Original Version*, in Percy's *Reliques*.—ED.

Now, I ask you, again, is it possible, by any process of legitimate analysis, to carry up the mingled feelings, some pleasurable, some painful, which are called up by this simple picture, into anything bearing the character of a knowledge, or a volition, or a desire? If we cannot do this, and if we cannot deny the reality of such feelings, we are compelled to recognise them as belonging to an order of phænomena, which, as they cannot be resolved into either of the other classes, must be allowed to constitute a third class by themselves.

Grounds on which objection has been taken to the Kantian classification of the mental phænomena.

But it is idle to multiply examples, and I shall now proceed to consider the grounds on which some philosophers, and among these, what is remarkable, a distinguished champion of the Kantian system, have endeavoured to discredit the validity of the classification.

Passing over the arguments which have been urged against the power of Feeling as a fundamental capacity of mind, in so far as these proceed merely on the ambiguities of language, I shall consider only the principal objections from the nature of the phænomena themselves, which have been urged by the three principal opponents of the classification in question, — Carus, Weiss, and Krug. The last of these is the philosopher by whom these objections have been urged most explicitly, and with greatest force. I shall, therefore, chiefly confine myself to a consideration of the difficulties which he proposes for solution.

I may premise that this philosopher (Krug), admitting only two fundamental classes of psychological phænomena,—the Cognitions and the Conations, —goes so far as not only to maintain, that what have obtained, from other psychologists, the name of *Feelings*, constitute no distinct and separate class of

LECT. XLI.

Krug quoted.

mental functions; but that the very supposition is absurd and even impossible. "That such a power of feeling," he argues,* "is not even conceivable, if by such is understood a power essentially different from the powers of Cognition and Conation," (thus I translate *Vorstellungs- und Bestrebungsvermögen*), "is manifest from the following consideration. . . . . The powers of cognition and the powers of conation are, in propriety, to be regarded as two different fundamental powers, only because the operation of our mind exhibits a twofold direction of its whole activity, —one inwards, another outwards; in consequence of which we are constrained to distinguish, on the one hand, an Immanent, ideal or theoretical, and, on the other, a Transeunt, real or practical, activity. Now, should it become necessary to interpolate between these two powers, a third; consequently, to convert the original duplicity of our activity into a triplicity; in this case, it would be requisite to attribute to the third power a third species of activity, the product of which would be, in fact, the Feelings. Now this activity of feeling must necessarily have either a direction inwards, or a direction outwards, or both directions at once, or finally neither of the two, that is, no direction at all; for apart from the directions inwards and outwards, there is no direction conceivable. But, in the first case, the activity of feeling would not be different from the cognitive activity, at least not essentially; in the second case, there is nothing but a certain appetency manifested under the form of a feeling; in the third, the activity of feeling

* This objection is given in substance, though not exactly in language, in Krug's *Philosophisches Lexikon*, art. *Seelenkräfte*. The author, in the same work, art. *Gefühl*, refers to his *Grundlage zu einer neuen Theorie der Gefühle und des sogenannten Gefühlsvermögens*, Königsberg, 1823, for a fuller discussion of the question. See also above, lect. xl. vol. I. p. 187.—ED.

LECT. XLI.

would be only a combination of theoretical and practical activity; consequently, there remains only the supposition that it has no direction. We confess, however, that an hypothetical activity of such a kind we cannot imagine to ourselves as a real activity. An activity without any determinate direction, would be in fact directed upon nothing, and a power conceived as the source of an activity, directed upon nothing, appears nothing better than a powerless power,—a wholly inoperative force,—in a word, a nothing."—So far our objectionist.

Criticised. 1. The suppositions on which the reasoning proceeds, are not exhaustive.

In answer to this reasoning, I would observe, that its cogency depends on this,—that the suppositions which it makes, and afterwards excludes, are exhaustive and complete. But this is not the case. "For, in place of two energies, an immanent and a transeunt, we may competently suppose three,—an ineunt, an immanent, and a transeunt. 1°, The Ineunt energy might be considered as an act of mind, directed upon objects in order to know them,—to bring them within the sphere of consciousness,—mentally to appropriate them; 2°, The Immanent energy might be considered as a kind of internal fluctuation about the objects, which had been brought to representation and thought,—a pleasurable or a painful affection caused by them,—in a word, a feeling; and, 3°, The Transeunt energy might be considered as an act tending towards the object in order to reach it, or to escape from it. This hypothesis is quite as allowable as that in opposition to which it is devised, and were it not merely in relation to an hypothesis, which rests on no valid foundation, it would be better to consider the feelings not as immanent activities, but as immanent passivities.

We may suppose three kinds of energy,—Ineunt, Immanent, and Transeunt.

"But, in point of fact, we are not warranted, by any analogy of our spiritual nature, to ascribe to the men-

2. But we are not warranted to

ascribe to the mental powers a direction either outwards or inwards.

tal powers a direction either outwards or inwards; on the contrary, they are rather the principles of our internal states, of which we can only improperly predicate a direction, and this only by relation to the objects of the states themselves. For directions are relations and situations of external things; but of such there are none to be met with in the internal world, except by analogy to outer objects. In our Senses, which have reference to the external world, there is an outward direction when we perceive, or when we act on external things; whereas, we may be said to turn inwards, when we occupy ourselves with what is contained within the mind itself, be this in order to compass a knowledge of our proper nature, or to elevate ourselves to other objects still more worthy of a moral intelligence. Rigorously considered, the feelings are in this meaning so many directions,—so many turnings of the mind on objects, internal or external; turnings towards those objects which determine the feelings, and which please or displease us. Take, for example, the respect, the reverence, we feel in the contemplation of the higher virtues of human nature; this feeling is an immanent conversion on its object.

3. The argument founded on the hypothesis, that what is true of inanimate, is true of animated nature; and would have no will or desire in the universe.

"The argument of the objectors is founded on the hypothesis, that as in the external world, all is action and reaction,—all is working and counterworking,—all is attraction and repulsion; so in the internal world, there is only one operation of objects on the mind, and one operation of the mind on objects; the former must consist in cognition, the latter in conation. But when this hypothesis is subjected to a scrutiny, it is at once apparent how treacherous is the reasoning which infers of animated, what is true of inanimate, nature; for, to say nothing of aught else that militates against it, this analogy would in truth leave no will or desire in the universe at all; for action and reaction are already compensated in cognition, or, to speak more correctly, in sensitive Perception itself."[a]

LECT. XLI.

Such is a specimen of the only argument of any moment, against the establishment of the Feelings as an ultimate class of mental phænomena.

I pass on to the second question,—What is the position of the Feelings by reference to the two other classes;—and, in particular, should the consideration of the Feelings precede, or follow, that of the Conations?

II. What is the position of the Feelings by reference to the two other classes of mental phænomena.

The answer to the second part of this question, will be given in the determination of the first part; for Psychology proposes to exhibit the mental phænomena in their natural consecution, that is, as they condition and suppose each other. A system which did not accomplish this, could make no pretension to be a veritable exposition of our internal life.

"To resolve this problem, let us take an example. A person is fond of cards. In a company where he beholds a game in progress, there arises a desire to join in it. Now the desire is here manifestly kindled by the pleasure, which the person had, and has, in the play. The feeling thus connects the cognition of the play with the desire to join in it; it forms the bridge, and contains the motive, by which we are roused from mere knowledge to appetency,—to conation, by reference to which we move ourselves so as to attain the end in view.

Resolved by an example.

"Thus we find, in actual life, the Feelings intermediate between the Cognitions and the Conations. And this relative position of these several powers is necessary; without the previous cognition, there could be neither feeling nor conation; and without the previous feeling there could be no conation. Without some

The Feelings intermediate between the Cognitions and Conations.

[a] Biunde, *Versuch der empirischen Psychologie*, II. § 207, p. 51-56.—Ed.

LECT. XLI.

kind or another of complacency with an object, there could be no tendency, no protension of the mind to attain this object as an end; and we could, therefore, determine ourselves to no overt action. The mere cognition leaves us cold and unexcited; the awakened feeling infuses warmth and life into us and our action; it supplies action with an interest, and, without an interest, there is for us no voluntary action possible. Without the intervention of feeling, the cognition stands divorced from the conation, and, apart from feeling, all conscious endeavour after anything would be altogether incomprehensible.

That the Conative Powers are determined by the Feelings further shown.

"That the manifestations of the Conative Powers are determined by the Feelings, is also apparent from the following reflection. The volition or desire tends towards a something, and this something is only given us in and through some faculty or other of cognition.

Mere cognition not sufficient to rouse Conation.

1. Because all objects known in the same manner and degree, are not equal objects of desire or will.

2. Because different individuals are desirous of different objects.

Now, were the mere cognition of a thing sufficient of itself to rouse our conation, in that case, all that was known in the same manner and in the same degree, would become an equal object of desire or will. But we covet one thing; we eschew another. On the supposition, likewise, that our conation was only regulated by our cognition, it behoved that every other individual besides should be desirous of the object which I desire, and be desirous of it also so long as the cognition of the object remained the same. But one person pursues what another person flies; the same person now yearns after something which anon he loathes. And why? It is manifest that here there lies hid some very variable quantity, which, when united with the cognition, is capable of rousing the powers of conation into activity. But such a quantity is given, and only given, in the feelings, that is, in our consciousness of the agreeable and disagreeable. If we take this element,—this influence,—this quantity,—into account, the whole anomalies are solved. We are able at once to understand why all that is thought or cognised with equal intensity, does not, with equal intensity, affect the desires or the will; why different individuals, with the same knowledge of the same objects, are not similarly attracted or repelled; and why the same individual does not always pursue or fly the same object. This is all explained by the fact, that a thing may please one person and displease another; and may now be pleasurable, now painful, and now indifferent, to the same person.

Importance of a correct understanding of the nature and influence of the Feelings.

"From these interests for different objects, and from these opposite interests which the same object determines in our different powers, are we alone enabled to render comprehensible the change and confliction of our desires, the vacillations of our volitions, the warfare of the sensual principle with the rational,—of the flesh with the spirit; so that, if the nature and influence of the feelings be misunderstood, the problems most important for man are reduced to insoluble riddles.

"According to this doctrine, the Feelings, placed in the midst between the powers of Cognition and the powers of Conation, perform the function of connecting principles to these two extremes; and thus the objection that has been urged against the feelings, as a class co-ordinate with the cognitions and the conations,—on the ground that they afford no principle of mediation,—is of all objections the most futile and erroneous. Our conclusion, therefore, is, that as, in our actual existence, the feelings find their place after the cognitions, and before the conations,—so, in the science of mind, the theory of the Feelings ought to follow that of our faculties of Knowledge, and to pre-

Place of the theory of the Feelings in the science of mind.

LECT. XLI.

cede that of our faculties of Will and Desire."[α] Notwithstanding this, various even of those psychologists who have adopted the Kantian trichotomy, have departed from the order which Kant had correctly indicated, and have inverted it in every possible manner,—some treating of the feelings in the last place, while others have considered them in the first.

III. Into what subdivisions are the Feelings to be distributed?

The last preliminary question which presents itself is,—Into what subdivisions are the Feelings themselves to be distributed? In considering this question, I shall first state some of the divisions which have been proposed by those philosophers, who have recognised the capacity of feeling as an ultimate, a fundamental, phænomenon of mind. This statement will be necessarily limited to the distributions adopted by the psychologists of Germany; for, strange to say, the Kantian reduction, though prevalent in the Empire, has remained either unknown to, or disregarded by, those who have speculated on the mind in France, Italy, and Great Britain.

Kant.

To commence with Kant himself. In the *Critique of Judgment*,[β] he enumerates three specifically different kinds of complacency, the objects of which are severally the Agreeable (*das Angenehm*), the Beautiful, and the Good. In his treatise of *Anthropology*,[γ] subsequently published, he divides the feelings of pleasure and pain into two great classes;—1°, The Sensuous; 2°, The Intellectual. The former of these classes is again subdivided into two subordinate kinds, inasmuch as the feeling arises either through the Senses (Sensual Pleasures), or through the Imagination (Pleasures of Taste.) The latter of

α Biunde, *Versuch der empirischen Psychologie*, ii. § 208, p. 60-64.—Ed.

β § 5. *Werke*, iv. p. 53.—Ed.

γ B. ii. *Werke*, vii. p. 143.—Ed.

these classes is also subdivided into subordinate kinds; for our Intellectual Feelings are connected either with the notions of the Understanding, or with the ideas of Reason. I may notice, that in his published manual of *Anthropology* the Intellectual Feelings of the first subdivision,—the feelings of the Understanding,—are not treated of in detail.

Schulze. Gottlob Schulze,[α] though a decided antagonist of the Kantian philosophy in general, adopts the threefold classification into the Cognitions, the Feelings, and the Conations; but he has preferred a division of the Feelings different from that of the philosopher of Königsberg. These he distributes into two classes,—the Corporeal and the Spiritual; to which he annexes a third class made up of these in combination,—the Mixed Feelings.

Hillebrand. Hillebrand[β] divides the Feelings, in a threefold manner, into those of States, those of Cognitions, and those of Appetency, (will and desire); and again into Real, Sympathetic, and Ideal.

Herbart. Herbart[γ] distributes them into three classes;—1°, Feelings which are determined by the character of the thing felt; 2°, Feelings which depend on the disposition of the feeling mind; 3°, Feelings which are intermediate and mixed.

Carus. Carus[δ] (of Leipzig,—the late Carus) thus distributes them. "Pure feeling," he says, "has relation either to Reason, and in this case we obtain the Intellectual Feelings; or it has relation to Desire and Will, and

α *Anthropologie*, § 144-146, p. 295 et seq., 3d edit. 1826.—Ed.

β *Anthropologie*, ii. 283.—Ed.

γ *Lehrbuch zur Psychologie*, § 98. *Werke*, vol. v. p. 72. On the divisions of the Feelings mentioned in the text, see Biunde, *Versuch einer systematischen Behandlung der empirischen Psychologie*, ii. § 210, p. 74, edit. 1831. Cf. Schmidler, *Psychologie*, § 61, p. 443, edit. 1833.—Ed.

δ *Psychologie*, *Werke*, i. 428, edit. Leipsic, 1808.—Ed.

LECT. XLI. in this case we have the Moral Feelings." Between these two classes, the Intellectual and the Moral Feelings, there are placed the Æsthetic Feelings, or Feelings of Taste, to which he also adds a fourth class, that of the Religious Feelings.

Such are a few of the more illustrious divisions of the Feelings into their primary classes. It is needless to enter at present into any discussion of the merits and demerits of these distributions. I shall hereafter endeavour to show you, that they may be divided, in the first place, into two great classes,—the Higher and the Lower,—the Mental and the Corporeal,—in a word, into Sentiments and Sensations.

LECTURE XLII.

THE FEELINGS.—THEORY OF PLEASURE AND PAIN.

LECT. XLII.

The Feelings.

In our last Lecture, we commenced the consideration of the Second Great Class of the Mental Phænomena,—the phænomena of Feeling,—the phænomena of Pleasure and Pain.

Cognitions, Feelings, and Conations,—their essential peculiarities.

Though manifestations of the same indivisible subject, and themselves only possible through each other, the three classes of mental phænomena still admit of a valid discrimination in theory, and require severally a separate consideration in the philosophy of mind. I formerly stated to you, that though knowledge, though consciousness, be the necessary condition not only of the phænomena of Cognition, but of the phænomena of Feeling, and of Conation, yet the attempts of philosophers to reduce the two latter classes to the first, and thus to constitute the faculty of Cognition into the one fundamental power of mind, had been necessarily unsuccessful; because, though the phænomena of Feeling and of Conation appear only as they appear in consciousness, and, therefore, in cognition; yet consciousness shows us in these phænomena certain qualities, which are not contained, either explicitly or implicitly, in the phænomena of Cognition itself. The characters by which these three

LECT. XLII.

classes are reciprocally discriminated are the following. —In the phænomena of Cognition, consciousness distinguishes an object known from the subject knowing. This object may be of two kinds:—it may either be the quality of something different from the ego; or it may be a modification of the ego or subject itself. In the former case, the object, which may be called for the sake of discrimination the *object-object*, is given as something different from the percipient subject. In the latter case, the object, which may be called the *subject-object*, is given as really identical with the conscious ego, but still consciousness distinguishes it, as an accident, from the ego,—as the subject of that accident, it projects, as it were, this subjective phænomenon from itself,—views it at a distance,—in a word, objectifies it. This discrimination of self from self, —this objectification,—is the quality which constitutes the essential peculiarity of Cognition.

Cognition.

Feeling,—how discriminated from Cognition.

In the phænomena of Feeling,—the phænomena of Pleasure and Pain,—on the contrary, consciousness does not place the mental modification or state before itself; it does not contemplate it apart,—as separate from itself,—but is, as it were, fused into one. The peculiarity of Feeling, therefore, is that there is nothing but what is subjectively subjective; there is no object different from self,—no objectification of any mode of self. We are, indeed, able to constitute our states of pain and pleasure into objects of reflection, but in so far as they are objects of reflection, they are not feelings, but only reflex cognitions of feelings.

Conation,—how discriminated from Cognition.

In the phænomena of Conation,—the phænomena of Desire and Will,—there is, as in those of Cognition, an object, and this object is also an object of knowledge. Will and desire are only possible through knowledge,—"Ignoti nulla cupido." But though both cognition and conation bear relation to an object, they are discriminated by the difference of this relation itself. In cognition, there exists no want; and the object, whether objective or subjective, is not sought for, nor avoided; whereas in conation, there is a want, and a tendency supposed, which results in an endeavour, either to obtain the object, when the cognitive faculties represent it as fitted to afford the fruition of the want; or to ward off the object, if these faculties represent it as calculated to frustrate the tendency, of its accomplishment.

Conation,—how discriminated from Feeling.

The feelings of Pleasure and Pain and the Conations are, thus, though so frequently confounded by psychologists, easily distinguished. It is, for example, altogether different to feel hunger and thirst, as states of pain, and to desire or will their appeasement; and still more different is it to desire or will their appeasement, and to enjoy the pleasure afforded in the act of this appeasement itself. Pain and pleasure, as feelings, belong exclusively to the present; whereas conation has reference only to the future, for conation is a longing,—a striving, either to maintain the continuance of the present state, or to exchange it for another. Thus, conation is not the feeling of pleasure and pain, but the power of overt activity, which pain and pleasure set in motion.

But although, in theory, the Feelings are thus to be discriminated from the Desires and Volitions, they are, as I have frequently observed, not to be considered as really divided. Both are conditions of perhaps all our mental states; and while the Cognitions go principally to determine our speculative sphere of exist-

LECT. XLII.

ence, the Feelings and the Conations more especially concur in regulating our practical.

What are the general conditions which determine the existence of Pleasure and Pain?

In my last Lecture, I stated the grounds on which it is expedient to consider the phænomena of Feeling prior to discussing those of Conation;—but before entering on the consideration of the several feelings, and before stating under what heads, and in what order, these are to be arranged, I think it proper, in the first place, to take up the general question,—What are the general conditions which determine the existence of Pleasure and Pain; for pleasure and pain are the phænomena which constitute the essential attribute of feeling, under all its modifications?

Order of discussion.

In the consideration of this question, I shall pursue the following order:—I shall, first of all, state the abstract Theory of Pleasure and Pain, in other words, enounce the fundamental law by which these phænomena are governed, in all their manifestations. I shall, then, take an historical retrospect of the opinions of philosophers in regard to this subject, in order to show in what relation the doctrine I would support stands to previous speculations. This being accomplished, we shall then be prepared to inquire, how far the theory in question is borne out by the special modifications of Feeling, and how far it affords us a common principle on which to account for the phænomena of Pleasure and Pain, under every accidental form they may assume.

1. The theory of Pleasure and Pain,—stated in the abstract.

I proceed, therefore, to deliver in somewhat abstruse formulæ, the theory of pleasure. The meaning of these formulæ I cannot expect should be fully apprehended, in the first instance,—far less can I expect that the validity of the theory should be recognised, before the universality of its application shall be illustrated in examples.

LECT. XLII.

First momentum.

I. Man exists only as he lives; as an intelligent and sensible being, he consciously lives, but this only as he consciously energises. Human existence is only a more general expression for human life, and human life only a more general expression for the sum of energies, in which that life is realised, and through which it is manifested in consciousness. In a word, life is energy, and conscious energy is conscious life.[a]

Comprehension of the term energy.

In explanation of this paragraph, and of those which are to follow, I may observe, that the term *energy*, which is equivalent to *act*, *activity*, or *operation*, is here used to comprehend also all the mixed states of action and passion, of which we are conscious; for, inasmuch as we are conscious of any modification of mind, there is necessarily more than a mere passivity of the subject; consciousness itself implying at least a reaction. Be this, however, as it may, the nouns *energy*, *act*, *activity*, *operation*, with the correspondent verbs, are to be understood to denote, indifferently and in general, all the processes of our higher and our lower life, of which we are conscious.[β] This being premised, I proceed to the second proposition.

Second.

II. Human existence, human life, human energy, is not unlimited, but, on the contrary, determined to a certain number of modes, through which alone it can possibly be exerted. These different modes of action are called, in different relations, *powers*, *faculties*, *capacities*, *dispositions*, *habits*.

In reference to this paragraph, it is only necessary to recall to your attention, that *power* denotes either

[a] Cf. Aristotle, *Eth. Nic.*, ix. 9; x. 4.—Ed. Lossius, *Lexikon*, v. *Vermögen*; theory of causation and activity; makes partly active, partly passive; partly tending to rest, partly to action.—*Memorandum*.

[β] Here a written interpolation,—*Occupation*, *exercise*, perhaps better [expressions than energy, as applying equally to all mental processes, whether active or passive.] See below p. 466.—Ed.

LECT. XLII.

Explication of terms,— power, faculty, &c.

a faculty or a capacity; *faculty* denotes a power of acting, *capacity* a power of being acted upon or suffering; *disposition*, a natural, and *habit*, an acquired, tendency to act or suffer.[a] In reference to habit, it ought however to be observed, that an acquired necessarily supposes a natural tendency. Habit, therefore, comprehends a disposition and something supervening on a disposition. The disposition, which, at first, was a feebler tendency, becomes, in the end, by custom, that is, by a frequent repetition of exerted energy, a stronger tendency. Disposition is the rude original, habit is the perfect consummation.

Third.

III. Man, as he consciously exists, is the subject of pleasure and pain; and these of various kinds: but as man only consciously exists in and through the exertion of certain determinate powers, so it is only through the exertion of these powers that he becomes the subject of pleasure and pain; each power being in itself at once the faculty of a specific energy, and a capacity of an appropriate pleasure or pain, as the concomitant of that energy.

Fourth.

IV. The energy of each power of conscious existence having, as its reflex or concomitant, an appropriate pleasure or pain, and no pain or pleasure being competent to man, except as the concomitant of some determinate energy of life, the all-important question arises,—What is the general law under which these counter-phænomena arise, in all their special manifestations?

Pleasure and Pain opposed as contraries, not as contradictories.

In reference to this proposition, I would observe that pleasure and pain are opposed to each other as contraries, not as contradictories, that is, the affirmation of the one implies the negation of the other, but

a See above, vol. I. lect. x. p. 177.—Ed.

the negation of the one does not infer the affirmation of the other ; for there may be a third or intermediate state, which is neither one of pleasure nor one of pain, but one of indifference. Whether such a state of indifference do ever actually exist ; or whether, if it do, it be not a complex state in which are blended an equal complement of pains and pleasures, it is not necessary, at this stage of our progress, to inquire. It is sufficient, in considering the quality of pleasure as one opposed to the quality of pain, to inquire, what are the proximate causes which determine them : or, if this cannot be answered, what is the general fact or law which regulates their counter-manifestation ; and if such a law can be discovered for the one, it is evident that it will enable us also to explain the other, for the science of contraries is one. I now proceed to the fifth proposition.

V. The answer to the question proposed is :—the more perfect, the more pleasurable, the energy ; the more imperfect, the more painful. Fifth.

In reference to this proposition, it is to be observed that the answer here given is precise, but inexplicit ; it is the enouncement of the law in its most abstract form, and requires at once development and explanation. This I shall endeavour to give in the following propositions.

VI. The perfection of an energy is twofold ; 1°, By relation to the power of which it is the exertion, and, 2°, By relation to the object about which it is conversant. The former relation affords what may be called its *subjective*, the latter what may be called its *objective*, condition. Sixth.

The explanation and development of the preceding proposition is given in the following.

LECT. XLII.

Seventh.

VII. By relation to its power:—An energy is perfect, when it is tantamount to the full, and not to more than the full, complement of free or spontaneous energy, which the power is capable of exerting; an energy is imperfect, either, 1°, When the power is restrained from putting forth the whole amount of energy it would otherwise tend to do, or, 2°, When it is stimulated to put forth a larger amount than that to which it is spontaneously disposed. The amount or quantum of energy in the case of a single power is of two kinds,—1°, An intensive, and, 2°, A protensive; the former expressing the higher degree, the latter the longer duration, of the exertion. A perfect energy is, therefore, that which is evolved by a power, both in the degree and for the continuance to which it is competent without straining; an imperfect energy, that which is evolved by a power in a lower or in a higher degree, for a shorter or for a longer continuance, than, if left to itself, it would freely exert. There are, thus, two elements of the perfection, and, consequently, two elements of the pleasure, of a simple energy:—its adequate degree and its adequate duration; and four ways in which such an energy may be imperfect, and, consequently, painful; inasmuch as its degree may be either too high, or too low; its duration either too long or too short.

When we do not limit our consideration to the simple energies of individual powers, but look to complex states, in which a plurality of powers may be called simultaneously into action, we have, besides the intensive and protensive quantities of energy, a third kind, to wit, the extensive quantity. A state is said to contain a greater amount of extensive energy, in proportion as it forms the complement of a greater number of simultaneously co-operating powers. This complement, it is evident, may be conceived as made up either of energies all intensively and protensively perfect and pleasurable, or of energies all intensively and protensively imperfect and painful, or of energies partly perfect, partly imperfect, and this in every combination afforded by the various perfections and imperfections of the intensive and protensive quantities. It may be here noticed, that the intensive and the two other quantities stand always in an inverse ratio to each other; that is, the higher the degree of any energy, the shorter is its continuance, and, during its continuance, the more completely does it constitute the whole mental state,—does it engross the whole disposable consciousness of the mind. The maximum of intensity is thus the minimum of continuance and of extension. So much for the perfection, and proportional pleasure, of an energy or state of energies, by relation to the power out of which it is elicited. This paragraph requires, I think, no commentary.

VIII. By relation to the object, (and by the term *object*, be it observed, is here denoted every objective cause by which a power is determined to activity), about which it is conversant, an energy is perfect, when this object is of such a character as to afford to its power the condition requisite to let it spring to full spontaneous activity; imperfect, when the object is of such a character as either, on the one hand, to stimulate the power to a degree, or to a continuance, of activity beyond its maximum of free exertion; or, on the other hand, to thwart it in its tendency towards this its natural limit. An object is, consequently, pleasurable or painful, inasmuch as it thus determines a power to perfect or to imperfect energy. Eighth.

LECT. XLII.

But an object, or complement of objects simultaneously presented, may not only determine one but a plurality of powers into coactivity. The complex state, which thus arises, is pleasurable, in proportion as its constitutive energies are severally more perfect; painful, in proportion as these are more imperfect; and in proportion as an object, or a complement of objects, occasions the average perfection or the average imperfection of the complex state, is it, in like manner, pleasurable or painful.

Ninth. Definitions of Pleasure and Pain.

IX. Pleasure is, thus, the result of certain harmonious relations,—of certain agreements; pain, on the contrary, the effect of certain unharmonious relations,—of certain disagreements. The pleasurable is, therefore, not inappropriately called *the agreeable*, the painful *the disagreeable*, and, in conformity to this doctrine, pleasure and pain may be thus defined :—

Pleasure is a reflex of the spontaneous and unimpeded exertion of a power, of whose energy we are conscious.[a] Pain, a reflex of the overstrained or repressed exertion of such a power.

The definition of Pleasure illustrated. 1. Pleasure the reflex of energy.

I shall say a word in illustration of these definitions. Taking pleasure,—pleasure is defined to be the reflex of energy and of perfect energy, and not to be either energy or the perfection of energy itself, —and why? It is not simply defined an energy, exertion, or act, because some energies are not pleasurable, —being either painful or indifferent. It is not simply

[a] This is substantially the definition of Aristotle, whose doctrine, as expounded in the 10th book of the *Nicomachean Ethics*, is more fully stated below, p. 450. In the less accurate dissertation, which occurs in the 7th book of the same treatise, and which perhaps properly belongs to the *Eudemian Ethics*, the pleasure is identified with the energy itself.—ED.

defined the perfection of an energy, because we can easily separate in thought the perfection of an act, a conscious act, from any feeling of pleasure in its performance. The same holds true, *mutatis mutandis*, of the definition of pain, as a reflex of imperfect energy.

LECT. XLIII.

2. Spontaneous and unimpeded.

Again, pleasure is defined the reflex of the spontaneous and unimpeded,—of the free and unimpeded, exertion of a power, of whose energy we are conscious. Here the term *spontaneous* refers to the subjective, the term *unimpeded* to the objective, perfection. Touching the term *spontaneous*, every power, all conditions being supplied, and all impediments being removed, tends, of its proper nature and without effort, to put forth a certain determinate maximum, intensive and protensive, of free energy. This determinate maximum of free energy, it, therefore, exerts spontaneously: if a less amount than this be actually put forth, a certain quantity of tendency has been forcibly repressed; whereas, if a greater than this has been actually exerted, a certain amount of nisus has been forcibly stimulated in the power. The term *spontaneously*, therefore, provides that the exertion of the power has not been constrained beyond the proper limit,—the natural maximum, to which, if left to itself, it freely springs.

Again, in regard to the term *unimpeded*,—this stipulates that the power should not be checked in the spring it would thus spontaneously make to its maximum of energy, that is, it is supposed that the conditions requisite to allow this spring have been supplied, and that all impediments to it have been removed. This postulates of course the presence of an object. The definition further states, that the exertion must be

LECT. XLII.

3. Of which we are conscious.

that of a power of whose energy we are conscious. This requires no illustration. There are powers in man, the activities of which lie beyond the sphere of consciousness. But it is of the very essence of pleasure and pain to be felt, and there is no feeling out of consciousness. What has now been said of the terms used in the definition of pleasure, renders all comment superfluous on the parallel expressions employed in that of pain.

Pleasure,—Positive and Negative.

On this doctrine it is to be observed, that there are given different kinds of pleasure, and different kinds of pain. In the first place, these are twofold, inasmuch as each is either Positive and Absolute, or Negative and Relative. In regard to the former, the mere negation of pain does, by relation to pain, constitute a state of pleasure. Thus, the removal of the toothache replaces us in a state which, though one really of indifference, is, by contrast to our previous agony, felt as pleasurable. This is negative or relative pleasure. Positive or absolute pleasure, on the contrary, is all that pleasure which we feel above a state of indifference, and which is, therefore, prized as a good in itself, and not simply as the removal of an evil.

Pain,—Positive and Negative.

On the same principle, pain is also divided into Positive or Absolute, and into Negative or Relative.

Positive pain, subdivided.

But, in the second place, there is, moreover, a subdivision of positive pain into that which accompanies a repression of the spontaneous energy of a power, and that which is conjoined with its effort, when stimulated to over-activity.[a]

[a] [With the foregoing theory compare Hutcheson, *System of Moral Philosophy*, i. p. 21 *et seq.* Lüders, *Kritik der Statistik*, p. 457-9. Tiedemann, *Psychologie*, p. 151, edit. 1804. Bonnet, *Essai Analytique sur l'Ame*, chaps. xvii. xx. Ferguson, *Principles of Moral and Political Science*, Part ii. c. i. § 2.—Ed.]

I proceed now to state certain corollaries, which flow immediately from the preceding doctrine. LECT. XLII.

Corollaries from preceding doctrine.

In the first place, as the powers which, in an individual, are either preponderantly strong by nature, or have become preponderantly strong by habit, have comparatively more perfect energies; so the pleasures which accompany these will be proportionally intense and enduring. But this being the case, the individual will be disposed principally, if not exclusively, to exercise these more vigorous powers, for their energies afford him the largest complement of purest pleasure. "Trahit sua quemque voluptas,"[a] each has his ruling passion.

1. The individual will be disposed to exercise his more vigorous powers.

But, in the second place, as the exercise of a power is the only mean by which it is invigorated, but as, at the same time, this exercise, until the development be accomplished, elicits imperfect, and, therefore, painful, or at least less pleasurable, energy,—it follows that those faculties which stand the most in need of cultivation, are precisely those which the least secure it; while, on the contrary, those which are already more fully developed, are precisely those which present the strongest inducements for their still higher invigoration.

2. Those faculties which most need cultivation, the least secure it.

[a] Virgil, *Ecl.* ii. 65.—Ed.

LECTURE XLIII.

THE FEELINGS.—HISTORICAL ACCOUNT OF THEORIES OF PLEASURE AND PAIN.

LECT. XLIII.

Recapitulation.

IN my last Lecture, I gave an abstract statement of that Theory of Pleasure and Pain, which, I think, is competent, and exclusively competent, to explain the whole multiform phænomena of our Feelings,—a theory, consequently, which those whole phænomena concur in establishing. It is, in truth, nothing but a generalisation of what is essential in the concrete facts themselves. Before, however, proceeding to show, by its application to particular cases, that this theory affords us a simple principle, on which to account for the most complicated and perplexing phænomena of Feeling, I shall attempt to give you a slight survey of the most remarkable opinions on this point. To do this, however imperfectly, is of the more importance, as there is no work in which any such historical deduction is attempted; but principally, because the various theories of philosophers on the doctrine of the pleasurable, are found, when viewed in connection, all to concur in manifesting the truth of that one which I have proposed to you,—a theory, in fact, which is the resumption and complement of them all. In attempting this survey, I by no means propose to furnish even an indication of all the opinions that have been held in regard to the pleasurable in general, nor even

General historical notices of Theories of the Pleasurable.

of all the doctrines on this subject that have been advanced by the authors to whom I specially refer. I can only afford to speak of the more remarkable theories, and, in these, only of the more essential particulars. But, in point of fact, though there is no end of what has been written upon pleasure and pain, considered in their moral relations and effects, the speculations in regard to their psychological causes and conditions are comparatively few. In general, I may also premise that there is apparent a remarkable gravitation in the various doctrines promulgated on this point, towards a common centre; and, however one-sided and insufficient the several opinions may appear, they are all substantially grounded upon truth, being usually right in what they affirm, and wrong only in what they deny; all are reflections, but only partial reflections, of the truth. These opinions, I may further remark, fall into two great classes; and at the head of each there is found one of the two great philosophers of antiquity,—Plato being the founder of the one general theory, Aristotle of the other. But though the distinction of these classes pervades the whole history of the doctrines, I do not deem it necessary to follow this classification in the following observations, but shall content myself with a chronological arrangement.

Three theories fall into two grand classes,—the Platonic and the Aristotelic.

Plato is the first philosopher who can be said to have attempted the generalisation of a law which regulates the manifestation of pleasure and pain; and it is but scanty justice to acknowledge that no subsequent philosopher has handled the subject with greater ingenuity and acuteness. For though the theory of Aristotle be more fully developed, and, as I am convinced, upon the whole the most complete and accurate

Plato the first to attempt the generalisation of a law of Pleasure and Pain.

LECT. XLIII.

which we possess, it is but fair to add, that he borrowed a considerable portion of it from Plato, whose doctrine he corrected and enlarged.

Plato's theory,—that a state of pleasure is always preceded by a state of pain.

The opinion of Plato regarding the source of pleasure is contained in the *Philebus*, and in the ninth book of the *Republic*, with incidental allusions to his theory in other dialogues. Thus, in the opening of the *Phædo*,* we have the following statement of its distinguishing principle,—that a state of pleasure is always preceded by a state of pain. Phædo, in describing the conduct of Socrates in the prison and on the eve of death, narrates, that, "sitting upright on the bed he (Socrates) drew up his leg, and stroking it with his hand, said at the same time,—'What a wonderful thing is this, my friends, which men call the pleasant and agreeable! and how wonderful a relation does it bear by nature to that which seems to be its contrary, the painful! For they are unwilling to be present with us both together; and yet, if any person pursues and obtains the one, he is almost always under a necessity of accepting also the other, as if both of them depended from a single summit. And it seems to me' (he continues), 'that if Æsop had perceived this, he would have written a fable upon it, and have told us that the Deity, being willing to reconcile the conflictive natures, but at the same time unable to accomplish this design, conjoined their summits in an existence one and the same; and that hence it comes to pass that whoever partakes of the one, is soon after compelled to participate in the other. And this, as it appears, is the case with myself at present; for the pain which was before in my leg, through the stricture of the fetter, is now succeeded by a pleasant sensation.'"

* P. 60.—ED.

The following extract from the *Philebus*[a] will, however, show more fully the purport and grounds of his opinion :—

Quotation from the Philebus.

"*Socrates.* I say then, that whenever the harmony in the frame of any animal is broken, a breach is then made in its constitution, and, at the same time, rise is given to pains.

"*Protarchus.* You say what is highly probable.

"*Soc.* But when the harmony is restored, and the breach is healed, we should say that then pleasure is produced; if points of so great importance may be despatched at once in so few words.

"*Prot.* In my opinion, O Socrates, you say what is very true; but let us try if we can show these truths in a light still clearer.

"*Soc.* Are not such things as ordinarily happen, and are manifest to us all, the most easy to be understood?

"*Prot.* What things do you mean?

"*Soc.* Want of food makes a breach in the animal system, and, at the same time, gives the pain of hunger.

"*Prot.* True.

"*Soc.* And food, in filling up the breach again, gives a pleasure.

"*Prot.* Right.

"*Soc.* Want of drink, also, interrupting the circulation of the blood and humours, brings on us corruption together with the pain of thirst: but the virtue of a liquid in moistening and replenishing the parts dried up, yields a pleasure. In like manner, unnatural suffocating heat, in dissolving the texture of the parts, gives a painful sensation; but a cooling again, a

a P. 31.—Ed.

LECT. XLIII. refreshment agreeable to nature, affects us with a sense of pleasure.

"*Prot.* Most certainly.

"*Soc.* And the concretion of the animal humours through cold, contrary to their nature, occasions pain; but a return to their pristine state of fluidity, and a restoring of the natural circulation, produce pleasure. See, then, whether you think this general account of the matter not amiss, concerning that sort of being which I said was composed of indefinite and definite,—that, when by nature any beings of that sort become animated with soul, their passage into corruption, or a total dissolution, is accompanied with pain; and their entrance into existence, the assembling of all those particles which compose the nature of such a being, is attended with a sense of pleasure.

"*Prot.* I admit your account of this whole matter; for, as it appears to me, it bears on it the stamp of truth."

And, in a subsequent part of the dialogue, Socrates is made to approve of the doctrine of the Eleatic School, in regard to the unreality of pleasure, as a thing always in generation, that is, always in progress towards existence, but never absolutely existent.

"*Soc.* But what think you now of this? Have we not heard it said concerning pleasure, that it is a thing always in generation, always produced anew, and which, having no stability of being, cannot properly be said to be at all? For some ingenious persons there are, who endeavour to show us that such is the nature of pleasure; and we are much obliged to them for this their account of it."[a]

Then, after an expository discourse on the Eleatic

[a] P. 53.—Ed.

doctrine, Socrates proceeds :[a]—"Therefore, as I said in the beginning of this argumentation, we are much obliged to the persons who have given us this account of pleasure,—that the essence of it consists in being always generated anew, but that never has it any kind of being. For it is plain that these persons would laugh at a man who asserted, that pleasure and good were the same thing.

"*Prot.* Certainly they would.

"*Soc.* And these very persons would undoubtedly laugh at those men, wherever they met with them, who place their chief good and end in a becoming, —an approximation to existence?

"*Prot.* How? what sort of men do you mean?

"*Soc.* Such as, in freeing themselves from hunger or thirst, or any of the uneasinesses from which they are freed by generation,—by tending towards being, are so highly delighted with the action of removing those uneasinesses, as to declare they would not choose to live without suffering thirst and hunger, nor without feeling all those other sensations which may be said to follow from such kinds of uneasiness."

Sum of Plato's doctrine of the Pleasurable.

The sum of Plato's doctrine on this subject is this, —that pleasure is nothing absolute, nothing positive, but a mere relation to, a mere negation of, pain. Pain is the root, the condition, the antecedent of pleasure, and the latter is only a restoration of the feeling subject, from a state contrary to nature to a state conformable with nature. Pleasure is the mere replenishing of a vacuum,—the mere satisfying of a want. With this principal doctrine,—that pleasure is only the negation of pain, Plato connects sundry collateral opinions in conformity to his general system. That

a P. 54.—ED.

pleasure, for example, is not a good, and that it is nothing real or existent, but something only in the progress towards existence,—never being, ever becoming (*ἀεὶ γιγνόμενον, οὐδέποτε ὄν*).

The doctrine of Aristotle proposed to correct and supplement the Platonic.

Aristotle saw the partiality and imperfection of this theory, and himself proposed another, which should supply its deficiencies. His speculations concerning the pleasurable are to be found in his Ethical Treatises, and, to say nothing of the two lesser works, the *Magna Moralia* and the *Eudemian Ethics*,[a] you will find the subject fully discussed in the seventh and tenth Books of the *Nicomachean Ethics*. I shall say nothing of Aristotle's arguments against Eudoxus, as to whether pleasure be the chief good, and against Plato, as to whether it be a good at all,—these are only ethical questions; I shall confine my observations to the psychological problem touching the law which governs its manifestation.

Aristotle refutes the Platonic doctrine,—that pleasure is only the removal of a pain.

Aristotle, in the first place, refutes the Platonic theory,—that pleasure is only the removal of a pain. "Since it is asserted," he says,[β] "that pain is a want, an indigence (ἔνδεια) contrary to nature, pleasure will be a repletion, a filling up (ἀναπλήρωσις) of that want in conformity to nature. But want and its repletion are corporeal affections. Now if pleasure be the repletion of a want contrary to nature, that which contains the repletion will contain the pleasure, and the faculty of being pleased. But the want and its repletion are in the body; the body, therefore, will be pleased,—the body will be the subject of this feeling. But the feeling of pleasure is

[a] The genuineness of these two works is questionable. The chapters on pleasure in the *Eudemian Ethics* are identical with those in the 7th book of the *Nicomachean*, being part of the three books which are common to both treatises.—Ed.

[β] *Eth. Nic.*, x. 3.—Ed.

an affection of the soul. Pleasure, therefore, cannot be merely a repletion. True it is, that pleasure is consequent on the repletion of a want, as pain is consequent on the want itself. For we are pleased when our wants are satisfied; pained when this is prevented.

"It appears," proceeds the Stagirite, "that this opinion has originated in an exclusive consideration of our bodily pains and pleasures, and more especially those relative to food. For when inanition has taken place, and we have felt the pains of hunger, we experience pleasure in its repletion. But the same does not hold good in reference to all our pleasures. For the pleasure we find, for example, in mathematical contemplations, and even in some of the senses, is wholly unaccompanied with pain. Thus the gratification we derive from the energies of hearing, smell, and sight, is not consequent on any foregone pain, and in them there is, therefore, no repletion of a want. Moreover, hope, and the recollection of past good, are pleasing; but are the pleasures from these a repletion? This cannot be maintained; for in them there is no want preceding, which could admit of repletion. Hence it is manifest, that pleasure is not the negation of a pain."

The theory of Aristotle.

Having disposed of Plato's theory, Aristotle proposes his own; and his doctrine, in as far as it goes, is altogether conformable to that I have given to you, as the one which appears to me the true.

Pleasure, according to Aristotle, is the concomitant of the unimpeded energy of a power.

Pleasure is maintained by Aristotle to be the concomitant of energy,—of perfect energy, whether of the functions of Sense or Intellect; and perfect energy he describes as that which proceeds from a power in health and vigour, and exercised upon an object rela-

LECT. XLIII. tively excellent, that is, suited to call forth the power into unimpeded activity. Pleasure, though the result, —the concomitant, of perfect action, he distinguishes from the perfect action itself. It is not the action, it is not the perfection, though it be consequent on action, and a necessary efflorescence of its perfection. Pleasure is thus defined by Aristotle to be the concomitant of the unimpeded energy of a natural power, faculty, or acquired habit.[a] "Thus when a sense, for example, is in perfect health, and it is presented with a suitable object of the most perfect kind, there is elicited the most perfect energy, which, at every instant of its continuance, is accompanied with pleasure. The same holds good with the function of Imagination, Thought, &c. Pleasure is the concomitant in every case where powers and objects are in themselves perfect, and between which there subsists a suitable relation. Hence arises the pleasure of novelty. For on the first presentation of a new object, the energy of cognition is intensely directed upon it, and the pleasure high; whereas when the object is again and again presented, the energy relaxes, and the pleasure declines. But pleasure is not merely the consequent of the most perfect exertion of power; for it reacts upon the power itself, by raising, invigorating, and perfecting its development. For we make no progress in a study, except we feel a pleasure in its pursuit.

Aristotle quoted.

"Every different power has its peculiar pleasure and its peculiar pain; and each power is as much corrupted by its appropriate pain as it is perfected by its appropriate pleasure. Pleasure is not something that arises,—that comes into existence, part after part;

[a] See above, p. 440.—Ed.

it is, on the contrary, complete at every indivisible instant of its continuance. It is not, therefore, as Plato holds, a change, a motion, a generation (γένεσις, κίνησις), which exists piecemeal as it were, and successively in time, and only complete after a certain term of endurance; but on the contrary something instantaneous, and, from moment to moment, perfect."[α]

Nothing added in antiquity to the two theories of Plato and Aristotle.

Such were the two theories touching the law of pleasure and pain, propounded by the two principal thinkers of antiquity. To their doctrines on this point we find nothing added, worthy of commemoration, by the succeeding philosophers of Greece and Rome; nay, we do not find that in antiquity these doctrines received any farther development or confirmation. Among the ancients, however, the Aristotelic theory seems to have soon superseded the Platonic; for, even among the lower Platonists themselves, there is no attempt to vindicate the doctrine of their master, in so far as to assert that all pleasure is only a relief from pain. Their sole endeavour is to reconcile Plato's opinion with that of Aristotle, by showing that the former did not mean to extend the principle in question to pleasure in general, but applied it only to the pleasures of certain of the senses. And in truth, various passages in the *Philebus* and in the ninth book of the *Republic*, afford countenance to this interpretation.[β] Be this, however, as it may, it

α See *Eth. Nic.*, x. 4, 5.—Ed. [On Aristotle's doctrine of the Pleasurable; see Tennemann, *Geschichte der Philosophie*, iii. p. 200.]

β [Plato, as well as Aristotle, seems to have made pleasure consist in a harmonious, pain in a disharmonious, energy. Every energy, both of Sense and Intellect, is, according to Plato, accompanied with a sensation of pleasure and pain. *Republic*, ix. p. 557. *Philebus*, p. 211, edit. Bipont. See Tennemann, *Geschichte der Philosophie*, ii. p. 290.]

was only in more recent times that the Platonic doctrine, in all its exclusive rigour, was again revived; and that too by philosophers who seem not to have been aware of the venerable authority in favour of the paradox which they proposed as new. I may add that the philosophers, who in modern times have speculated upon the conditions of the pleasurable, seem, in general, unaware of what had been attempted on this problem by the ancients; and it is indeed this circumstance alone that enables us to explain, why the modern theories on this subject, in principle the same with that of Aristotle, have remained so inferior to his in the great virtues of a theory,—comprehension and simplicity.

The theories of Plato and Aristotle reduced to unity.

Before, however, proceeding to the consideration of subsequent opinions, it may be proper to observe that the theories of Plato and Aristotle, however opposite in appearance, may easily be reduced to unity, and the theory of which I have given you the general expression, will be found to be the consummated complement of both. The two doctrines differ only essentially in this:—that the one makes a previous pain the universal condition of pleasure; while the other denies this condition as a general law, and holds that pleasure is a positive reality, and more than the mere alternative of pain. Now, in regard to this difference, it must be admitted, on the one hand, that in so far as the instances are concerned, on which Plato attempts to establish his principle, Aristotle is successful in showing, that these are only special cases, and do not warrant the unlimited conclusion in support of which they are adduced.

But, on the other hand, it must be confessed that

Aristotle has not shown the principle to be false,—that all pleasure is an escape from pain. He shows, indeed, that the analogy of hunger, thirst, and other bodily affections, cannot be extended to the gratification we experience from the energies of intellect,—cannot be extended even to that which we experience in the exercise of the higher senses. It is true, that the pleasure I experience in this particular act of vision, cannot be explained from the pain I had felt in another particular act of vision, immediately preceding; and if this example were enough, it would certainly be made out that pleasure is not merely the negation of a foregoing pain. But let us ascend a step higher and inquire,—would it not be painful if the faculty of vision, (to take the same example), were wholly restrained from operation? Now it will not be denied, that the repression of any power in its natural *nisus*,—*conatus*, to action, is positively painful; and, therefore, that the exertion of a power, if it afforded only a negation of that positive pain, and were, in its own nature, absolutely indifferent, would, by relation to the pain from which it yields us a relief, appear to us a real pleasure. We may, therefore, I think, maintain, with perfect truth, that as the holding back of any power from exercise is positively painful, so its passing into energy is, were it only the removal of that painful repression, negatively pleasurable; on this ground, consequently, and to this extent, we may rightly hold with Plato,—that every state of pleasure and free energy is, in fact, the escape from an alternative state of pain and compulsory inaction.

In what sense the Platonic dogma is true.

So far we are warranted in going. But we should be wrong were we to constitute this partial truth into

LECT. XLIII.

The doctrine that the whole pleasure of activity arises from the negation of the pain of forced inertion,—erroneous.

an unlimited,—an exclusive principle; that is, were we to maintain that the whole pleasure we derive from the exercise of our powers, is nothing more than a negation of the pain we experience from their forced inertion. This I say would be an erroneous, because an absolute, conclusion. For the pleasure we find in the free play of our faculties is, as we are most fully conscious, far more than simply a superseding of pain. That philosophy, indeed, would only provoke a smile which would maintain, that all pleasure is in itself a zero,—a nothing, which becomes a something only by relation to the reality of pain which it annuls.

After compulsory inertion, pleasure higher than in ordinary circumstances,—explained.

It is true, indeed, that after a compulsory inertion, our pleasure, in the first exertion of our faculties, is frequently far higher than that which we experience in their ordinary exercise, when left at liberty. But this does not, at least does not exclusively, arise from the contrast of the previous and subsequent states of pain and pleasure, but principally because the powers are in excessive vigour,—at least in excessive erethism or excitation, and have thus a greater complement of intenser energy suddenly to expend. On the principle, therefore, that the degree of pleasure is always in the ratio of the degree of spontaneous activity, the pleasure immediately consequent on the emancipation of a power from thraldom, would, if the power remain uninjured by the constraint, be naturally greater, because the energy would in that case be, for a season, more intense. At the same time, the state of pleasure would in this case appear to be higher than what it absolutely is; because it would be set off by proximate contrast with a previous state of pain. Thus it is that a basin of water of ordinary blood heat, appears hot, if we plunge in it a hand which had previously been dipped in snow; and cold, if we immerse in it another which had previously been placed in water of a still higher temperature. But it is unfair to apply this magnifying effect of contrast to the one relative and not to the other; and any argument drawn from it against the positive reality of pleasure, applies equally to disprove the positive reality of pain. The true doctrine I hold to be this,—that pain and pleasure are, as I have said, each to be considered both as Absolute and as Relative:—absolute, that is, each is something real, and would exist were the other taken out of being; relative, that is, each is felt as greater or less by immediate contrast to the other. I may illustrate this by the analogy of a scale. Let the state of indifference,—that is, the negation of both pain and pleasure,—be marked as zero, let the degrees of pain be denoted by a descending series of numbers below zero, and the degrees of pleasure by an ascending series of numbers above zero. Now, suppose the degree of pain we feel from a certain state of hunger, to be six below zero; in this case our feeling, in the act of eating, will not merely rise to zero, that is, to the mere negation of pain, as the Platonic theory holds, but to some degree of positive pleasure, say six. And here I may observe, that, were the insufficiency of the Platonic theory shown by nothing else, this would be done by the absurd consequences it implies, in relation to the function of nutrition alone; for if its principle be true, then would our gratification from the appeasement of hunger, be equally great by one kind of viand as by another.

Unfair to apply the magnifying effect of contrast to disprove the positive reality of pleasure more than of pain.

Pleasure and pain both Absolute and Relative.

Thus, then, the counter theories of Plato and Aris-

LECT. XLIII.

totle are, as I have said, right in what they affirm, wrong in what they deny; each contains the truth, but not the whole truth. By supplying, therefore, to either that in which it was defective, we reduce their apparent discord to real harmony, and show that they are severally the partial expressions of a theory which comprehends and consummates them both. But to proceed in our historical survey.

The counter theories of Plato and Aristotle the partial expressions of the true.

Passing over a host of commentators in the Lower Empire, and during the middle ages, who were content to repeat the doctrines of Aristotle and Plato; in modern times, the first original philosopher I am aware of, who seems to have turned his attention upon the phænomena of pain and pleasure, is the celebrated Cardan; and the result of his observation was a theory identical with Plato's, though of Plato's speculation he does not seem to have been aware. In the sixth chapter of his very curious autobiography, *De Vita Propria Liber*, he tells us, that it was his wont to anticipate the causes of disease, because he was of opinion that pleasure consisted in the appeasement of a pre-existent pain, (quod arbitrarer voluptatem consistere in dolore procedenti sedato). But in the thirteenth book of his great work *De Subtilitate*, this theory is formally propounded. This, however, was not done in the earlier editions of the work; and the theory was, therefore, not canvassed by the ingenuity of his critic, the elder Scaliger, whose *Exercitationes contra Cardanum* are totally silent on the subject. It is only in the editions of the *De Subtilitate* of Cardan, subsequent to the year 1560, that a statement of the theory in question is to be found. The following is a summary of his reasoning:—"All pleasure has its root in a preceding

Historical notices of theories of the Pleasurable, resumed.

Cardan,—held a theory identical with Plato's.

Summary of his doctrine.

pain. Thus it is that we find pleasure in rest after hard labour; in meat and drink after hunger and thirst; in the sweet after the bitter; in light after darkness; in harmony after discord. Such are the facts in confirmation of this doctrine, which simple experience affords. But philosophy supplies, likewise, a reason from the nature of things themselves. Pleasure and pain exist only as they are states of feeling; but feeling is a change, and change always proceeds from one contrary to another; consequently, either from the good to the bad, or from the bad to the good. The former of these alternatives is painful, and, therefore, the other, when it takes place, is pleasing; a state of pain must thus always precede a state of pleasure." Such are the grounds on which Cardan thinks himself entitled to reject the Aristotelic theory of pleasure, and to substitute in its place the Platonic. It does not, however, appear from anything he says, that he was aware of the relative speculations of these two philosophers.

His theory criticised.

But the reasoning of Cardan is incompetent: for if it proves anything, it proves too much, seeing that it would follow from his premises, that a pleasurable feeling cannot gradually, continually, uninterruptedly, rise in intensity; for it behoves that every new degree of pleasure should be separated from the preceding by an intermediate state of higher pain; a conclusion which is contradicted by the most ordinary and manifest experience. This theory remained, therefore, in Cardan's, as in Plato's, hands, destitute of the necessary proof.

Montaigne, —held a similar doctrine.

The same doctrine,—that pleasure is only the alternation and consequent of pain,—was adopted, likewise, by Montaigne. In the famous twelfth chapter of the

LECT. XLIII.

second book of his *Essays*, he says:—"Our states of pleasure are only the privation of our states of pain;" but this universal inference he, like his predecessors, deduces only from the special phænomena given in certain of the senses.

Descartes.

The philosopher next in order is Descartes;[a] and his opinion is deserving of attention, not so much from its intrinsic value, as from the influence it has exerted upon those who have subsequently speculated upon the causes of pleasure. These philosophers seem to have been totally ignorant of the far profounder theories of the ancients; and while the regular discussions of the subject by Aristotle and Plato were, for our modern psychologists, as if they had never been, the incidental allusion to the matter by Descartes, originated a series of speculations which is still in progress.

His doctrine of the pleasurable.

Descartes' philosophy of the pleasurable is promulgated in one short sentence of the sixth letter of the First Part of his *Epistles*, which is addressed to the Princess Elizabeth. It is as follows:—"All our pleasure is nothing more than the consciousness of some one or other of our perfections."—"Tota nostra voluptas posita est tantum in perfectionis alicujus nostræ conscientia." It is curious to hear the praises that have been lavished upon this definition of the pleasurable. It has been lauded for its novelty; it

a Before Descartes, Vives held a positive theory of the pleasurable. His definition of pleasure and its illustration, are worthy of a passing notice: "Delectatio sita est in congruentia, quam invenire non est sine proportionis ratione aliqua inter facultatem et objectum, ut quædam sit quasi similitudo inter illa; tum ne notabiliter sit majus, quod affert delectationem: nec notabiliter minus, quam ea vis quæ recipit voluptatem, ea utique parte qua recipitur. Ideo mediocris lux gratior est oculis, quam ingens; et subobscura gratiora sunt hebeti visui: eundem in modum de sonis." *De Anima*, lib. iii. p. 202, edit. 1555.—ED.

Groundlessly lauded for its novelty and importance.

has been lauded for its importance. "Descartes," says Mendelssohn in his *Letters on the Sensations*, (*Briefe über die Empfindungen*), "was the first who made the attempt to give a real explanation of the pleasurable."[α] The celebrated Kaestner thus opens his *Réflexions sur l'Origine du Plaisir*.[β]—"I shall not pretend decidedly to assert that no one before Descartes has said, that pleasure consisted in the feeling of some one of our perfections. I confess, however, that I have not found this definition in any of the dissertations, sometimes tiresome, and frequently uninstructive, of the ancient philosophers on the nature and effects of pleasure. I am, therefore, disposed to attribute a discovery which has occasioned so many controversies, to that felicitous genius, which has disencumbered metaphysics of the confused chaos of disputes, as unintelligible as vain, in order to render it the solid and instructive science of God and of the human soul." And M. Bertrand, another very intelligent philosopher, in his *Essai sur le Plaisir*[γ] says, "Descartes is probably the first who has enounced, that all pleasure consists in the inward feeling we have of some of our perfections, and, in these few words, he has unfolded a series of great truths."

The doctrine of Descartes, a vague version of that of Aristotle.

Now what is the originality, what is the importance, of this celebrated definition? This is easily answered, —in so far as it has any meaning, it is only a statement, in vague and general terms, of the truth which Aristotle had promulgated, in precise and proximate

α Anmerkung, 6.—Ed.

β The *Réflexions sur l'Origine du Plaisir*, is appended to the *Nouvelle Théorie des Plaisirs, par M. Sulzer* (1767.) The *Nouvelle Théorie* is a French version of Sulzer's treatise, *Untersuchung über den Ursprung der angenehmen und unangenehmen Empfindungen*. See above, p. 416.—Ed.

γ Sect. i. ch. i. p. 3. Neuchatel, 1777.—Ed.

LECT. XLIII.

expressions. Descartes says, that pleasure is the consciousness of one or other of our perfections. This is not false; but it is not instructive. We are not conscious of any perfection of our nature, except in so far as this is the perfection of one or other of our powers; and we are not conscious of a power at all, far less of its perfection, except in so far as we are conscious of its operation. It, therefore, behoved Descartes to have brought down his definition of pleasure from the vague generality of a consciousness of perfection, to the precise and proximate declaration, that pleasure is a consciousness of the perfect energy of a power. But this improvement of his definition would have stripped it of all novelty. It would then have appeared to be, what it truly is, only a version, and an inadequate version, of Aristotle's. These are not the only objections that could be taken to the Cartesian definition; but for our present purpose it would be idle to advance them.

Leibnitz,—adopted both the counter theories.

Leibnitz is the next philosopher to whose opinion I shall refer; and this you will find stated in his *Nouveaux Essais*,[a] and other works latterly published. Like Descartes, he defines pleasure the feeling of a perfection, pain the feeling of an imperfection; and, in another part of the work,[β] he adopts the Platonic theory, that all pleasure is grounded in pain, which he ingeniously connects with his own doctrine of latent modifications, or, as he calls them, obscure perceptions. As this work, however, was not published till long after not only his own death, but that of his great disciple Wolf, the indication, (for it is nothing more), of his opinion on this point had little influence on

a Liv. ii. ch. xxi. § 41. *Opera*, ed. Erdmann, p. 261.—Ed.

β Liv. ii. ch. xx. § 6. *Opera*, ed. Erdmann, p. 248.—Ed.

subsequent speculations; indeed I do not remember to have seen the doctrine of Leibnitz upon pleasure even alluded to by any of his countrymen.

Wolf. Wolf, with whose doctrine that of Baumgarten[α] nearly coincides, defines pleasure, the intuitive cognition, (that is, in our language, the perception or imagination), of any perfection whatever, either true or apparent.—"Voluptas est intuitus, seu cognitio intuitiva, perfectionis cujuscunque, sive veræ sive apparentis."[β] His doctrine you will find detailed in his *Psychologia Empirica*, and in his *Horæ Subsecivæ*. It was manifestly the offspring, but the degenerate offspring, of the doctrine of Descartes, which, as we have seen, was itself only a corruption of that of Aristotle. Descartes rightly considered pleasure as a quality of the subject, in defining it a consciousness of some perfection in ourselves. Wolf, on the contrary, wrongly considers pleasure more as an attribute of the object, in defining it a cognition of any perfection whatever. Now in their definitions of pleasure, as Descartes was inferior to Aristotle, so Wolf falls far below Descartes, and in the same quality,—in want of precision and proximity.

His doctrine criticised.

1. Wrongly considers pleasure as an attribute of the object.

Pleasure is a feeling, and a feeling is a merely subjective state, that is, a state which has no reference to anything beyond itself,—which exists only as we are conscious of its existence. Now, then, the perfection or imperfection of an object, considered in itself, and as out of relation to our subjective states, is thought,—is judged, but is not felt; and this judgment is not pleasure or pain, but approbation or disapprobation,

α See his *Metaphysik*, § 482 *et seq.*, p. 233, edit. 1783. Cf. Platner, *Phil. Aphorismen*, ii. § 366, p. 218.—Ed.

β *Psychologia Empirica*, § 511, where he expressly refers to Descartes as the author of the definition.—Ed.

that is, an act of the cognitive faculties, but not an affection of the capacities of feeling. In this point of view, therefore, the definition of pleasure, as the cognition of any sort of perfection, is erroneous. It may, indeed, be true that the perfection of an object can determine the cognitive faculty to a perfect energy; and the concomitant of this perfect energy will be a feeling of pleasure. But, in this case, the objective perfection, as cognised, is not itself the pleasure; but the pleasure is the feeling which we have of the perfection, that is, of the state of vigorous and unimpeded energy of the cognitive faculty, as exercised on that perfection. Wolf ought, therefore, to have limited his definition, like Descartes, to the consciousness of subjective perfection; as Descartes should have explicated his consciousness of subjective perfection into the consciousness of full, spontaneous, and unimpeded activity.

2. Limits pleasure to the cognition of perfection by the Intuitive Faculties.

But there is another defect in the Wolfian definition:—it limits the pleasure from the cognition of perfection to the Intuitive Faculties, that is, to Sense and Imagination, denying it to the Understanding,—the faculty of Relations,—Thought Proper. This part of his theory was, accordingly, assailed by Moses Mendelssohn,—one of the best writers and most ingenious philosophers of the last century,—who, in other respects, however, remained faithful to the objective point of view, from whence Wolf had contemplated the phænomenon of pleasure. This was done in his *Briefe über die Empfindungen*, 1755.[a] A reaction was, however, inevitable; and other German philosophers were soon found who returned to the subjective

This part of Wolf's doctrine assailed by Mendelssohn.

a See Anmerkung, 6; and Reinhold, *Über die bisherigen Begriffe vom Vergnügen*, § 2.—*Vermischte Schriften*, i. p. 281 *et seq.*—ED.

point of view, from which Wolf, Baumgarten, and Mendelssohn had departed. LECT. XLIII.

Du Bos and Pouilly,—considered pleasure in its subjective aspect.

But before passing to these, it would be improper to overlook the doctrine of two French philosophers, who had already explained pleasure in its subjective aspect, and who prepared the way for the profounder theories of the German speculators,—I mean Du Bos and Pouilly. As their doctrines nearly coincide, I shall consider them as one. The former treats of this subject in his *Réflexions Critiques sur la Peinture*,[a] &c.; the latter in his *Théorie des Sentimens Agréables*.[β] The following are the principal moments of their inquiries:—

Their theory stated.

"1. Considering pleasure only in relation to the subject, the question they propose to answer is, What takes place in the state which we call pleasurable?

"2. The gratification of a want causes pleasure. If the want be natural, the result is a natural pleasure, and an unnatural pleasure, if the want be unnatural.

"3. The fundamental want,—the want to which all others may be reduced,—is the occupation of the mind. All that we know of the mind is that it is a thinking, a knowing power. We desire objects only for the sake of intellectual occupation.

"4. The activity of mind is either occupied or

[a] See tom. I. partie I. §§ 1, 2. First published in 1719, Paris.—Ed.

[β] See chaps. i. iii. iv. v. First published in 1743. To these should be added the valuable treatise of the Père André,—the *Essai sur le Beau*, which was first published in 1741. There is also, previously to Sulzer, another French æsthetical writer of merit,—Batteux, whose treatise *Les Beaux Arts réduits à un même Principe*, first appeared in 1746. This work, along with two relative treatises, was republished in 1774, under the title of *Principes de la Littérature*. All these authors consider pleasure, more or less, from the subjective point of view, and are, in principle, Aristotelic. For a collection of treatises, in whole and part, on pleasure in its psychological and moral aspects, see *Le Temple du Bonheur, ou Recueil des plus Excellens Traités sur le Bonheur*; in 4 vols. New edition, 1770.—Ed.

LECT. XLIII. occupies itself. The matters which afford the objects of our faculties of knowledge are either sensible impressions, which are delivered over to the understanding—this is the case in perception of sense; or this matter is furnished by the cognitive faculty itself—as is the case in thinking.

"5. If this activity meets with impediments in its prosecution,—be this in the functions either of thought or sense,—there results a feeling of restraint; and this of two kinds, positive and negative.

"6. When the activity, whether in perception or thinking, is prevented from being brought to its conclusion, there emerges the feeling of straining,—of effort,—the feeling of positive limitation of our powers. This is painful.

"7. If the mind be occupied less than usual in all its functions, there arises a feeling of unsatisfied want; this constitutes that state of negative restraint,—the state of ennui, of tedium. This is painful.

"8. The stronger and at the same time the easier the activity of mind in any of its functions, the more agreeable."[a]

This theory is evidently only that of Aristotle; to whom, however, the French philosophers make no allusion. What they call *occupation* or *exercise*, he calls *energy*. The former expressions are, perhaps, preferable on this account, that they apply equally well to the mental processes, whether active or passive, whereas the terms *energy*, *act*, *activity*, *operation*, &c., only properly denote these processes as they are considered in the former character.

Sulzer,—his theory a reaction

Subsequently to the French philosophers, and as a reaction against the partial views of the school of

a Abridged from Reinhold, *Über die bisherigen Begriffe vom Vergnügen*, § 1. *Vermischte Schriften*, p. 275.—Ed.

Against the views of Wolf.

Wolf, there appeared the theory of Sulzer, the Academician of Berlin,—a theory which was first promulgated in his *Enquiry into the Origin of our Agreeable and Disagreeable Feelings*,[a] in 1752. This is one of the ablest discussions upon the question, and though partial, like the others, it concurs in establishing the truth of that doctrine of which Aristotle has left, in a short compass, the most complete and satisfactory exposition. The following are the leading principles of Sulzer's theory :—

His theory stated.

"1. We must penetrate to the essence of the soul, if we would discover the primary source of pleasure.

"2. The essence of the soul consists in its natural activity, and this activity again consists in the production of ideas." [By that he means the faculty in general of Cognition or Thought. I may here observe, by the way, that he adopts the opinion that the faculty of thought or cognition is the one fundamental power of mind; and in this he coincides with Wolf, whose theory of pleasure, however, he rejects.]

"3. In this essential tendency to activity are grounded all our pleasurable and painful feelings.

"4. If this natural activity of the soul, or this ceaseless tendency to think, encounters an impediment, pain is the result; whereas if it be excited to a lively activity, the result is pleasure.

"5. There are two conditions which regulate the degree of capacity and incapacity in the soul for pleasurable and painful feelings, the habitude of reflection, and the natural vivacity of thought; and both together constitute the perfect activity of mind.

a *Untersuchung über den Ursprung der angenehmen und unangenehmen Empfindungen.* Published in the Memoirs of the Royal Academy of Berlin for the years 1751, 1752. See *Vermischte philosophische Schriften*, vol. i. p. 1. Leipsic, 1773. See above, p. 416. Ed.

"6. Pleasurable feelings, consequently, can only be excited by objects which at once comprise a variety of constituent qualities or characters, and in which these characters are so connected that the mind recognises in them materials for its essential activity. An object which presents to the mental activity no exercise, remains altogether indifferent.

"7. No object which moves the mind in a pleasurable or in a painful manner is simple;[a] it is necessarily composite or multiplex. The difference between agreeable and disagreeable objects can only lie in the connection of the parts of this multiplicity. Is there order in this connection, the object is agreeable; is there disorder, it is painful.

"8. Beauty is the manifold, the various, recalled to unity. The mere multitude of parts does not constitute an object beautiful; for there is required that an object should have at once such multiplicity and connection as to form a whole.

"9. This is the case in intellectual beauty; that is, in the beauty of those objects which the understanding contemplates in distinct notions. The beauty of geometrical theorems, of algebraic formulæ, of scientific principles, of comprehensive systems, consists, no less than the beauty of objects of Imagination and Sense, in the unity of the manifold, and rises in proportion to the quantity of the multiplicity and the unity.

"10. All these objects present a multitude of constituent characters,—of elementary ideas, at once; and these are so connected, so bound together by a principle of unity, that the mind is, in consequence thereof, enabled to unfold and then to bring back the different

a [But see Tiedemann's *Psychologie*, p. 152.]

parts to a common centre, that is, reduce them to unity,—to totality,—to system.

"11. From this it is evident, that the Beautiful only causes pleasure through the principle of activity. Unity, multiplicity, correspondence of parts, render an object agreeable to us, only inasmuch as they stand in a favourable relation to the active power of the mind.

"12. The relation in which beauty stands to the mind is thus necessary, and, consequently, immutable. A single condition is alone required in order that what is in itself beautiful should operate on us; it is necessary that we should know it; and to know it, it is necessary that, to a certain extent, we be conversant with the kind to which it belongs; for otherwise we should not be competent to apprehend the beauty of an object. (!)

"13. A difference of tastes is found only among the ignorant or the half-learned; and taste is a necessary consequence of knowledge."[α]

I shall not pursue this theory in the explanation it attempts of the pleasures of the Senses and of the Moral Powers, in which it is far less successful than in those of the Intellect. This was to be expected in consequence of the one-sided view Sulzer had taken of the mental phænomena, in assuming the Cognitive Faculty as the elementary power out of which the Feelings and Conations are evolved.[β]

The theory of Sulzer criticised.

The theory of Sulzer is manifestly only a one-sided modification of the Aristotelic; but it does not appear that he was himself aware how completely he had

α See Reinhold [*Ueber die bisherigen Begriffe vom Vergnügen*, § 3. *Verm. Schriften*, p. 290 *et seq.*—ED.]

β For Sulzer's doctrines on these points, see Reinhold, as above, p. 301 *et seq.*—ED.

LECT. XLIII.

been anticipated by the Stagirite. "On the contrary, he once and again denominates his explanation of the pleasurable a discovery. This can, however, hardly be allowed him, even were the Aristotelic theory out of the question; for it required no mighty ingenuity for a philosopher who was well acquainted with the works of his immediate predecessors, in France and Germany, by whom pleasure had been explained as the vigorous and easy exercise of the faculties,—as the feeling of perfection in ourselves, and as the apprehension of perfection in other things, that is, their unity in variety:—I say, after these opinions of his precursors, it required no such uncommon effort of invention to hit upon the thought,—that pleasure is determined when the variety in the object calls forth the activity of the subject, and when this activity is rendered easy by the unity in which the variety is contained. His explanation is more explicit, but, except a change of expression, it is not easy to see what Sulzer added to Du Bos and Pouilly, to say nothing of Wolf and Mendelssohn. . . . .

Summary of the theory.

"The theory of Sulzer is summed up in the following result:—Every variety of pleasure may, subjectively considered, be carried into the prompt and vigorous activity of the cognitive faculty; and, objectively considered, be explained as the product of objects which, in consequence of their variety in unity, intensely occupy the mind without fatiguing it. The peculiar merit of the theory of Sulzer, in contrast to those of his immediate predecessors, is that it combines both the subjective and objective points of view. In this respect, it is favourably contrasted with the opinion of Wolf and Mendelssohn. But it takes a one-sided view of the character of the subject. In the first place, the essence of the mind in general, and

Its merit.

Its defect.

the essence of the cognitive faculty in particular, does not consist of activity exclusively, but of activity and receptivity in correlation. But receptivity is a passive power, not an active, and thus the theory in its fundamental position is only half true. This one-sided view by Sulzer, in which regard is had to the active or intellectual element of our constitution to the exclusion of the passive or sensual, is precisely the opposite to that other, and equally one-sided, view which was taken by Helvetius[α] and the modern Epicureans and Materialists; but their theory of the pleasurable may be passed over as altogether without philosophical importance. In the second place, it is erroneous to assert that pleasure is nothing else than the consciousness of the unimpeded activity of mind. The activity of mind is manifested principally in thinking, whereas the state of pleasure consists wholly of a consciousness of feeling. In the enjoyment of pleasure we do not think, but feel; and in an intenser enjoyment there is almost a suspension of thought."[β]

Genovesi and Verri adopted the Platonic theory.

It is not necessary to say much of the speculations upon pleasure subsequent to Sulzer, and prior to Kant. In Italy I find that two philosophers of the last century had adopted the Platonic opinion,—of pleasure being always an escape from pain,—Genovesi and Verri; the former in a chapter of his *Metaphysics*,[γ] the latter in a chapter of his *Dissertation on the Nature of Pleasure and Pain.*[δ] This opinion, however, reacquires importance from having been

α *De l'Esprit*, disc. i. ch. i. Cf. *De l'Homme*, sect. ii. ch. x.—Ed.

β See Reinhold, as above, pp. 308, 316, 317.—Ed.

γ Cap. vi. t. ii. p. 213, edit. 1753.—Ed.

δ *Discorso sull' Indole del Piacere, e del Dolore*, §§ iii. iv. *Opere Filosofiche*, t. p. 20 *et seq.*, edit. 1784. This treatise is translated into German by Meiners,—*Gedanken über die Natur des Vergnügens*. Leipsic, 1777.—Ed.

LECT. XLIII.

adopted from Verri by the philosopher of Königsberg. In his *Manual of Anthropology*, Kant briefly and generally states his doctrine on this point; but in the notes which have been recently printed of his Lectures on this subject, we have a more detailed view of the character and grounds of his opinion. The Kantian doctrine is as follows:—

Kant adopted the Platonic theory.

His doctrine stated.

"Pleasure is the feeling of the furtherance, (*Beförderung*), pain of the hindrance of life. Under pleasure is not to be understood the feeling of life; for in pain we feel life no less than in pleasure, nay, even perhaps more strongly. In a state of pain, life appears long, in a state of pleasure, it seems brief; it is only, therefore, the feeling of the promotion,—the furtherance, of life, which constitutes pleasure. On the other hand, it is not the mere hindrance of life which constitutes pain; the hindrance must not only exist, it must be felt to exist." (Before proceeding further, I may observe, that these definitions of pleasure and pain are virtually identical with those of Aristotle, only far less clear and explicit.)

But to proceed—"If pleasure be a feeling of the promotion of life, this presupposes a hindrance of life; for there can be no promotion, if there be no foregoing hindrance to overcome. Since, therefore, the hindrance of life is pain, pleasure must presuppose pain. . . . .

"If we intend our vital powers above their ordinary degree, in order to go out of the state of indifference or equality, we induce an opposite state; and when we intend the vital powers above the suitable degree we occasion a hindrance, a pain. The vital force has a degree along with which a state exists, which is one neither of pleasure nor of pain, but of content, of comfort, (*das Wohlbefinden*). When this state is reduced to a lower pitch by any hindrance, then, a promotion,—a furtherance, of life is useful in order to overcome this impediment. Pleasure is thus always a consequent of pain. When we cast our eyes on the progress of things, we discover in ourselves a ceaseless tendency to escape from our present state. To this we are compelled by a physical stimulus, which sets animals, and man, as an animal, into activity. But in the intellectual nature of man, there is also a stimulus, which operates to the same end. In thought, man is always dissatisfied with the actual; he is ever looking forward from the present to the future; he is incessantly in a state of transition from one state to another, and is unable to continue in the same. But what is it that thus constrains us to be always passing from one state to another, but pain? And that it is not a pleasure which entices us to this, but a kind of discontent with present suffering, is shown by the fact that we are always seeking for some object of pleasure, without knowing what that object is, merely as an aid against the disquiet,—against the complement of petty pains, which in a moment irritate and annoy us. It is thus apparent that man is urged on by a necessity of his nature to go out of the present as a state of pain, in order to find in the future one less irksome. Man thus finds himself in a never-ceasing pain; and this is the spur for the activity of human nature. Our lot is so cast that there is nothing enduring for us, but pain; some indeed have less, others more, but all, at all times, have their share; and our enjoyments at best are only slight alleviations of pain. Pleasure is nothing positive; it is only a liberation of pain, and, therefore, only something negative. Hence

LECT. XLIII. it follows, that we never begin with pleasure but always with pain; for while pleasure is only an emancipation from pain, it cannot precede that of which it is only a negation. Moreover, pleasure cannot endure in an unbroken continuity, but must be associated with pain, in order to be always suddenly breaking through this pain,—in order to realise itself. Pain, on the contrary, may subsist without interruption in one pain, and be only removed through a gradual remission; in this case, we have no consciousness of pleasure. It is the sudden,—the instantaneous, removal of the pain, which determines all that we can call a veritable pleasure. We find ourselves constantly immersed, as it were, in an ocean of nameless pains, which we style disquietudes or desires, and the greater the vigour of life an individual is endowed with, the more keenly is he sensible to the pain. Without being in a state of determinate corporeal suffering, the mind is harrassed by a multitude of obscure uneasinesses, and it acts, without being compelled to act, for the mere sake of changing its condition. Thus men run from solitude to society, and from society to solitude, without having much preference for either, in order merely, by the change of impressions, to obtain a suspension of their pain. It is from this cause that so many have become tired of their existence, and the greater number of such melancholic subjects have been urged to the act of suicide in consequence of the continual goading of pain,—of pain from which they found no other means of escape.[a]

"It is certainly the intention of Providence that, by the alternation of pain, we should be urged on to activity. No one can find pleasure in the continual

[a] Cf. *Anthropologie*, § 60.—Ed.

enjoyment of delights; these soon pall upon us,—pall upon us in fact the sooner, the more intense was their enjoyment. There is no permanent pleasure to be reaped except in labour alone. The pleasure of toil consists in a reaction against the pain to which we should be a victim, did we not exert a force to resist it. Labour is irksome, labour has its annoyances, but these are fewer than those we should experience were we without labour. As man, therefore, must seek even his recreation in toil itself, his life is at best one of vexation and sorrow; and as all his means of dissipation afford no alleviation, he is left always in a state of disquietude, which incessantly urges him to escape from the state in which he actually is." [This is the doom of man,—to be born to sorrow as the sparks fly upwards, and to eat his bread in the sweat of his brow.]

"Men think that it is ungrateful to the Creator to say, that it is the design of Providence to keep us in a state of constant pain; but this is a wise provision in order to urge human nature on to exertion. Were our joys permanent, we should never leave the state in which we are, we should never undertake aught new. That life we may call happy, which is furnished with all the means by which pain can be overcome; we have in fact no other conception of human happiness. Contentment is when a man thinks of continuing in the state in which he is, and renounces all means of pleasure; but this disposition we find in no man."[a]

[a] *Menschenkunde*, p. 248 *et seq.*; published by Starke, 1831. This is not included in Kant's collected works by Rosenkranz and Schubert. *Cf. Anthropologie*, § 59. *Werke*, vii. part ii. p. 144.—ED. [For further historical notices of theories of the Pleasurable, see Lossius, *Lexikon*, v. *Vergnügen*.

LECTURE XLIV.

THE FEELINGS.—APPLICATION OF THE THEORY OF PLEASURE AND PAIN TO THE PHÆNOMENA.

LECT. XLIV.

Feelings,—their principle of classification internal.

THE Feelings being mere subjective states, involving no cognition or thought, and, consequently, no reference to any object, it follows, that they cannot be classified by relation to aught beyond themselves. The differences in which we must found all divisions of the Feelings into genera and species, must be wholly internal, and must be sought for and found exclusively in the states of Feeling themselves.

Admit of a twofold classification,—as Causes and as Effects.

Now, in considering these states, it appears to me, that they admit of a classification in two different points of view;—we may consider these states either as Causes or as Effects. As causes, they are viewed in relation to their product,—their product either of pleasure or of pain. As effects, they are viewed as themselves products,—products of the action of our different constitutive functions. In the former of these points of view, our states of Feeling will be divided simply into the three classes—1°, The Pleasurable; 2°, The Painful; and, 3°, The partly Pleasurable partly Painful,—without considering what kind of pleasure and what kind of pain it is which they involve; and here, it only behoves us to inquire,—what are the general conditions which determine in a feeling one or other of these counter qualities. In the latter of these points of view, our states of Feeling will be divided according as the energy, of which they are concomitant, be that of a power of one kind or of another,—a distinction, which affords a divison of our pleasures and pains, taken together, into various sorts. I shall take these points of view in their order.

The Feelings as Causes,—divided into Pleasurable and Painful.

In the former point of view, these feelings are distributed simply into the Pleasurable and the Painful; and it remains, on the theory I have proposed, to explain, in general, the causes of these opposite affections, without descending to their special kinds.

Application of foregoing theory to explain in general the causes of Pleasurable and Painful feeling.

Now, it has been stated, that a feeling of pleasure is experienced, when any power is consciously exerted in a suitable manner; that is, when we are neither, on the one hand, conscious of any restraint upon the energy which it is disposed spontaneously to put forth, nor, on the other, conscious of any effort in it, to put forth an amount of energy greater, either in degree or in continuance, than what it is disposed freely to exert. In other words, we feel positive pleasure, in proportion as our powers are exercised, but not over-exercised; we feel positive pain, in proportion as they are compelled either not to operate, or to operate too much. All pleasure, thus, arises from the free play of our faculties and capacities; all pain from their compulsory repression or compulsory activity.

Apparent contradictions of the doctrine prove real confirmations.

The doctrine meets with no contradiction from the facts of actual life; for the contradictions which, at first sight, these seem to offer, prove, when examined, to be real confirmations. Thus it might be thought, that the aversion from exercise,—the love of idleness,

The *dolce far niente*.

—in a word, the *dolce far niente*,—is a proof that the inactivity, rather than the exertion, of our powers, is

LECT. XLIV.

the condition of our pleasurable feelings. This objection, from a natural proneness to inertion in man, is superficial; and the very examples on which it proceeds, refute it, and, in refuting it, concur in establishing our theory of pleasure and pain.

This not a negation of activity, but the opposite.

Now, is the *far niente*,—is that doing nothing, in which so many find so sincere a gratification, in reality a negation of activity, and not in truth itself an activity intense and varied? To do nothing in this sense, is simply to do nothing irksome,—nothing difficult,—nothing fatiguing,—especially to do no outward work. But is the mind internally, the while, unoccupied and inert? This, on the contrary, may be vividly alive, —may be intently engaged in the spontaneous play of imagination; and so far, therefore, in this case, from pleasure being the concomitant of inactivity, the activity is, on the contrary, at once vigorous and unimpeded; and such, accordingly, as, on our theory, would be accompanied by a high degree of pleasure.[a]

Ennui—what.

Ennui is the state in which we find nothing on which to exercise our powers; but ennui is a state of pain.

All occupation either play or labour.

We must recollect, that all energy, all occupation, is either play or labour. In the former, the energy appears as free or spontaneous; in the latter, as either compulsorily put forth, or its exertion so impeded by difficulties, that it is only continued by a forced and painful effort, in order to accomplish certain ulterior ends. Under certain circumstances, indeed, play may become a labour, and labour may become a play. A play is, in fact, a labour, until we have acquired the dexterity requisite to allow the faculties exerted to operate with ease; and, on the other hand, a labour is said to become a play, when a person has by nature,

a [See Krug, *Geschmackslehre oder Aesthetik*, p. 89, note.]

or has acquired by custom, such a facility in the relative operations, as to energise at once vigorously and freely.[α] In point of fact, as man by his nature is determined to pursue happiness, (happiness is only another name for a complement of pleasures), he is determined to that spontaneous activity of his faculties, in which pleasure consists. The love of action is, indeed, signalised, as a fact in human nature, by all who have made man an object of observation, though few of them have been able to explain its true rationale. "The necessity of action," says Samuel Johnson,[β] "is not only demonstrable from the fabric of the body, but evident from observation of the universal practice of mankind, who, for the preservation of health," (he should have said for pleasure), "in those whose rank or wealth exempts them from the necessity of lucrative labour, have invented sports and diversions, which, though not of equal use to the world with manual trades, are yet of equal fatigue to those who practise them."

The love of action signalised as a fact in human nature by all observers.

Samuel Johnson.

It is finely observed by another eloquent philosopher,[γ] in accounting, on natural principles, for man's love of war:—"Every animal is made to delight in the exercise of his natural talents and forces: the lion and the tiger sport with the paw; the horse delights to commit his mane to the wind, and forgets his pasture to try his speed in the field; the bull, even before his brow is armed, and the lamb, while yet an emblem of innocence, have a disposition to strike with the forehead, and anticipate in play the conflicts they are doomed to sustain. Man, too, is disposed to opposition, and to employ the forces of his nature

Adam Ferguson.

α Cf. Krug, *Geschmackslehre oder Aesthetik*, § 21. pp. 89, 90.—ED.

β *Rambler*, No. 85.—ED.

γ Adam Ferguson, *Essay on the History of Civil Society*, part i. section iv.—ED.

LECT. XLIV. against an equal antagonist; he loves to bring his reason, his eloquence, his courage, even his bodily strength, to the proof. His sports are frequently an image of war; sweat and blood are freely expended in play; and fractures or death are often made to terminate the pastime of idleness and festivity. He was not made to live for ever, and even his love of amusement has opened a way to the grave."

Paley. "The young of all animals," says Paley,* "appear to me to receive pleasure simply from the exercise of their limbs and bodily faculties, without reference to any end to be attained, or any use to be answered by the exertion. A child, without knowing anything of the use of language, is in a high degree delighted with being able to speak. Its incessant repetition of a few articulate sounds, or, perhaps, of the single word which it has learnt to pronounce, proves this point clearly. Nor is it less pleased with its first successful endeavours to walk, or rather to run, (which precedes walking), although entirely ignorant of the importance of the attainment to its future life, and even without applying it to any present purpose. A child is delighted with speaking, without having anything to say, and with walking, without knowing where to go. And, prior to both these, I am disposed to believe, that the waking hours of infancy are agreeably taken up with the exercise of vision, or perhaps, more properly speaking, with learning to see.

"But it is not for youth alone that the great Parent of creation hath provided. Happiness is found with the purring cat, no less than with the playful kitten; in the arm-chair of dozing age, as well as in either the sprightliness of the dance, or the animation of the

* *Natural Theology.* Works, vol. iv. chap. xxvi. p. 359.

LECT. XLIV.

chase. To novelty, to acuteness of sensation, to hope, to ardour of pursuit, succeeds, what is, in no inconsiderable degree, an equivalent for them all, 'perception of ease.' Herein is the exact difference between the young and the old. The young are not happy, but when enjoying pleasure; the old are happy, when free from pain. And this constitution suits with the degrees of animal power which they respectively possess. The vigour of youth was to be stimulated to action by impatience of rest; whilst to the imbecility of age, quietness and repose become positive gratifications. In one important respect, the advantage is with the old. A state of ease is, generally speaking, more attainable than a state of pleasure. A constitution, therefore, which can enjoy ease, is preferable to that which can taste only pleasure. This same perception of ease oftentimes renders old age a condition of great comfort, especially when riding at its anchor after a busy or tempestuous life."

The theory confirmed by the phænomena presented by the Painful Affections.

A strong confirmation of the doctrine, that all pleasure is a reflex of activity, and that the free energy of every power is pleasurable, is derived from the phænomena presented by those affections which we emphatically denominate the Painful. This fact is too striking, from its apparent inconsistency, not to have soon attracted attention:—

"Non tantum sanctis instructae legibus urbes,
  Tectaque divitiis luxuriosa suis
Mortalem alliciunt pulcra ad spectacula visum,
  Sed placet annoso squalida terra situ.
Oblectat pavor ipse animum; sunt gaudia curis,
  Et stupuisse juvat, quem doluisse piget." [a]

---

a Virginius Cæsarinus [*Poemata Virginii Cæsarini, Urbani viii. Pont. Opt. Max. Cubiculo Præfecti.* Printed in *Septem Illustrium Virorum Poemata.* Amstelodami, apud Dan. Elzevirium, 1672, p. 465.—Ed.]

LECT. XLIV.

Grief accompanied with pleasure.

Take, for example, in the first place, the affection of Grief,—the sorrow we feel in the loss of a beloved object. Is this affection unaccompanied with pleasure? So far is this from being the case, that the pleasure so greatly predominates over the pain as to produce a mixed emotion, which is far more pleasurable than any other of which the wounded heart is susceptible.

Noticed by Pliny.

It is expressly stated by the younger Pliny, in a passage which commences with these words:—"Est quædam etiam dolendi voluptas," &c.[α] This has also been frequently signalised by the poets:—

Ovid.

Thus Ovid[β]:—

> "Fleque meos casus: est quædam flere voluptas;
> Expletur lacrymis egeriturque dolor."

Lucan.

Thus Lucan[γ]; of Cornelia after the murder of Pompey:—

> "Caput ferali obduxit amictu,
> Decrevitque pati tenebras, puppisque cavernis
> Delituit: sævumque arcte complexa dolorem,
> Perfruitur lachrymis, et amat pro conjuge luctum."

Statius.

Thus Statius[δ]:—

> "Nemo vetat, satiare malis; ægrumque dolorem
> Libertate doma, jam flendi expleta voluptas."

Seneca.

Thus Seneca, the tragedian[ε]:—

> "Mœror lacrymas amat assuetas,
> Flendi miseris dira cupido est."

Petrarch.

Thus Petrarch[ζ]:—

> "Non omnia terræ
> Obruta; vivit amor, vivit dolor; ora negatur
> Regia conspicere, at flere et meminisse relictum est."

α Lib. viii. ep. 10: "Est quædam etiam dolendi voluptas; præsertim si in amici sinu defleas, apud quem lacrymis tuis vel laus sit parata, vel venia."—Ed.

β *Tristia*, iv. iii. 37.—Ed.

γ *Pharsalia*, ix. 108.—Ed.

δ *Sylvæ*, ii. i. 14.—Ed.

ε *Thyestes*, l. 952.—Ed.

ζ *Epist.* lib. i., *Barbato Sulmonensi*.—Ed.

Thus Shenstone[a]:—

"Heu quanto minus est cum reliquis versari, quam tui meminisse."

Shenstone.

Finally, Lord Pembroke[β]:—

Lord Pembroke.

"I would not give my dead son for the best living son in Christendom."

Fear, not simply painful.

In like manner, Fear is not simply painful. It is a natural disposition; has a tendency to act; and there is, consequently, along with its essential pain, a certain pleasure, as the reflex of its energy. This is finely expressed by Akenside[γ]:—

Akenside quoted.

"Hence, finally, by night
The village matron round the blazing hearth
Suspends the infant audience with her tales,
Breathing astonishment! of witching rhymes,
And evil spirits; of the deathbed call
Of him who robb'd the widow, and devour'd
The orphan's portion; of unquiet souls
Ris'n from the grave to ease the heavy guilt
Of deeds in life conceal'd; of shapes that walk
At dead of night and clank their chains, and wave
The torch of hell around the murd'rer's bed.
At every solemn pause, the crowd recoil,
Gazing each other speechless, and congeal'd
With shiv'ring sighs till, eager for th' event,
Around the beldame all erect they hang,
Each trembling heart with grateful terrors quell'd."

Pity.

In like manner, Pity, which, being a sympathetic passion, implies a participation in sorrow, is yet confessedly agreeable. The poet even accords to the energy of this benevolent affection a preference over the enjoyments of an exclusive selfishness:—

"The broadest mirth unfeeling folly wears,
Is not so sweet as virtue's very tears."[δ]

---

a Inscription on an urn. See Dodsley's *Description of the Leasowes*, in Shenstone's *Works* (1777), vol. ii. p. 307.—Ed.

β The anecdote is told in a somewhat different form of the Duke of Ormond. See Carte's *Life*, b. viii. Anno 1680. Hume, chap. lxix., tells the story of the Duke of Ormond, but as in the text.—Ed.

γ *Pleasures of Imagination*, b. i. 255.—Ed.

δ Pope, *Essay on Man*, iv. 319. The correct reading of the second line is,—"Less pleasing far than virtue's very tears."—Ed.

Energetic emotions painful in themselves still pleasurable.

On the same principle is to be explained the enjoyment which men have in spectacles of suffering,—in the combats of animals and men, in executions, in tragedies, &c.,—a disposition which not unfrequently becomes an irresistible habit, not only for individuals, but for nations. The excitation of energetic emotions painful in themselves is, however, also pleasurable. St Austin affords curious examples of this in his own case, and in that of his friend Alypius.

Illustrated in the case of St Augustin.

Speaking of himself in his *Confessions*,[α] he says:—"Theatrical spectacles were to me irresistible, replete as they were with the images of my own miseries, and the fuel of my own fire. What is the cause why a man chooses to grieve at scenes of tragic suffering, which he would have the utmost aversion himself to endure? And yet the spectator wishes to derive grief from these; in fact, the grief itself constitutes his pleasure. For he is attracted to the theatre, not to succour, but only to condole."

Also in the case of his friend Alypius.

In another part of the same work,[β] he gives the following account of his friend Alypius, who had been carried by his fellow-students, much against his inclination, to the amphitheatre, where there was to be a combat of gladiators. At first, unable to regard the atrocious spectacle, he closed his eyes, but, to give you the result of the story in the words of St Austin, "Abstulit inde secum insaniam qua stimularetur redire, non tantum cum illis a quibus prius abstractus est, sed etiam præ illis, et alios trahens."

General Causes which contribute to raise or

I now proceed to consider the General Causes which contribute to raise or to lower the intensity of our energies, and, consequently, to determine the correspond-

α Lib. iii. cap. 2.—ED.

β *Confessiones*, lib. vi. cap. 8.—ED. [See Purchot, *Physica*, pars ii. § iii. c. v. *Institut. Phil.*, iii. p. 416.]

ing degree of pleasure or pain. These may be reduced to Four; for an object rouses the activity of our powers, 1°, In proportion as it is New or Unexpected; 2°, In proportion as it stands in a relation of Contrast; 3°, In proportion as it stands in a relation of Harmony; and, 4°, In proportion as it is Associated with more, or more interesting objects.

LECT. XLIV.

lower the intensity of our energies.

I. The principle on which Novelty determines a higher energy, and, consequently, a higher feeling of pleasure, is twofold; and of these the one may be called the Subjective, the other the Objective.

I. Novelty.

In a subjective relation,—the new is pleasurable, inasmuch as this supposes that the mind is determined to a mode of action, either from inactivity, or from another state of energy. In the former case, energy, (the condition of pleasure), is caused: in the latter, a change of energy is afforded, which is also pleasurable; for powers energise less vigorously in proportion to the continuance of the same exertion, consequently, a new activity being determined, this replaces a strained or expiring exercise, that is, it replaces a painful, indifferent, or unpleasurable feeling, by one of comparatively vivid enjoyment. Hence all that the poets, from Homer downward, have said of the satiety consequent on our enjoyments, and of the charms of variety and change; but if I began to give quotations on these heads there would be no end. In an objective relation,—a novel object is pleasing, because it affords a gratification to our desire of knowledge; for to learn, as Aristotle has observed,[a] is to man naturally pleasing. But the old is already known,—it has been learned,—has been referred to its place, and, therefore, no longer occupies the cognitive faculties;

Twofold,-subjective and objective.

[a] *Rhet.*, i. 11, 21; iii. 10, 2.—ED.

LECT. XLIV.

whereas, the new, as new, is still unknown, and rouses to energy the powers by which it is to be brought within the system of our knowledge.

II. Contrast.

II. The second general principle is Contrast. Contrast operates in two ways; for it has the effect both of enhancing the real or absolute intensity of a feeling, and of enhancing the apparent or relative. As an instance of the former, the unkindness of a person from whom we expect kindness, rouses to a far higher pitch the emotions consequent on injury. As an instance of the latter, the pleasure of eating appears proportionally great, when it is immediately connected and contrasted with the removal of the pangs of hunger. It is on this principle, that the recollection of our past suffering is agreeable,—"hæc olim meminisse juvabit."[α] To the same purport Seneca,[β] the tragedian:—

Subordinate applications of this principle. 1. Recollection of past suffering.

> "Quæ fuit durum pati
> Meminisse dulce est."

Cowley.

And Cowley[γ]:—

> "Things which offend, when present, and affright,
> In memory, well painted, move delight."

Whereas the remembrance of a former happiness only augments the feeling of a present misery.

Southern.

> "Could I forget
> What I have been, I might the better bear
> What I am destin'd to. I'm not the first
> That have been wretched: but to think how much
> I have been happier."[δ]

It is, likewise, on this principle, that whatever recalls

α Virgil, *Æneid*, i. 203.—Ed.

β *Hercules Furens*, act. iii. 656.—Ed.

γ *Ode upon his Majesty's Restoration*.—Ed.

δ Southern, *Innocent Adultery*, act ii.

LECT. XLIV.

us to a vivid consciousness of our own felicity, by contrasting it with the wretchedness of others, is, though not unaccompanied with sympathetic pain, still predominantly pleasurable. Hence, in part, but in part only, the enjoyment we feel from all representations of ideal suffering. Hence, also in part, even the pleasure we have in witnessing real suffering:—

2. Consciousness of our own felicity as contrasted with wretchedness of others.

Lucretius quoted.

> "Suave, mari magno turbantibus æquora ventis,
> E terra magnum alterius spectare laborem:
> Non quia vexari quemquam est jucunda voluptas,
> Sed quibus ipse malis careas, quia cernere suave est.
> Suave etiam belli certamina magna tueri
> Per campos instructa, tua sine parte pericli."[a]

But on this, and other subjects, I can only touch.

III. Harmony and Discord.

III. The third general principle on which our powers are roused to a perfect and pleasurable, or to an imperfect and painful energy, is the relation of Harmony, or Discord, in which one coexistent activity stands to another.

Illustrated.

It is sufficient merely to indicate this principle, for its influence is manifest. At different times, we exist in different complex states of feeling, and these states are made up of a number of constituent thoughts and affections. At one time,—say during a sacred solemnity,—we are in a very different frame of mind from what we are at another,—say during the representation of a comedy. Now, then, in such a state of mind, if anything occurs to awaken to activity a power previously unoccupied, or to occupy a power previously in energy in a different manner, this new mode of activity is either of the same general character and tendency with the other constituent elements of the complex state, or it is not. In the former case, the new energy

a Lucretius, ii. 1.—Ed.

LECT. XLIV. chimes in with the old; each operates without impediment from the other, and the general harmony of feeling is not violated: in the latter case, the new energy jars with the old, and each severally counteracts and impedes the other. Thus, in the sacred solemnity, and when our minds are brought to a state of serious contemplation, everything that operates in unison with that state,—say a pious discourse, or a strain of solemn music,—will have a greater effect, because all the powers which are thus determined to exertion, go to constitute one total complement of harmonious energy. But suppose that, instead of the pious discourse or the strain of solemn music, we are treated to a merry tune or a witty address;—these, though at another season they might afford us considerable pleasure, would, under the circumstances, cause only pain; because the energies they elicited, would be impeded by those others with which the mind was already engrossed, while those others would, in like manner, be impeded by them. But, as we have seen, pleasure is the concomitant of unimpeded energy.

IV. Association. IV. The fourth and last general principle by which the activity of our powers is determined to pleasurable or painful activity, is Association. With the Its nature. nature and influence of association you are familiar, and are aware that, a determinate object being present in consciousness with its proper thought, feeling, or desire, it is not present, isolated and alone, but may draw after it the representation of other objects, with their respective feelings and desires.

And influence. Now it is evident, in the first place, that one object, considered simply and in itself, will be more pleasing than another, in proportion as it, of its proper nature, determines the exertion of a greater amount of free energy. But, in the second place, the amount of free energy which an object may itself elicit, is small, when compared to the amount that may be elicited by its train of associated representations. Thus, it is evident, that the object which in itself would otherwise be pleasing, may, through the accident of association, be the occasion of pain; and, on the contrary, that an object naturally indifferent or even painful may, by the same contingency, be productive of pleasure.

Association supposes as its condition pains and pleasures not founded on itself.

This principle of Association accounts for a great many of the phænomena of our intellectual pleasures and pains; but it is far from accounting for everything. In fact, it supposes, as its condition, that there are pains and pleasures not founded on Association. Association is a principle of pleasure and pain, only as it is a principle of energy of one character or another; and the attempts that have been made to resolve all our mental pleasures and pains into Association, are guilty of a twofold vice. For, in the first place, they convert a partial into an exclusive law; and, in the second, they elevate a subordinate into a supreme principle. The influence of Association, by which Mr Alison[a] and Lord Jeffrey,[β] among others, have attempted to explain the whole phænomena of our intellectual pleasures, was more properly, I think, appreciated by Hutcheson,—a philosopher whose works are deserving of more attention than has latterly been paid to them. "We shall see hereafter," he says, and Aristotle said the same thing, "that associations of ideas make objects pleasant and delightful, which are not naturally apt to give any such pleasures; and, in the same way, the casual conjunction of ideas may give a disgust

The attempt to resolve all our pleasures and pains into Association, vicious in a twofold way.

Hutcheson more properly appreciated the influence of Association.

a See his *Essays on Taste*. 6th edit, Edinburgh, 1825.—Ed.

β See *Encyclopædia Britannica*, art. *Beauty*, 7th edit. p. 487.—Ed.

LECT. XLIV. where there is nothing disagreeable in the form itself. And this is the occasion of many fantastic aversions to figures of some animals, and to some other forms. Thus swine, serpents of all kinds, and some insects really beautiful enough, are beheld with aversion, by many people who have got some accidental ideas associated with them. And for distastes of this kind no other account can be given."[a]

[a] *Inquiry into the Origin of our Ideas of Beauty and Virtue*, treatise I. sect. vi., 4th edition, p. 73.—ED.

LECTURE XLV.

THE FEELINGS.—THEIR CLASSES.

LECT. XLV.

The Feelings,—considered as Effects.

Having thus terminated the consideration of the Feelings considered as Causes,—causes of Pleasure and Pain,—I proceed to consider them as Effects,—as products of the action of our different powers. Now, it is evident, that, since all Feeling is the state in which we are conscious of some of the energies or processes of life, as these energies or processes differ, so will the correlative feelings. In a word, there will be as many different Feelings as there are distinct modes of mental activity. In the Lecture in which I commenced the discussion of the Feelings, I stated to you various distributions of these states by different philosophers.[a] To these I do not think it necessary again to recur, and shall simply state to you the grounds of the division I shall adopt.

As many different feelings as there are distinct modes of mental activity.

Two grand classes of Feelings. I. Sensations.

As the Feelings, then, are not primitive and independent states, but merely states which accompany the exertion of our faculties, or the excitation of our capacities, they must, as I have said, take their differences from the differences of the powers which they attend. Now, though all consciousness and all feeling be only mental, and, consequently, to say that any feeling is corporeal, would, in one point of view, be inaccurate, still it is manifest that there is a consider-

a See above, vol. ii. lect. xli. p. 420.—Ed.

LECT. XLV.

able number of mental functions, cognitive as well as appetent, clearly marked out as in proximate relation to the body; and to these functions we give the name of *Sensitive, Sensible, Sensuous,* or *Sensual.* Now, the feelings which accompany the exertion of these Sensitive or Corporeal Powers, whether cognitive or appetent, will constitute a distinct class, and to these we may, with great propriety, give the name of *Sensations;* whereas, on the Feelings which accompany the energies of all our higher powers of mind, we may, with equal propriety, bestow the name of *Sentiments.*

II. Sentiments.

The first grand distribution of our feelings will, therefore, be into the Sensations,—that is, the Sensitive or External Feelings; and into the Sentiments,—that is, the Mental or Internal Feelings. Of these in their order.

Sensations. Two classes— 1. Of the Five Senses. 2. Of the Sensus Vagus.

I. Of the Sensations.—The Sensations may be divided into two classes. The first class will contain those which accompany our perceptions through the five determinate Senses, — of Touch, Taste, Smell, Hearing, and Sight,—the *Sensus Fixus.* The second class will comprise those sensations which are included under what has been called the *Cœnæsthesis* or *Sensus Communis,—the Common Sense,—Vital Sense,—Sensus Vagus,*—such as the feelings of Heat and Cold, of Shuddering, the feeling of Health, of Muscular Tension and Lassitude, of Hunger and Thirst, the Visceral Sensations, &c., &c.[a]

The first class considered.

In regard to the determinate senses, each of these organs has its specific action, and its appropriate pleasure and pain; for there is a pleasure experienced in each of these, when an object is presented which determines it to suitable activity; and a pain or dis-

a See above, vol. ii. lect. xxvii. p. 157. — Ed.

LECT. XLV.

satisfaction experienced, when the energy elicited is either inordinately vehement or too remiss. This pleasure and pain, which is that alone belonging to the action of the living organ, and which, therefore, may be styled *organic*, we must distinguish from that higher feeling, which, perhaps, results from the exercise of Imagination and Intellect upon the phænomena delivered by the senses. Thus, I would call *organic* the pleasure we feel in the perception of green or blue, and the pain we feel in the perception of a dazzling white; but I would be, perhaps, disposed to refer to some other power than the External Sense, the enjoyment we experience in the harmony of colours, and certainly that which we find in the proportions of figure. The same observation applies to Hearing. I would call *organic* the pleasure we have in single sounds; whereas the satisfaction we receive from the harmony, and, still more, from the melody of tones, seems to require a higher faculty. This, however, is a very obscure and difficult problem; but, in whatever manner it be determined, the Aristotelic theory of pleasure and pain is still the only one which can account for the phænomena. Limiting, however, the organic pleasure of which a sense is capable, to that from the activity determined in it by its elementary objects,—this will be competent to every sense, but in very different degrees. In treating of the Cognitive Powers, I formerly noticed that in all the senses we could discriminate two phænomena,—the phænomenon of Perception Proper, and the phænomenon of Sensation Proper.[a] By *perception* is understood the objective relation of the sense, that is, the information obtained through it of the qualities of external

Organic pleasure and pain discriminated and illustrated.

The degree of organic pleasure determined by the objectivity and subjectivity of the Senses.

[a] See above, vol. ii. lect. xxiv. p. 93.—ED.

LECT. XLV.

existences in their action on the organ; by *sensation* is understood the subjective relation of the sense, that is, our consciousness of the affection of the organ itself, as acted on,—as affected by an object. I stated that these phænomena were in an inverse ratio to each other,—that is, the greater the perception the less always the sensation, the greater the sensation the less always the perception. I further observed, that, of the senses, some were more objective, others more subjective;—that in some the phænomenon of perception predominated, in others the phænomenon of sensation; that is, some gave us much information in regard to the qualities of their object and little in regard to their own affection in the act; whereas the information we received from others, was almost limited exclusively to their own modification, when at work.

Sight and Hearing objective; Taste and Smell subjective; hence in the two former, organic pleasure and pain feeble, in the two latter, strong.

Thus the two higher senses of Sight and Hearing might be considered as pre-eminently objective, the two lower senses of Taste and Smell might be considered as pre-eminently subjective; while the sense of Touch might be viewed as that in which the two phænomena are, as it were, *in æquilibrio*. Now, according to this doctrine, we ought to find the organic pleasure and pain in the two higher senses comparatively feeble, in the two lower, comparatively strong. And so it is. The satisfaction or dissatisfaction we receive from certain single colours and certain single sounds, in determining the organs of Sight and Hearing to perfect or imperfect activity, is small in proportion to the pleasure or the displeasure we are conscious of from the application of certain single objects to the organs of Taste and Smell.+

How far the theory of pleasure

So far we may safely go. But when it is required of us to explain, particularly and in detail, why the rose, for example, produces this sensation of smell, assafœtida that other, and so forth, and to say in what peculiar action does the perfect or pleasurable, and the imperfect or painful, activity of an organ consist, we must at once profess our ignorance. But it is the same with all our attempts at explaining any of the ultimate phænomena of creation. In general, we may account for much; in detail, we can rarely account for anything; for we soon remount to facts which lie beyond our powers of analysis and observation.

and pain affords an explanation of the phænomena.

All that we can say in explanation of the agreeable in sensation, is, that, on the general analogy of our being, when the impression of an object on a sense is in harmony with its amount of power, and thus allows it the condition of springing to full spontaneous energy, the result is pleasure; whereas, when the impression is out of harmony with the amount of power, and thus either represses it or stimulates it to over-activity, the result is pain.

The theory applicable to the Vital Sense.

The same explanation, drawn from the observation of the phænomena within our reach, must be applied to the sensations which belong to the Vital Sense, but in regard to these it is not necessary to say anything in detail.

II. Sentiments,—divided into Contemplative and Practical.

II. The Mental or Internal Feelings,—the Sentiments,—may be divided into Contemplative and Practical. The former are the concomitants of our Cognitive Powers, the latter of our Powers of Conation. Of these in their order.

Contemplative Feelings divided into those of the Subsidiary Faculties; and of the Elaborative. The first

The Contemplative Feelings are again distributed into two classes,—into those of the Subsidiary Faculties, and those of the Elaborative; and the Feelings accompanying the subsidiary faculties may be again subdivided into those of Self-Consciousness or

LECT. XLV.

Sentiments classified into those of Self-Consciousness, and of Imagination.

Internal Perception, and into those of Imagination,—*Imagination* being here employed to comprehend its relative faculty, the faculty of Reproduction. Of these in their order; and first of the Feelings or Sentiments attending the faculty of Reflex Perception or Self-Consciousness.

a. Sentiments attending Self-Consciousness.

By this faculty we become aware of our internal states; that is, in other words, that we live. Now we are conscious of our life only as we are conscious of our activity, and we are conscious of activity only as we are conscious of a change of state,—for all activity is the going out of one state into another; while, at the same time, we are only conscious of one state by contrast to, or as discriminated from, a preceding.

Tedium or Ennui.

Now pleasure, we have also seen, is the consciousness of a vigorous and unimpeded energy; pain, the consciousness of repressed or impeded tendency to action. This being the case, if there be nothing which presents to our faculties the objects on which they may exert their activity, in other words, if there be no cause whereby our actual state may be made to pass into another, there results a peculiar irksome feeling of a want of excitement, which we denominate *tedium* or *ennui*. This feeling is like that of being unable to die, and not being allowed to live; and sometimes becomes so oppressive that it leads to suicide or madness.

Arises from a repressed tendency to action.

The pain we experience in the feeling of Tedium, arises from the feeling of a repressed tendency to action; and it is intense in proportion as this feeling is lively and vigorous. An inability to thought is a security against this feeling, and, therefore, tedium is far less felt by the uncultivated than by the educated. The more varied the objects presented to our thought,

LECT. XLV.

—the more varied and vivacious our activity, the intenser will be our consciousness of living, and the more rapidly will the time appear to fly. But when we look back upon the series of thoughts, with which our mind was occupied the while, we marvel at the apparent length of its duration. Thus it is that, in travelling, a month seems to pass more rapidly than a week; but cast a retrospect upon what has occurred, and occupied our attention during the interval, and the month appears to lengthen to a year. Hence we explain why we call our easy occupations *pastimes*; and why play is so engaging when it is at all deep. Games of hazard determine a continual change,—now we hope, and now we fear; while in games of skill, we experience also the pleasure which arises from the activity of the understanding, in carrying through our own, and in frustrating the plan of our antagonist.

The more varied and vivacious our activity, the intenser our consciousness of life, and the more rapidly does time appear to fly.

Pastimes.

Games of chance and skill.

All that relieves tedium, by affording a change and an easy exercise for our thoughts, causes pleasure. The best cure of tedium is some occupation which, by concentrating our attention on external objects, shall divert it from a retortion on ourselves. All occupation is either labour or play; labour when there is some end ulterior to the activity, play when the activity is for its own sake alone. In both, however, there must be ever and anon a change of object, or both will soon grow tiresome. Labour is thus the best preventive of tedium, for it has an external motive which holds us steadfast to the work; while after the completion of our task, the feeling of repose, as the change from the feeling of a constrained to that of a spontaneous state, affords a vivid and peculiar pleasure. Labour must alternate with repose, or we shall never know what is the true enjoyment of life.

Tedium, how cured.

LECT. XLV.

The change of our perceptions and thoughts to be pleasing must not be too rapid.

Thus it appears that a uniform continuity in our internal states is painful, and that pleasure is the result of their commutation. It is, however, to be observed, that the change of our perceptions and thoughts to be pleasing must not be too rapid; for as the intervals, when too long, produce the feeling of Tedium, so, when too short, they cause that of Giddiness or Vertigo. The too rapid passing, for example, of visible objects or of tones before the Senses, of images before the Phantasy, of thoughts before the Understanding, occasions the disagreeable feeling of confusion or stupefaction, which, in individuals of very sensitive temperament, results in Nausea,—Sickness.[a]

Giddiness.

Nausea.

b. Sentiments concomitant of Imagination.

I proceed now to the Speculative Feelings which accompany the energies of Imagination. It has already been frequently stated, that whatever affords to a power the mean of full spontaneous energy is a cause of pleasure; and that whatever either represses the free exertion of a power, or stimulates it into strained activity, is the cause of pain.

Condition of the pleasurable applicable to Imagination, both as Reproductive and as Plastic.

I shall now apply this law to the Imagination. Whatever, in general, facilitates the play of the Imagination, is felt as pleasing; whatever renders it more difficult is felt as displeasing. And this applies equally to Imagination considered as merely reproductive of the objects presented by sense, or as combining these in the phantastic forms of its own productive, or rather plastic, activity. Considering the Phantasy merely as reproductive, we are pleased with the portrait of a person whose face we know, if like, because it enables us to recall the features into consciousness easily and freely; and we are displeased with it if

A. Reproductive.

a See Marcus Herz, *Über den Schwindel*, 1791.

unlike, because it not only does not assist, but thwarts us in our endeavour to recall them; while after this has been accomplished, we are still farther pained by the disharmony we experience between the portrait on the canvass and the representation in our own imagination. A short and characteristic description of things which we have seen, pleases us, because, without exacting a protracted effort of attention, and through a few striking traits, it enables the imagination to place the objects vividly before it. On the same principle, whatever facilitates the reproduction of the objects which have been consigned to memory, is pleasurable; as for example, resemblances, contrasts, other associations with the passing thought, metre, rhyme, symmetry, appropriate designations, &c. To realise an act of imagination, it is necessary that we grasp up,—that we comprehend, the manifold as a single whole: an object, therefore, which does not allow itself, without difficulty, to be thus represented in unity, occasions pain; whereas an object which can easily be recalled to system, is the cause of pleasure. The former is the case when the object is too large or too complex to be perceived at once; when the parts are not prominent enough to be distinctly impressed upon the memory. Order and symmetry facilitate the acts of Reproduction and Representation, and, consequently, afford us a proportional gratification. But, on the other hand, as pleasure is in proportion to the amount of free energy, an object which gives no impediment to the comprehensive energy of Imagination, may not be pleasurable, if it be so simple as not to afford to this faculty a sufficient exercise. Hence it is, that not variety alone, and not unity alone, but variety combined with unity, is that

An act of Imagination involves the comprehension of the manifold as a single whole.

The Beautiful in objects constituted by variety in unity.

LECT. XLV.

quality in objects, which we emphatically denominate *beautiful*.

Office of the Plastic Imagination to reconstruct and rearrange.

As to what is called the Productive or Creative Imagination,—this is dependent for its materials on the Senses and on the Reproductive Imagination. The Imagination produces, the Imagination creates, nothing; it only rearranges parts,—it only builds up old materials into new forms; and in reference to this act, it ought, therefore, to be called, not the *productive* or *creative*, but the *plastic*.[a] Now this reconstruction of materials by the Plastic Imagination is twofold; for it either arranges them in one representation, or in a series of representations. Of the pleasure we receive from single representations, I have already spoken; it, therefore, only remains to consider the enjoyment we find in the activity of imagination, in so far as this is excited in concatenating a series of representations. I do not at present speak of any pleasure or pain which the contents of these concatenated representations may produce; these are not feelings of imagination, but of appetency or conation; I have here exclusively in view the feelings which accompany the facilitated, or impeded, energy of this function of the phantasy.

This reconstruction twofold.

Conditions of the pleasurable, as regards the Understanding.

Now it is manifest that a series of representations are pleasing:—1°, In proportion as they severally call up in us a more varied and harmonious image; and, 2°, In proportion as they stand to each other in a logical dependence. This latter is, however, a condition not of the Imagination, but of the Understanding or Elaborative Faculty; and, therefore, before speaking of those feelings which accompany the joint energies of these faculties, it will be proper to consider those which arise from the opera-

[a] See above, vol. ii. lect. xxxiii. p. 262.—Ed.

tions of the Understanding by itself. To these, therefore, I now pass on.

Function of the Understanding.

The function of the Understanding may, in general, be said to bestow on the cognitions which it elaborates, the greatest possible compass, (comprehension and extension), the greatest possible clearness and distinctness, the greatest possible certainty and systematic order; and in as much as we approximate to the accomplishment of these ends, we experience pleasure, in as much as we meet with hindrances in our attempts, we experience pain. The tendency, the desire we have, to amplify the limits of our knowledge, is one of the strongest principles of human nature. To learn is thus pleasurable; to be frustrated in our attempted knowledge, painful.

Obscure and confused cognitions, how disagreeable.

Obscurity and confusion in our cognitions we feel as disagreeable; whereas their clearness and distinctness affords us sincere gratification. We are pained by a hazy and perplexed discourse; but rejoice in one perspicuous and profound. Hence the pleasure we experience in having the cognitions we possessed, but darkling and confused, explicated into life and order; and, on this account, there is hardly a more pleasing object than a tabular conspectus of any complex whole. We are soothed by the solution of a riddle; and the wit which, like a flash of lightning, discovers similarities between objects which seemed contradictory, affords a still intenser enjoyment.

Wit,—how pleasing.

Cognitions divided into two classes,—Empirical and Rational.

Our cognitions may be divided into two classes,—the Empirical or Historical, and the Rational. In the former we only apprehend the fact that they are; in the latter, we comprehend the reason why they are. The Understanding, therefore, does not for each demand the same kind or degree of knowledge; but

LECT. XLV.

in each, if its demand be successful, we are pleased; if unsuccessful, we are chagrined.

Sentiment of Truth,—what, and how pleasurable.

From the tendency of men towards knowledge and certainty, there arises a peculiar feeling which is commonly called the Feeling or Sentiment of Truth, but might be more correctly styled the Feeling or Sentiment of Conviction. For we must not mistake this feeling for the faculty by which we discriminate truth from error; this feeling, as merely subjective, can determine nothing in regard to truth and error, which are, on the contrary, of an objective relation; and there are found as many examples of men who have died the confessors of an error they mistook for truth, as of men who have laid down their lives in testimony of the real truth. "Every opinion," says Montaigne,[a] "is strong enough to have had its martyrs." Be this, however, as it may, the feeling of conviction is a pleasurable sentiment, because it accompanies the consciousness of an unimpeded energy; whereas the counter-feeling,—that of doubt or uncertainty, is a painful sentiment, because it attends a consciousness of a thwarted activity. The uneasy feeling which is thus the concomitant of doubt, is a powerful stimulus to the extension and perfecting of our knowledge.

Generalization and Specification,—how pleasurable.

The multitude,—the multifarious character, of the objects presented to our observation, stands in signal contrast with the very limited capacity of the human intellect. This disproportion constrains us to classify; that is, by a comparison of the objects of sense to reduce these to notions; on these primary notions we repeat the comparison, and thus carry them up into higher, and these higher into highest, notions. This

[a] *Essais*, liv. i. ch. xl.—ED.

process is performed by that function of the Understanding, which apprehends resemblances; and hence originate *species* and *genera* in all their gradations. In this detection of the similarities between different objects, an energy of the understanding is fully and freely exerted; and hence results a pleasure. But as in these classes,—these general notions, the knowledge of individual existences loses in precision and completeness, we again endeavour to find out differences in the things which stand under a notion, to the end that we may be able to specify and individualise them. This counter-process is performed by that function of the Understanding, which apprehends dissimilarities between resembling objects, and in the full and free exertion of this energy there is a feeling of pleasure.

Science,—how pleasing.

The Intellect further tends to reduce the piecemeal and fragmentary cognitions it possesses, to a systematic whole, in other words, to elevate them to a Science; hence the pleasure we derive from all that enables us with ease and rapidity to survey the relation of complex parts, as constituting the members of one organic whole.

Deduction from first principles.

The Intellect, from the necessity it has of thinking everything as the result of some higher reason, is thus determined to attempt the deduction of every object of cognition from a simple principle. When, therefore, we succeed or seem to succeed in the discovery of such a principle, we feel a pleasure; as we feel a pain, when the intellect is frustrated in this endeavour.

Apprehension of adaptation of Means to Ends,—

To the feelings of pleasure which are afforded by the unimpeded energies of the Understanding, belongs, likewise, the gratification we find in the apprehension

LECT. XLV.

How pleasurable.

Ends of two kinds,—external and internal. Hence the Useful and the Perfect.

of external or internal adaptation of Means to Ends. Human intelligence is naturally determined to propose to itself an end; and, in the consideration of objects, it thus necessarily thinks them under this relation. If an object, viewed as a mean, be fitted to effect its end, this end is either an external, that is, one which lies beyond the thing itself, in some other existence; or an internal, that is, one which lies within the thing itself, and consummates its own existence. If the end be external, an object suited to accomplish it is said to be *useful*. If, again, the end be internal, and all the parts of the object be viewed in relation to their whole as to their end, an object, as suited to effect this end, is said to be *perfect*. If, therefore, we consider an object in reference either to an external or to an internal end, and if this object be recognised to fulfil the conditions which this relation implies, the act of thought in which this is accomplished is an unimpeded, and, consequently, pleasurable, energy; whereas the act of cognising that these conditions are awanting, and the object therefore ill adapted to its end, is a thwarted, and therefore a painful, energy of thought.

LECTURE XLVI.

THE FEELINGS.—THEIR CLASSES.—THE BEAUTIFUL AND SUBLIME.

LECT. XLVI.

Recapitulation.

AFTER terminating the consideration of the Feelings viewed as Causes,—causes of Pleasure and Pain, we entered, in our last Lecture, on their discussion regarded as Effects,—effects of the various processes of conscious life. In this latter relation, I divided them into two great classes,—the Sensations and the Sentiments. The Sensations are those feelings which accompany the vital processes more immediately connected with the corporeal organism. The Sentiments are those feelings which accompany the mental processes, which, if not wholly inorganic, are at least less immediately dependent on the conditions of the nervous system. The Sensations I again subdivided into two orders,—into those which accompany the action of the five Determinate Senses, and into those which accompany, or, in fact, constitute the manifestations of the Indeterminate or Vital Sense. After a slight consideration of the Sensations, I passed on to the Sentiments. These I also subdivided into two orders, according as they accompany the energies of the Cognitive, or the energies of the Conative, Powers. The former of these I called the Contemplative,—the latter, the Practical Feelings or Sentiments. Taking the former,—the Contemplative,—into discussion, I further subdivided

LECT. XLVI.

these into two classes, according as they are the concomitants of the lower or Subsidiary, or of the higher or Elaborative, Faculty of Cognition. The sentiments which accompany the lower or Subsidiary Faculties, by a final subdivision, I distributed into those of the Faculty of Self-consciousness and into those of the Imagination,—referring to the Imagination the relative faculty of Reproduction. I ought also to have observed, that, as the Imagination always co-operates in every act of complex perception, and, in fact, bestows on such a cognition its whole unity, under the Feelings of Imagination (or of Imagination and the Understanding in conjunction) would fall to be considered those sentiments of pleasure which, in the perceptions of sense, we receive from the relations of the objects presented. Under the Feelings connected with the energies of the Elaborative Faculty or Understanding, I comprehended those which arise from the gratification of the Regulative Faculty,—Reason or Intelligence,—because it is only through the operations of the former that the laws of the latter are carried into effect. In relation to Feelings, the two faculties may, therefore, be regarded as one. I then proceeded to treat of the several kinds of Contemplative Feeling in detail; and, before the conclusion of the Lecture, had run rapidly through those of Self-consciousness, those of Imagination, considered apart from the Understanding, and those of the Understanding, considered apart from Imagination. We have now, therefore, in the first place, to consider the feelings which arise from the acts of Imagination and Understanding in conjunction.

Feelings that arise from the Imagination and Understanding in conjunction.

Beauty and Sublimity.

The feelings of satisfaction which result from the joint energy of the Understanding and Phantasy, are principally those of Beauty and Sublimity; and the judgments which pronounce an object to be *sublime*, *beautiful*, &c., are called, by a metaphorical expression, *Judgments of Taste*. These have also been styled *Æsthetical Judgments*; and the term *æsthetical* has now, especially among the philosophers of Germany, nearly superseded the term *taste*. Both terms are unsatisfactory.

The gratification we feel in the beautiful, the sublime, the picturesque, &c., is purely contemplative, that is, the feeling of pleasure which we then experience, arises solely from the consideration of the object, and altogether apart from any desire of, or satisfaction in, its possession. In the following observations, it is almost needless to observe, that I can make no attempt at more than a simple indication of the origin of the pleasure we derive from the contemplation of those objects, which, from the character of the feelings they determine, are called *beautiful*, *sublime*, &c.

Beauty distinguished as Absolute and Relative.

In relation to the Beautiful, this has been distinguished into the Free or Absolute, and into the Dependent or Relative.[a] In the former case, it is not necessary to have a notion of what the object ought to be, before we pronounce it beautiful or not; in the latter case, such a previous notion is required. Flowers, shells, arabesques, &c., are freely or absolutely beautiful. We judge, for example, a flower to be beautiful, though unaware of its destination, and that it contains a complex apparatus of organs all admirably adapted to the propagation of the plant. When we are made cognisant of this, we obtain, indeed, an additional gratification, but one wholly different from that which we experience in the contemplation of the flower itself,

a See Hutcheson, *Inquiry*, treatise i. sects. 2, 4.—ED.

LECT. XLVI.

apart from all consideration of its adaptations. A house, a pillar, a piece of furniture, are dependently or relatively beautiful; for here the object is judged beautiful by reference to a certain end, for the sake of which it exists. This distinction, which is taken by Kant[a] and others, appears to me unsound. For Relative Beauty is only the confusion of two elements, which ought to have been kept distinct. There is no doubt, I think, that certain objects please us directly and of themselves, that is, no reference being had to aught beyond the form itself which they exhibit. These are things of themselves beautiful. Other things, again, please us not directly and of themselves; that is, their form presents nothing, the cognition of which results in an agreeable feeling. But these same things may please indirectly and by relation; that is, when we are informed that they have a purpose, and are made aware of their adaptation to its accomplishment, we may derive a pleasure from the admirable relation which here subsists between the end and means. These are things Useful. But the pleasure which results from the contemplation of the useful, is wholly different from that which results from the contemplation of the beautiful, and, therefore, they ought not to be confounded. It may, indeed, happen that the same object is such as affords us both kinds of pleasure, and it may at once be beautiful and useful. But why, on such a ground, establish a second series of beauty? In this respect, St Augustin shows himself superior to our great modern analyst. In his *Confessions*, he informs us that he had written a book, (unfortunately lost),

This distinction unsound.

The Useful and the Beautiful distinct.

St Augustin's doctrine on this point superior to the modern.

a Partially, perhaps; see *Kritik der Urtheilskraft*, §§ 6, 10. But Kant distinguishes Beauty from Adaptation to an End, though he refers both to the faculty of Judgment.—Ed.

addressed to Hierius, the Roman rhetorician, under the title *De Apto et Pulcro*, in which he maintained, that the beautiful is that which pleases absolutely and of itself, the well-adapted that which pleases from its accommodation to something else,—"Pulcrum esse, quod per se ipsum; aptum, autem, quod ad aliquid accommodatum deceret."[a]

LECT. XLVI.

Relative Beauty is only a beautified utility, or utilised beauty.

Now what has been distinguished as Dependent or Relative Beauty, is nothing more than a beautified utility, or a utilised beauty. For example, a pillar taken by itself and apart from all consideration of any purpose it has to serve, is a beautiful object; and a person of good taste, and ignorant of its relations, would at once pronounce it so. But when he is informed that it is also a mean towards an end, he will then find an additional satisfaction in the observation of its perfect adaptation to its purpose; and he will no longer consider the pillar as something beautiful and useless; his taste will desiderate its application, and will be shocked at seeing, as we so often see, a set of columns stuck on upon a building, and supporting nothing. Be this, however, as it may, our pleasure, in both cases, arises from a free and full play being allowed to our cognitive faculties.

The theory of Free or Absolute Beauty.

In the case of Beauty,—Free Beauty,—both the Imagination and the Understanding find occupation; and the pleasure we experience from such an object, is in proportion as it affords to these faculties the opportunity of exerting fully and freely their respective energies. Now, it is the principal function of the Understanding, out of the multifarious presented to it, to form a whole. Its entire activity is, in fact, a tendency towards unity; and it is only satisfied when this object is so constituted as to afford

a Lib. iv. cap. xv.—Ed.

the opportunity of an easy and perfect performance of this its function. In this case, the object is judged beautiful or pleasing.

The greater the number of the parts of any object given by the Imagination, which the Understanding has to bind up into a whole, and the shorter the time in which it is able to bring this process to its issue, the more fully and the more easily does the understanding energise, and, consequently, the greater will be the pleasure afforded as the reflex of its energy.[a]

The theory explains the differences of individuals in the apprehension of the Beautiful.

This not only affords us the rationale of what the Beautiful is, but it also enables us to explain the differences of different individuals in the apprehension of the beautiful. The function of the Understanding is in all men the same; and the understanding of every man binds up what is given as plural and multifarious into the unity of a whole. But as it is only the full and facile accomplishment of this function, which has pleasure for its concomitant, it depends wholly on the capacity of the individual understanding, whether this condition shall be fulfilled. If an understanding, by natural constitution, by cultivation and exercise, be vigorous enough to think up rapidly into a whole what is presented in complexity,—multiplicity,—the individual has an enjoyment in the exertion, and he regards the object as beautiful; whereas, if an intellect perform this function slowly and with effort, if it succeed in accomplishing the end at all, the individual can feel no pleasure, (if he does not experience pain), and the object must to him appear as one destitute of beauty, if not positively ugly. Hence it is that children, boors, in a word, persons of

a [Cf. Mendelssohn, *Philosophische Schriften*, ii. p. 74. Hemsterhuis, *Lettre sur la Sculpture*, *Œuvres Philosophiques*, t. i. p. 12.]

a weak or uncultivated mind, may find the parts of a building beautiful, while unable to comprehend the beauty of it as a whole. On the other hand, we may also explain why the pleasure we have in the contemplation of an object is lessened, if not wholly annihilated, if we mentally analyse it into its parts. The fairest human head would lose its beauty were we to sunder it in thought, and consider how it is made up of integuments, of cellular tissue, of muscular fibres, of bones, of brain, of blood-vessels, &c. It is no longer a whole; it is the multifarious without unity.

And affords the reason why our pleasure in the contemplation of an object is lessened, when we analyse it into its parts.

In reference to Taste, it is quite a different thing to sunder a whole into its parts, and a whole into its lesser wholes. In the one case, we separate only to separate, and not again to connect. In the other, we look to the parts, in order to be able in a shorter time more perfectly to survey the whole. This must enhance the gratification, and it is a process always requisite when the whole comprises a more multiplex plurality than our understanding is competent to embrace at the first attempt. When a whole head is found too complex to be judged at once, out of the brow, eyes, nose, cheeks, mouth, &c., we make so many lesser wholes, in order, in the first place, to comprehend them by the intellect as wholes together; we then bind up these petty wholes into one great whole, which, in a shorter or longer time, we overlook, and award to it, accordingly, a greater or a less amount of beauty.

Difference between sundering a whole into its parts, and into its lesser wholes.

In the case of Relative or Dependent Beauty, we must distinguish the pleasure we receive into two, combined indeed, but not identical. The one of these pleasures is that from the beauty which the object contains, and the principle of which we have been

Relative Beauty, from the conformity of Means to End.

LECT. XLVI.

just considering. The other of these pleasures is that which, in our last Lecture, we showed was attached to a perfect energy of the Understanding, in thinking an object under the notion of conformity as a mean adapted to an end.

Judgments of Taste either Pure or Mixed.

A judgment of Taste may be called *pure*, when the pleasure it enounces is one exclusively derived from the Beautiful, and *mixed*, when with this pleasure there are conjoined feelings of pain or pleasure from other sources. Such, for example, are the organic excitations of particular colours, tones, &c., emotions, the moral feeling, the feeling of pleasure from the sublime, &c. It requires a high cultivation of the taste in order to find gratification in a pure beauty, and also to separate from our judgment of an object, in this respect, all that is foreign to this source of pleasure. The uncultivated man at first finds gratification only in those qualities which stimulate his organs; and it is only gradually that he can be educated to pay attention to the form of objects, and to find pleasure in what lightly exercises his faculties of Imagination and Thought,—the Beautiful. The result, then, of what has now been said is, that a thing beautiful is one whose form occupies the Imagination and Understanding in a free and full, and, consequently, in an agreeable, activity: and to this definition of the Beautiful all others may without difficulty be reduced; for these, like the definitions of the pleasurable, are never absolutely false, but, in general, only partial expressions of the truth. On these it is, however, at present impossible to touch.

The Beautiful defined.

The Sublime,—the feeling partly pleasurable,

The feeling of pleasure in the Sublime is essentially different from our feeling of pleasure in the Beautiful. The beautiful awakens the mind to a soothing contemplation; the sublime rouses it to strong emotion. The beautiful attracts without repelling; whereas the sublime at once does both; the beautiful affords us a feeling of unmingled pleasure, in the full and unimpeded activity of our cognitive powers; whereas our feeling of sublimity is a mingled one of pleasure and pain,—of pleasure in the consciousness of the strong energy, of pain in the consciousness that this energy is vain.[a]

partly painful.

Theory of the Sublime.

But as the amount of pleasure in the sublime is greater than the amount of pain, it follows, that the free energy it elicits must be greater than the free energy it repels. The beautiful has reference to the form of an object, and the facility with which it is comprehended. For beauty, magnitude is thus an impediment. Sublimity, on the contrary, requires magnitude as its condition; and the formless is not unfrequently sublime. That we are at once attracted and repelled by sublimity, arises from the circumstance that the object which we call *sublime*, is proportioned to one of our faculties, and disproportioned to another; but as the degree of pleasure transcends the degree of pain, the power whose energy is promoted must be superior to that power whose energy is repressed.

The Sublime,—divided into that of Extension, Protension, and Intension.

The Sublime has been divided into two kinds, the Theoretical and the Practical, or, as they are also called, the Mathematical and the Dynamical.[β] A preferable division would be according to the three quantities,—into the sublime of Extension, the sublime of

a [That the sublime has a painful feeling with it, see Fracastorius, *De Sympathia et Antipathia*, c. xx., *Opera* (edit. 1584), f 73[b]; Mendelssohn, *Recherches sur les Sentiments Moraux, traduit par M. Abbt* (1764), p. 6 *et seq.*; Kant, *Kritik der Urtheilskraft*, § 23; Burke, *On the Sublime and Beautiful*, part i. § 7: part ii. §§ 1, 2; part iii. § 27; part iv. § 5-8.]

β Kant, *Kritik der Urtheilskraft*, § 24 *et seq.*—Ed.

LECT. XLVI.

Protension, and the sublime of Intension; or, what comes to the same thing,—the sublime of Space, the sublime of Time, and the sublime of Power. In the two former the cognitive, in the last the conative, powers come into play. An object is extensively, or protensively sublime, when it comprises so great a multitude of parts that the Imagination sinks under the attempt to represent it in an image, and the Understanding to measure it by reference to other quantities. Baffled in the attempt to reduce the object within the limits of the faculties by which it must be comprehended, the mind at once desists from the ineffectual effort, and conceives the object not by a positive, but by a negative, notion; it conceives it as inconceivable, and falls back into repose, which is felt as pleasing by contrast to the continuance of a forced and impeded energy. Examples of the sublime,—of this sudden effort, and of this instantaneous desisting from the attempt, are manifested in the extensive sublime of Space, and in the protensive sublime of Eternity.

These divisions illustrated.

The sublime of Extension and Protension.

The sublime of Intension.

An object is intensively sublime, when it involves such a degree of force or power that the Imagination cannot at once represent, and the Understanding cannot bring under measure, the quantum of this force; and when, from the nature of the object, the inability of the mind is made at once apparent, so that it does not proceed in the ineffectual effort, but at once calls back its energies from the attempt. It is thus manifest that the feeling of the sublime will be one of mingled pain and pleasure; pleasure, from the vigorous exertion and from the instantaneous repose; pain, from the consciousness of limited and frustrated activity. This mixed feeling in the contemplation of a sublime object is finely expressed by Lucretius when he says:—

> "Me quædam divina voluptas,
> Percipit atque horror."[α]

I do not know a better example of the sublime, in all its three forms, than in the following passage of Kant:[β]—

The Sublime, in its three forms, exemplified in a passage from Kant.

"Two things there are, which, the oftener and the more steadfastly we consider, fill the mind with an ever new, an ever rising admiration and reverence;—*the* STARRY HEAVEN *above, the* MORAL LAW *within.* Of neither am I compelled to seek out the reality, as veiled in darkness, or only to conjecture the possibility, as beyond the hemisphere of my knowledge. Both I contemplate lying clear before me, and connect both immediately with my consciousness of existence. The one departs from the place I occupy in the outer world of sense; expands, beyond the bounds of imagination, this connection of my body with worlds lying beyond worlds, and systems blending into systems; and protends it also into the illimitable times of their periodic movement,—to its commencement and continuance. The other departs from my invisible self, from my personality; and represents me in a world, truly infinite indeed, but whose infinity can be tracked out only by the intellect, with which also my connection, unlike the fortuitous relation I stand in to all worlds of sense, I am compelled to recognise as universal and necessary. In the former, the first view of a countless multitude of worlds annihilates, as it were, my importance as an *animal product*, which, after a brief and that incomprehensible endowment with the powers of life, is compelled to refund its constituent matter to the

α Lib. iii. 28.—ED. β *Kritik der practischen Vernunft*, Beschluss.—ED.

LECT. XLVI.

planet—itself an atom in the universe—on which it grew. The aspect of the other, on the contrary, elevates my worth as an *intelligence* even without limit; and this through my personality, in which the moral law reveals a faculty of life independent of my animal nature, nay, of the whole material world:—at least, if it be premitted to infer as much from the regulation of my being, which a conformity with that law exacts; proposing, as it does, my moral worth for the absolute end of my activity, conceding no compromise of its imperative to a necessitation of nature, and spurning, in its infinity, the conditions and boundaries of my present transitory life."

> "Spirat enim majora animus seque altius effert
> Sideribus, transitque vias et nubila fati,
> Et momenta premit pedibus quæcunque putantur
> Figere propositam natali tempore sortem."[a]

Here we have the extensive sublime in the heavens and their interminable space, the protensive sublime in their illimitable duration, and the intensive sublime in the omnipotence of the human will, as manifested in the unconditional imperative of the moral law.

The Picturesque,—wherein it consists, and how it differs from the Sublime and Beautiful.

The Picturesque, however opposite to the Sublime, seems, in my opinion, to stand to the Beautiful in a somewhat similar relation. An object is positively ugly, when it is of such a form that the Imagination and Understanding cannot help attempting to think it up into unity, and yet their energies are still so impeded that they either fail in the endeavour, or accomplish it only imperfectly, after time and toil. The cause of this continuance of effort is, that the object does not present such an appearance of incon-

[a] Prudentius, *Contra Sym.*, II. 479. Quoted in *Discussions*, p. 311.—Ed.

gruous variety as at once to compel the mind to desist from the attempt of reducing it to unity; but, on the contrary, leads it on to attempt what it is yet unable to perform,—its reduction to a whole. But variety, —variety even apart from unity,—is pleasing; and if the mind be made content to expatiate freely and easily in this variety, without attempting painfully to reduce it to unity, it will derive no inconsiderable pleasure from this exertion of its powers. Now a picturesque object is precisely of such a character. It is so determinately varied and so abrupt in its variety, it presents so complete a negation of all rounded contour, and so regular an irregularity of broken lines and angles, that every attempt at reducing it to an harmonious whole is at once found to be impossible. The mind, therefore, which must forego the energy of representing and thinking the object as a unity, surrenders itself at once to the energies which deal with it only in detail.

The Practical Feelings.

I proceed now to those feelings which I denominated Practical,—those, namely, which have their root in the powers of Conation, and thus have reference to overt action.

Their divisions.

The Conative, like the Cognitive, powers are divided into a higher and a lower order, as they either are, or are not, immediately relative to our bodily condition. The former may be called the Pathological, the latter the Moral. Neglecting this distribution, the Practical Feelings are relative either—1°, To our Self-preservation; or, 2°, To the Enjoyment of our Existence; or, 3°, To the Preservation of the Species; or, 4°, To our Tendency towards Development and Perfection; or, 5°, To the Moral Law. Of these in their order.

LECT. XLVI.

Those relative—1. To Self-preservation.

In the first place, of the feelings relative to Self-preservation:—these are the feelings of Hunger and Thirst, of Loathing, of Sorrow, of Bodily Pain, of Repose, of Fear at danger, of Anxiety, of Shuddering, of Alarm, of Composure, of Security, and the nameless feeling at the Representation of Death. Several of these feelings are corporeal, and may be considered, with equal propriety, as modifications of the Vital Sense.

2. Enjoyment of existence.

In the second place, man is determined not only to exist, but to exist well; he is, therefore, determined also to desire whatever tends to render life agreeable, and to eschew whatever tends to render it disagreeable. All, therefore, that appears to contribute to the former, causes in him the feeling of Joy; whereas all that seems to threaten the latter, excites in him the repressed feelings of Fear, Anxiety, Sorrow, &c., which we have already mentioned.

3. Preservation of the species.

In the third place, man is determined, not only to preserve himself, but to preserve the species to which he belongs, and with this tendency various feelings are associated. To this head belong the feelings of Sexual Love; and the sentiment of Parental Affection. But the human affections are not limited to family connections. "Man," says Aristotle, "is the sweetest thing to man."[α] "Man is more political than any bee or ant."[β] We have thus a tendency to social intercourse, and society is at once the necessary condition of our happiness and our perfection. "The solitary," says Aristotle again, "is either above or below humanity; he is either a god or a beast."[γ]

Sympathy.

In conformity with his tendency to social existence,

α *Eth. Eud.*, vii. 2, 25.—Ed.

β *Polit.*, i. 2, 10.—Ed.

γ *Polit.*, i. 2, 9, 14.—Ed.

man is endowed with a Sympathetic Feeling, that is, he rejoices with those that rejoice, and grieves with those that grieve. Compassion,—Pity,—is the name given to the latter modification of sympathy; the former is without a definite name. Besides sympathetic sorrow and sympathetic joy, there are a variety of feelings which have reference to our existence in a social relation. Of these there is that connected with Vanity, or the wish to please others from the desire of being respected by them; with Shame, or the fear and sorrow at incurring their disrespect; with Pride, or the overweening sentiment of our own worth. To the same class we may refer the feelings connected with Indignation, Resentment, Anger, Scorn, &c.

Vanity. Shame. Pride.

4. Tendency to development.

In the fourth place, there is in man implanted a desire of developing his powers,—there is a tendency towards perfection. In virtue of this, the consciousness of all comparative inability causes pain; the consciousness of all comparative power causes pleasure. To this class belong the feelings which accompany Emulation,—the desire of rising superior to others; and Envy,—the desire of reducing others beneath ourselves.

5. The Moral Law.

In the fifth place, we are conscious that there is in man a Moral Law,—a Law of Duty, which unconditionally commands the fulfilment of its behests. This supposes, that we are able to fulfil them, or our nature is a lie; and the liberty of human action is thus, independently of all direct consciousness, involved in the datum of the Law of Duty. Inasmuch also as Moral Intelligence unconditionally commands us to perform what we are conscious to be our duty, there is attributed to man an absolute worth,—an absolute dignity. The feeling which the manifestation of this

LECT. XLVI. worth excites, is called Respect. With the consciousness of the lofty nature of our moral tendencies, and our ability to fulfil what the law of duty prescribes, there is connected the feeling of Self-respect; whereas, from a consciousness of the contrast between what we ought to do, and what we actually perform, there arises the feeling of Self-abasement. The sentiment of respect for the law of duty is the Moral Feeling, which has by some been improperly denominated the Moral Sense; for through this feeling we do not take cognisance whether anything be morally good or morally evil, but when, by our intelligence, we recognise aught to be of such a character, there is herewith associated a feeling of pain or pleasure, which is nothing more than our state in reference to the fulfilment or violation of the law.

Man, as conscious of his liberty to act, and of the law by which his actions ought to be regulated, recognises his personal accountability, and calls himself before the internal tribunal which we denominate Conscience. Here he is either acquitted or condemned. The acquittal is connected with a peculiar feeling of pleasurable exultation, as the condemnation with a peculiar feeling of painful humiliation,—Remorse.

APPENDIX.

I. PERCEPTION.—FRAGMENTS.—(See Vol. II. p. 29.)

(Written in connection with proposed MEMOIR OF MR STEWART. On Desk, May 1856; written Autumn 1855.—ED.)

THERE are three considerations which seem to have been principally effective in promoting the theory of a Mediate or Representative Perception, and by *perception* is meant the apprehension, through sense, of external things. These might operate severally or together.

The first is, that such a hypothesis is necessary to render possible the perception of distant objects. It was taken as granted that certain material realities, (as a sun, stars, &c.), not immediately present to sense, were cognised in a perceptive act. These realities could not be known immediately, or in themselves, unless known as they existed, and they existed only as they existed in their place in space. If, therefore, the perceptive mind did not sally out to them, (which, with the exception of one or two theorists, was scouted as an impossible hypothesis), an immediate perception behoved to be abandoned, and the sensitive cognition we have of them must be vicarious; that is, not of the realities themselves, as present to our organs, and presented to apprehension, but of something different from the realities externally existing, through which, however, they are mediately represented. Various theories in regard to the nature of this medium or vicarious object may be entertained; but these may be overpassed. This first consideration alone was principally effectual among materialists: on them the second had no influence.

A second consideration was the opposite and apparently inconsistent nature of the object and subject of cognition; for here the reality to be known is material, whereas the mind knowing is immaterial; while it was long generally believed, that what is known must be of an analogous essence, (the same or similar), to what knows. In consequence of this persuasion, it was deemed impossible that the immaterial unextended mind could apprehend in itself, as extended, a material reality. To explain the fact of sensitive perception, it was therefore supposed requisite to attenuate,—to immaterialise the immediate object of perception, by dividing the object known from the reality existing. Perception thus became a vicarious or mediate cognition, in which the corporeal was said to be represented by the incorporeal.

Perception—Positive Result.

1. We perceive only through the senses.

2. The senses are corporeal instruments,—parts of our bodily organism.

3. We are, therefore, percipient only through, or by means of, the body. In other words, material and external things are to us only not as zero, inasmuch as they are apprehended by the mind in their relation with the material organ which it animates, and with which it is united.

4. An external existence, and an organ of sense, as both material, can stand in relation only according to the laws of matter. According to these laws, things related,—connected, must act and be acted on; but a thing can act only where it is. Therefore the thing perceived, and the percipient organ, must meet in place,—must be contiguous. The consequence of this doctrine is a complete simplification of the theory of perception, and a return to the most ancient speculation on the point. All sensible cognition is, in a certain acceptation, reduced to Touch, and this is the very conclusion maintained by the venerable authority of Democritus.

According to this doctrine, it is erroneous, in the first place, to affirm that we are percipient of distant, &c. objects.

It is erroneous, in the second place, to say that we perceive external things in themselves, in the signification that we perceive them as existing in their own nature, and not in relation to the living organ. The real, the total, the only object perceived has, as a relative, two phases. It may be described either as the idiopathic affection of the sense, (*i. e.* the sense in relation to an external reality), or as the quality of a thing actually determining such or such an affection of the sentient organ, (*i. e.* an external reality in correlation to the sense).

A corollary of the same doctrine is, that what have been denominated the Primary Qualities of body, are only perceived through the Secondary; in fact, Perception proper cannot be realised except through Sensation Proper. But synchronous.

The object of perception is an affection, not of the mind as apart from body, not of the body as apart from mind, but of the composite formed by union of the two; that is, of the animated or living organism (Aristotle).

In the process of perception there is required both an act of the conscious mind and a passion of the affected body; the one without the other is null. Galen has, therefore, well said, "Sensitive perception is not a mere passive or affective change, but the discrimination of an affective change."[α] (Aristotle,—judgment.)

Perception supposes Consciousness, and Consciousness supposes Memory and Judgment; for, abstract Consciousness, and there is no Perception; abstract Memory, or Judgment, and Consciousness is abolished. (Hobbes,—Memory; Aristotle,—Judgment of Sense.) Memory, Recollection; for change is necessary to Consciousness, and change is only to be apprehended through the faculty of Remembrance. Hobbes has, therefore, truly said of Perception,—"Sentire semper idem, et non sentire, ad idem recidunt."[β] But there could be no discriminative apprehension, supposing always memory without an act whereby difference was affirmed or sameness denied; that is, without an act of Judgment. Aristotle[γ] is, therefore, right in making Perception a Judgment.

II. LAWS OF THOUGHT.—(See Vol. II. p. 308.)

(Written in connection with proposed MEMOIR OF MR STEWART. On Desk, May 1856; written Autumn 1855.—ED.)

The doctrine of Contradiction, or of Contradictories, (ἀξίωμα τῆς ἀντιφάσεως), that Affirmation or Negation is a necessity of thought, whilst Affirmation and Negation are incompatible, is developed into three sides or phases, each of which implies both the others,—

α See *Reid's Works*, p. 878.—ED. β See *Ibid.*—ED. γ See *Ibid.*—ED.

phases which may obtain, and actually have received, severally, the name of *Law*, *Principle*, or *Axiom*. Neglecting the historical order in which these were scientifically named and articulately developed, they are :—

1°, The Law, Principle, or Axiom, of *Identity*, which, in regard to the same thing, immediately or directly enjoins the affirmation of it with itself, and mediately or indirectly prohibits its negation : (*A is A*).

2°, The Law, &c. of *Contradiction*, (properly *Non-contradiction*), which, in regard to contradictories, explicitly enjoining their reciprocal negation, implicitly prohibits their reciprocal affirmation : (*A is not Not-A.*) In other words, contradictories are thought as existences incompatible at the same time,—as at once mutually exclusive.

3°, The Law, &c. of *Excluded Middle* or *Third*, which declares that, whilst contradictories are only two, everything, if explicitly thought, must be thought as of these either the one or the other : (*A is either B or Not-B.*) In different terms :—Affirmation and Negation of the same thing, in the same respect, have no conceivable medium ; whilst anything actually may, and virtually must, be either affirmed or denied of anything. In other words :—Every predicate is true or false of every subject ; or, contradictories are thought as incompossible, but, at the same time, the one or the other as necessary. The argument from Contradiction is omnipotent within its sphere, but that sphere is narrow. It has the following limitations :—

1°, It is negative, not positive ; it may refute, but it is incompetent to establish. It may show what is not, but never, of itself, what is. It is exclusively Logical or Formal, not Metaphysical or Real ; it proceeds on a necessity of thought, but never issues in an Ontology or knowledge of existence.

2°, It is dependent ; to act it presupposes a counter-proposition to act from.

3°, It is explicative, not ampliative ; it analyses what is given, but does not originate information, or add anything, through itself, to our stock of knowledge.

4°, But, what is its principal defect, it is partial, not thoroughgoing. It leaves many of the most important problems of our knowledge out of its determination ; and is, therefore, all too narrow in its application as a universal criterion or instrument of judgment. For were we left, in our reasonings, to a dependence on the principle of Contradiction, we should be unable competently to attempt any argument with regard to some of the most interesting and important questions. For there are many problems in the philosophy of mind where the solution necessarily lies between what are, to us, the one or the other of two counter and, therefore, incompatible alternatives, neither of which are we able to conceive as possible, but of which, by the very conditions of thought, we are compelled to acknowledge that the one or the other cannot but be; and it is as supplying this deficiency, that what has been called the argument from Common Sense becomes principally useful.

The principle of Contradiction, or rather of Non-Contradiction, appears in two forms, and each of these has a different application.

In the first place, (what may be called the *Logical* application), it declares that, of Contradictories, two only are possible in thought; and that of these alternatives the one or the other, exclusively, is thought as necessarily true. This phasis of the law is unilateral; for it is with a consciousness or cognition that the one contradictory is necessarily true, and the other contradictory necessarily false. This, the logical phasis of the law, is well known, and has been fully developed.

In the second place, (what may be called the *Psychological* application), while it necessarily declares that, of Contradictories, both cannot, but one must, be, still bilaterally admits that we may be unable positively to think the possibility of either alternative. This, the psychological phasis of the law, is comparatively unknown, and has been generally neglected. Thus, *Existence* we cannot but think,—cannot but attribute in thought; nevertheless we can actually conceive neither of these contradictory alternatives,—the absolute commencement, the infinite non-commencement, of being. As it is with Existence, so is it with *Time*. We cannot think time beginning; we cannot think time not beginning. So also with *Space*. We are unable to conceive an existence out of space; yet we are equally unable to compass the notion of illimitable or infinite space. Our capacity of thought is thus peremptorily proved incompetent to what we necessarily think about; for, whilst what we think about must be thought to Exist,—to exist in Time,—to exist in Space,—we are unable to realise the counter-notions of Existence commencing or not commencing, whether in Time or in

Space. And thus, whilst Existence, Time, and Space, are the indispensable conditions, forms, or categories of actual thought, still are we unable to conceive either of the counter-alternatives, in one or other of which we cannot but admit that they exist. These and such like impotences of positive thought have, however, as I have stated, been strangely overlooked.

III. THE CONDITIONED.

(a.) KANT'S ANALYSIS OF JUDGMENTS.—(See Vol. II. p. 375.)

(Fragment from Early Papers, probably before 1830.—ED.)

Kant analysed judgments (*a priori*) into *analytic* or *identical* [or *explicative*], and *synthetical*, or [*ampliative*, *non-identical*]. Great fame from this. But he omitted a third kind,—those that the mind is compelled to form by a law of its nature, but which can neither be reduced to analytic judgments, because they cannot be subordinated to the law of Contradiction, nor to synthetical, because they do not seem to spring from a positive power of mind, but only arise from the inability of the mind to conceive the contrary.

In Analytic judgments,—(principle of contradiction),—we conceive the one alternative as necessary, and the other as impossible. In Synthetic judgments, we conceive the affirmative as necessary, but not [its negation as self-contradictory].

Would it not be better to make the synthetic of two kinds,—a positive and negative? Had Kant tried whether his synthetic judgments *a priori* were positive or negative, he would have reached the law of the Conditioned, which would have given a totally new aspect to his Critique,—simplified, abolished the distinction of *Verstand* and *Vernunft*, which only positive and negative, (at least as a faculty conceiving the Unconditioned, and left it only, as with Jacobi, the Νοῦς, the *locus principiorum*,—the faculty,—revelation, of the primitive facts or faiths of consciousness,—the Common Sense of Reid), the distinction of *Begriffe* and *Ideen*, and have reduced his whole Categories and Ideas to the category of the Conditioned and its subordinates.

* * * * * * * * * *

(1853, November).—There are three degrees or epochs which

we must distinguish in philosophical speculation touching the Necessary.

In the first, which we may call the Aristotelic or Platonico-Aristotelic, the Necessary was regarded, if not exclusively, principally and primarily, in an objective relation;—at least the objective and subjective were not discriminated; and it was defined that of which the existence of the opposite,—contrary,—is impossible,—what could not but be.

In the second, which we may call the Leibnitzian or Leibnitio-Kantian, the Necessary was regarded primarily in a subjective respect, and it was defined that of which the thought of the opposite,—contrary,—is impossible,—what we cannot but think. It was taken for granted, that what we cannot think, cannot be, and what we must think, must be; and from hence there was also inferred, without qualification, that this subjective necessity affords the discriminating criterion of our native or *a priori* cognitions,—notions and judgments.

But a third discrimination was requisite; for the necessity of thought behoved to be again distinguished into two kinds.—(See *Discussions*, 2d edit., Addenda.)

(*b.*) Contradictions proving the Psychological Theory of the Conditioned.—(July 1852.)

1. Finite cannot comprehend, contain the Infinite.—Yet an inch or minute, say, are finites, and are divisible *ad infinitum*, that is, their terminated division incogitable.

2. Infinite cannot be terminated or begun.—Yet eternity *ab ante* ends *now;* and eternity *a post* begins *now.*—So apply to Space.

3. There cannot be two infinite maxima.—Yet eternity *ab ante* and *a post* are two infinite maxima of time.

4. Infinite maximum if cut into two, the halves cannot be each infinite, for nothing can be greater than infinite, and thus they could not be parts; nor finite, for thus two finite halves would make an infinite whole.

quantities

5. What contains infinite extensions, protensions, intensions, cannot be passed through,—come to an end. An inch, a minute, a degree contains these; *ergo*, &c. Take a minute. This contains an infinitude of protended quantities, which must follow one

after another; but an infinite series of successive protensions can, *ex termino*, never be ended; *ergo*, &c.

6. An Infinite maximum cannot but be all inclusive. Time *ab ante* and *a post* infinite and exclusive of each other; *ergo*, &c.

7. An infinite number of quantities must make up either an infinite or a finite whole. I. The former.—But an inch, a minute, a degree, contain each an infinite number of quantities; therefore, an inch, a minute, a degree, are each infinite wholes; which is absurd. II. The latter.—An Infinite number of quantities would thus make up a finite quantity; which is equally absurd.

8. If we take a finite quantity, (as an inch, a minute, a degree), it would appear equally that there are, and that there are not, an equal number of quantities between these and a greatest, and between these and a least.[α]

9. An absolutely quickest motion is that which passes from one point to another in space in a minimum of time. But a quickest motion from one point to another, say a mile distance, and from one to another, say a million million of miles, is thought the same; which is absurd.

10. A wheel turned with quickest motion; if a spoke be prolonged, it will therefore be moved by a motion quicker than the quickest. The same may be shown using the rim and the nave.[β]

11. Contradictory are Boscovich Points, which occupy space, and are unextended.[γ] Dynamism, therefore, inconceivable. *E contra*,

12. Atomism also inconceivable; for this supposes atoms,—minima extended but indivisible.

13. A quantity, say a foot, has an infinity of parts. Any part of this quantity, say an inch, has also an infinity. But one infinity is not larger than another. Therefore, an inch is equal to a foot.[δ]

14. If two divaricating lines are produced *ad infinitum* from a point where they form an acute angle, like a pyramid, the base will be infinite and, at the same time, not infinite; 1°, Because terminated by two points; and, 2°, Because shorter than the sides[ε]; 3°, Base could not be drawn, because sides infinitely long.[ζ]

α See Boscovich on Stay, *Philosophia Recentior*, I. p. 281, edit. 1755.

β See Leibnitz, *Meditationes de Cognitione, Veritate, et Ideis*.—Ed.

γ See Boscovich on Stay, as above, I. p. 304.

δ See Tellez, quoted by F. Bonæ Spei, [*Physica*, pars I. tract. iii. disp. I. dub. 4, p. 154, edit. 1652.—Ed.]

ε See Bonæ Spei, *Physica*, [pars I. tract. iii. disp. I. dub. 2, p. 130.—Ed.]

ζ See Carleton, [*Philosophia Universa, Auctore Thoma Compton Carleton*, Antverpiæ, 1649, p. 392.—Ed.]

15. An atom, as existent, must be able to be turned round. But if turned round, it must have a right and left hand, &c., and these its signs must change their place; therefore, be extended.*

(c.) Philosophy of Absolute—Distinctions of Mode of Reaching it.

I. Some carry the Absolute by assault,—by a single leap,—place themselves at once in the absolute,—take it as a datum; others climb to it by degrees,—mount to the absolute from the conditioned,—as a result.

Former—Plotinus, Schelling; latter—Hegel, Cousin, are examples.

II. Some place cognition of Absolute above, and in opposition to consciousness,—conception,—reflection, the conditions of which are difference, plurality, and, in a word, condition, limitation. (Plotinus, Schelling.) Others do not, but reach it through consciousness, &c.—the consciousness of difference, contrast, &c.; giving, when sifted, a cognition of identity (absolute). (Hegel, Cousin.)

III. Some, to realise a cognition of Absolute, abolish the logical laws of Contradiction and Excluded Middle, (as Cusa, Schelling, Hegel. Plotinus is not explicit). Others do not, (as Cousin).

IV. Some explicitly hold that as the Absolute is absolutely one, cognition and existence must coincide;—to know the absolute is to be the absolute,—to know the absolute is to be God. Others do not explicitly assert this, but only hold the impersonality of reason,—a certain union with God; in holding that we are conscious of eternal truths as in the divine mind. (Augustin, Malebranche, Price, Cousin.)

V. Some carry up man into the Deity, (as Schelling). Others

* See Kant in Krug's *Metaphysik*, p. 193.

bring down the Deity to man; in whose philosophy the latter is the highest manifestation of the former,—man apex of Deity.

VI°. Some think Absolute can be known as an object of knowledge,—a notion of absolute competent; others that to know the absolute we must *be* the absolute, (Schelling, Plotinus?)

° Some [hold] that unconditioned is to be believed, not known; others that it can be known.[a]

(*d.*) SIR W. HAMILTON TO MR HENRY CALDERWOOD.

MY DEAR SIR, *Cordale, 26th Sept.* 1854.

I received a few days ago your *Philosophy of the Infinite*, and beg leave to return you my best thanks, both for the present of the book itself, and for the courteous manner in which my opinions are therein controverted. The ingenuity with which your views are maintained, does great credit to your metaphysical ability; and however I may differ from them, it gives me great satisfaction to recognise the independence of thought by which they are distinguished, and to acknowledge the candid spirit in which you have written.

At the same time, I regret that my doctrines, (briefly as they are promulgated on this abstract subject), have been, now again, so much mistaken, more especially in their theological relations. In fact, it seems to me, that your admissions would, if adequately developed, result in establishing the very opinions which I maintain, and which you so earnestly set yourself to controvert.

In general, I do not think that you have taken sufficiently into account the following circumstances:—

1°, That the Infinite which I contemplate is considered only as *in thought;* the Infinite beyond thought being, it may be, an object of belief, but not of knowledge. This consideration obviates many of your objections.

2°, That the sphere of our belief is much more extensive than the sphere of our knowledge; and, therefore, when I deny that the Infinite can by us be *known*, I am far from denying that by us it

[a] Cf. *Discussions*, p. 12 *et seq.*—ED.

is, must, and ought to be, *believed.* This I have indeed anxiously evinced, both by reasoning and authority. When, therefore, you maintain, that in denying to man any positive cognisance of the Infinite, I virtually extenuate his belief in the infinitude of Deity, I must hold you to be wholly wrong, in respect both of my opinion, and of the theological dogma itself.

Assuredly, I maintain that an infinite God cannot be by us (positively) comprehended. But the Scriptures, and all theologians worthy of the name, assert the same. Some indeed of the latter, and, among them, some of the most illustrious Fathers, go the length of asserting, that "an understood God is no God at all," and that, "if we maintain God to be as we can think that he is, we blaspheme." Hence the assertion of Augustin: "Deum potius ignorantia quam scientia attingi."

3°, That there is a fundamental difference between *The Infinite*, (τὸ Ἓν καὶ Πᾶν), and a relation to which we may apply the term *infinite.* Thus, Time and Space must be excluded from the supposed notion of *The Infinite;* for The Infinite, if positively thought it could be, must be thought as under neither Space nor Time.

But I would remark specially on some essential points of your doctrine; and these I shall take up without order, as they present themselves to my recollection.

You maintain (*passim*) that thought, conception, knowledge, is and must be finite, whilst the *object of thought*, etc., may be infinite. This appears to me to be erroneous, and even contradictory. An existence can only be an object of thought, conception, knowledge, inasmuch as it is an object thought, conceived, known; as such only does it form a constituent of the circle of thought, conception, knowledge. A thing may be partly known, conceived, thought, partly unknown, &c. But that part of it only which is thought, can be an object of thought, &c.; whereas the part of it not thought, &c., is, as far as thought, &c. is concerned, only tantamount to zero. The infinite, therefore, in this point of view, can be *no object* of thought, &c.; for nothing can be more self-repugnant than the assertion, that we know the infinite through a finite notion, or have a finite knowledge of an infinite object of knowledge.

But you assert (*passim*) that we have a knowledge, a notion of the infinite; at the same time asserting (*passim*) that this knowledge or notion is "inadequate,"—"partial,"—"imperfect,"—

"limited,"—"not in all its extent,"—"incomplete,"—"only to some extent,"—"in a certain sense,"—"indistinct," &c. &c.

Now, in the first place, this assertion is in contradiction of what you also maintain, that "the infinite is one and indivisible" (pp. 25, 26, 226); that is, that having *no parts*, it cannot be *partially* known. But, in the second place, this also subverts the possibility of conceiving, of knowing, the Infinite; for as partial, inadequate, not in all its extent, &c., our conception includes *some part* only of the object supposed infinite, and *does not include* the rest. Our knowledge is, therefore, by your own account, limited and finite; consequently, you implicitly admit that we have no knowledge, at least no positive knowledge, of the infinite.

Neither can I surmise how we should ever come to know that the object thus partially conceived *is* in itself infinite; seeing that we are denied the power of knowing it *as* infinite, that is, not partially, not inadequately, not in some parts only of its extent, &c., but totally, adequately, in its whole extent, &c.; in other words, under the criteria compatible with the supposition of infinitude. For, as you truly observe, "everything *short* of the infinite is limited" (p. 223).

Again, as stated, you describe the infinite to be "one and indivisible." But, to conceive as inseparable into *parts*, an entity which, not excluding, in fact includes, the worlds of mind and matter, is for the human intellect utterly improbable. And does not the infinite contain the finite? If it does, then it contains what has parts, and is divisible; if it does not, then is it exclusive: the finite is out of the infinite; and the infinite is conditioned, limited, restricted,—*finite*.

You controvert, (p. 233, *alibi*), my assertion, that to conceive a thing *in relation*, is, *ipso facto*, to conceive it as finite, and you maintain that the relative is not incompatible with infinity, unless it be also restrictive. But restrictive I hold the relative always to be, and, therefore, incompatible with *The Infinite* in the more proper signification of the term, though infinity, in a looser signification, may be applied to it. My reasons for this are the following:—A relation is always a *particular* point of view; consequently, the things thought as relative and correlative are always thought restrictively, in so far as the thought of the one discriminates and excludes the other, and likewise all things not conceived in the same special or relative point of view. Thus, if we think of Socrates and Xanthippe under the matrimonial relation, not only do the thoughts of Socrates and Xanthippe exclude each other as separate existences, and, *pro tanto*, therefore are restrictive; but thinking of Socrates *as husband*, this excludes our conception of him as citizen, &c. &c. Or, to take an example from higher relatives: what is thought as the *object*, excludes what is viewed as the *subject*, of thought, and hence the necessity which compelled Schelling and other absolutists to place *The Absolute* in the indifference of subject and object, of knowledge and existence. Again: we conceive God in the relation of Creator, and in so far as we merely conceive Him as Creator, we do not conceive Him as unconditioned, as infinite; for there are many other relations of the Deity under which we may conceive Him, but which are not included in the relation of Creator. In so far, therefore, as we conceive God only in this relation, our conception of Him is manifestly restrictive. Further, the created universe is, and you assert it to be, (pp. 175, 180, 229), finite. The creation is, therefore, an act, of however great, of finite power; and the Creator is thus thought only in a finite capacity. God, in His own nature, is infinite, but we do not positively think Him as infinite, in thinking Him under the relation of the Creator of a finite creation. Finally, let us suppose the created universe, (which you do not), to be infinite; in that case we should be reduced to the dilemma of asserting *two* infinites, which is contradictory, or of asserting, the supernal absurdity, that God the Creator is finite, and the universe created by Him is infinite.

In connection with this, you expressly deny Space and Time to be restrictions, whilst you admit them to be necessary conditions of thought (p. 103-117). I hold them both to be restrictive.

In the first place, take *Space*, or Extension. Now, what is conceived as extended, does it not exclude the unextended? Does it not include body, to the exclusion of mind? *Pro tanto*, therefore, space is a limitation, a restriction.

In the same way *Time*,—is it not restrictive in excluding the Deity, who must be held to exist above or beyond the condition of time or succession? This, His existence, we must believe as real, though we cannot positively think, conceive, understand its possibility. Time, like Space, thus involving limitation, both must be excluded, as has been done by Schelling, from the sphere,—from the supposed notion of the infinite-absolute,—

"Whose kingdom is where Time and Space are not."

You ask, if we had not a positive notion of the thing, how such a name as *Infinite* could be introduced into language (p. 58). The answer to this is easy. In the first place, the word Infinite, (*infinitum*, *ἄπειρον*), is negative, expressing the negation of limits; and I believe that this its negative character holds good in all languages. In the second place, the question is idle; for we have many words which, more directly and obtrusively expressing a negation of thought, are extant in every language, as *incogitable*, *unthinkable*, *incomprehensible*, *inconceivable*, *unimaginable*, *nonsense*, &c. &c.; whilst the term *infinite* directly denotes only the negation of limits, and only indirectly a negation of thought.

I may here notice what you animadvert on, (p. 60, 76), the application of the term *notion*, &c., to what cannot be positively conceived. At best this is merely a verbal objection against an abuse of language; but I hardly think it valid. The term *notion* can, I think, be not improperly applied to what we are unable positively to construe in thought, and which we understand only by a problematic supposition. *A round square* cannot certainly be represented; but, understanding what is hypothetically required, the union of the attribute *round* with the attribute *square*, I may surely say, "the notion round-square is a representative impossibility."

You misrepresent, in truth reverse, my doctrine, in saying, (p. 160), that I hold "God *cannot* act as a cause, for the unconditioned cannot exist in relation." I never denied, or dreamed of denying, that the Deity, though infinite, though unconditioned, *could* act in a finite relation. I only denied, in opposition to Cousin, that so He *must*. True it is, indeed, that in thinking God under relation, we do not *then* think Him, even negatively, as infinite; and in general, whilst always believing Him to be infinite, we are ever unable to construe to our minds,—positively to conceive,—His attribute itself of infinity. This is "unsearchable." This is "past finding out." What I have said as to the infinite being (subjectively) inconceivable, does not at all derogate from our belief of its (objective) reality. In fact, the main scope of my speculation is to show articulately, that we *must believe*, as actual, much that we are unable (positively) *to conceive*, as even possible.

I should have wished to make some special observations on your seventh chapter, in relation to Causality; for I think your objections to my theory of causation might be easily obviated. Assuredly that theory applies equally to mind and matter. These, however, I must omit. But what can be more contradictory than your assertion, "that creation is conceived, and is by us conceivable, only as *the origin of existence*, by the fiat of the Deity?" (p. 156.) Was the *Deity not existent before the creation?* or did the *non-existent Deity at the creation originate existence?* I do not dream of imputing to you such absurdities. But you must excuse me in saying, that there is infinitely less ground to wrest my language, (as you seem to do), to the assertion of a material Pantheism, than to suppose you guilty of them.

Before concluding, I may notice your denial, (p. 108), of my statement, that time present is conceivable only as a line in which the past and future limit each other. As a position of time, (time is a protensive quantity), the present, if positively conceived, must have a certain duration, and that duration can be measured and stated. Now, does the present endure for an hour, a minute, a second, or for any part of a second? If you state what length of duration it contains, you are lost. So true is the observation of St Augustin.

These are but a few specimens of the mode in which I think your objections to my theory of the infinite may be met. But, however scanty and imperfect, I have tired myself in their dictation, and must, therefore, now leave them, without addition or improvement, to your candid consideration.—Believe me, my dear sir, very truly yours,

(Signed) W. HAMILTON.

(*e*.) Doctrine of Relation.

(Written in connection with proposed Memoir of Mr Stewart. On Desk, May 1856; written Autumn 1855.—Ed.)

I. Every Relation, (*Quod esse habet ad aliud,—unius accidens,—σχέσις,—respectivum,—ad aliquid,—ad aliud,—relatum,—*

comparatum,—sociale), supposes at least two things, or, as they are called, terms thought as relative; that is, thought to exist only as thought to exist in reference to each other: in other words, Relatives, (τὰ πρός τι σχέσιν ἔχοντα,—*relativa sunt, quorum esse est ad aliud*), are, from the very notion of relativity, necessarily plural. Hence Aristotle's definition is not of Relation but of things relative. Indeed, a relation of one term,—a relative not referred,—not related (πρός τι οὐ πρός τι), is an overt contradiction, —a proclaimed absurdity. The Absolute, (the one, the not-relative,—not-plural), is diametrically opposed to the relative,—these mutual negatives.

II. A relation is a unifying act,—a synthesis; but it is likewise an antithesis. For even when it results in denoting agreement, it necessarily proceeds through a thought of difference; and thus relatives, however they may in reality coincide, are always mentally contrasted. If it be allowed, even the relation of identity,—of the sameness of a thing to itself, in the formula A = A, involves the discrimination and opposition of the two terms. Accordingly, in the process of relation, there is no conjunction of a plurality in the unity of a single notion, as in a process of generalisation; for in the relation there is always a division, always an antithesis of the several connected and constituent notions.

III. Thus relatives are severally discriminated; inasmuch as the one is specially *what is referred*, the other specially *what is referred to*. The former, opening the relation, retains the generic name of *the Relative*, (and is sometimes called exclusively *the Subject*); whilst the latter, closing it, is denominated *the Correlative*, (and to this the word *Term* is not unfrequently restricted). Accordingly, even the relation of the thing to itself in the affirmation of identity, distinguishes a Relative and a Correlative. Thus in the judgment, "God is just," God is first posited as subject and Relative, and then enounced as predicate and Correlative.

IV. The Relative and the Correlative are mutually referred, and can always be reciprocated or converted, (πρὸς ἀντιστρέφοντα λέγεσθαι,—*reciproce, ad convertentiam dici*); that is, we can view in thought the Relative as the Correlative, and the Correlative as the Relative. Thus, if we think the Father as the Relative of the Son as Correlative, we can also think the Son as Relative of the Father as Correlative. But, in point of fact, there are here always, more or less obtrusive, two different, though not independent, relations: for the relation, in which the Father is relative and the Son correlative, is that of Paternity; while the relation, in which the Son is relative and the Father correlative, is that of Filiation; relations, however, which mutually imply each other. Thus, also, Cause and Effect may be either Relative or Correlative. But where Cause is made the Relative, the relation is properly styled *Causation;* whereas we ought to denominate it *Effectuation*, when the Effect becomes the relative term. To speak of the relation of Knowledge; we have here Subject and Object, either of which we may consider as the Relative or as the Correlative. But, in rigid accuracy, under Knowledge, we ought to distinguish two reciprocal relations,—the relation of *knowing*, and the relation of *being known*. In the former, the Subject, (that *known as knowing*), is the Relative, the Object, (that *known as being known*), is the Correlative; in the latter, the terms are just reversed.

V. The Relatives, (the things relative and correlative), as relative, always coexist in nature (ἅμα τῇ φύσει), and coexist in thought (ἅμα τῇ γνώσει). To speak now only of the latter simultaneity;—we cannot conceive, we cannot know, we cannot define the one relative, without, *pro tanto*, conceiving, knowing, defining also the other. Relative and Correlative are each thought through the other; so that in enouncing Relativity as a condition of the thinkable, in other words, that thought is only of the Relative; this is tantamount to saying that we think one thing only as we think two things mutually and at once; which again is equivalent to a declaration that the Absolute (the non-Relative) is for us incognisable, and even incogitable.

In these conditions of Relativity, all philosophers are at one; so far there is among them no difference or dispute.

Note.—No part of philosophy has been more fully and more accurately developed, or rather no part of philosophy is more determinately certain than the doctrine of Relation; insomuch that in this, so far as we are concerned, there is no discrepancy of opinion among philosophers. The only variation among them is merely verbal; some giving a more or less extensive meaning to the words employed in the nomenclature. For whilst all agree in calling by the generic name of *relative* both what are specially denominated the *Relative* and the *Correlative;* some limit the expression, *Term*, (*terminus*), to the latter, and others the expression, *Subject*, (*subjectum*), to the former; whilst the greater number of recent philosophers, (and these I follow), apply these expressions indifferently to both Relative and Correlative.

IV. CAUSATION.—LIBERTY AND NECESSITY.

(See Vol. II. p. 413.)

(*a.*) CAUSATION.

(Written in connection with proposed MEMOIR OF MR STEWART. On Desk, May 1856; written Autumn 1855.—ED.)

My doctrine of Causality is accused of neglecting the phænomenon of *change*, and of ignoring the attribute of *power*. This objection precisely reverses the fact. Causation is by me proclaimed to be identical with change,—change of power into act, ("omnia mutantur"); change, however, only of appearance,—we being unable to realise in thought either existence (substance) apart from phænomena, or existence absolutely commencing, or absolutely terminating. And specially as to power; power is the property of an existent something, (for it is thought only as the essential attribute of what is able so or so to exist); power is, consequently, the correlative of existence, and a necessary supposition, in this theory, of causation. Here the cause, or rather the complement of causes, is nothing but powers capable of producing the effect; and the effect is only that now existing actually, which previously existed potentially, or in the causes. We must, in truth, define:—a cause, the power of effectuating a change; and an effect, a change actually caused. Let us make the experiment.

And, first, of Causation at its highest extremity: Try to think creation. Now, all that we can here do is to think the existence of a creative power,—a Fiat; which creation, (unextended or mental, extended or material), must be thought by us as the evolution, the incomprehensible evolution, by the exertion or putting forth of God's attribute of productive power, into energy. This Divine power must always be supposed as pre-existent. Creation excludes the commencement of being: for it implies creative God as prior;

and the existence of God is the negation of nonentity.[a] We cannot, indeed, compass the thought of what has no commencement; we cannot, therefore, positively conceive, (what, however, we firmly believe), the eternity of a Self-existent,—of God: but still less can we think, or tolerate the supposition, of something springing out of nothing,—of an absolute commencement of being.

Again, to think Causation at its lowest extremity: As it is with Creation, so it is with Annihilation. The thought of both supposes a Deity and Divine power; for as the one is only the creative power of God exerted or put forth into act, so the other is only the withdrawal of that exerted energy into power. We are able to think no complete annihilation,—no absolute ending of existence; ("omnia mutantur, nihil interit"); as we cannot think a creation from nothing, in the sense of an origination of being without a previously existing Creator,—a prior creative power. Causation is, therefore, necessarily *within* existence; for we cannot think of a change either from non-existence to existence, or from existence to non-existence. The thought of power, therefore, always precedes that of creation, and follows that of annihilation; and as the thought of power always involves the thought of existence, therefore, in so far as the thoughts of creation and annihilation go, the necessity of thinking a cause for these changes exemplifies the facts,—that change is only from one form of existence to another, and that causation is simply our inability to think an absolute commencement or an absolute termination of being. The sum of being (actual and potential) now extant in the mental and material worlds, together with that in their Creator, and the sum of being (actual and potential) in the Creator alone, before and after these worlds existed, is necessarily thought as precisely the same. Take the instance of a neutral salt. This is an effect, the product of various causes,—and all are necessarily powers. We have here, 1°, An acid involving its power (active or passive) of combining with the alkali; 2°, An alkali, involving its power (active or passive) of combining with the acid; 3°, (Since, as the chemical brocard has

[a] I have seen an attempt at the correction of my theory of creation, in which the Deity is made to originate or create existence. That is, either existence is created by an existent God, on which alternative the definition is stultified by self-contradiction; or existence is created by a non-existent God,—an alternative, if deliberately held, at once absurd and impious.

it, "corpora non agunt nisi soluta"), a fluid, say water, with its power of dissolving and holding in solution the acid and alkali; 4°, A translative power, say the human hand, capable of bringing the acid, the alkali, and the water, into correlation, or within the sphere of mutual affinity. These, (and they might be subdivided), are all causes of the effect; for, abstract any one, and the salt is not produced. It wants a coefficient cause, and the concurrence of every cause is requisite for an effect.[a]

But all the causes or coefficient powers being brought into reciprocal relation, the salt is the result; for an effect is nothing but the actual union of its constituent entities,—concauses or coefficient powers. In thought, causes and effects are thus, *pro tanto*, tautological: an effect always pre-existed potentially in its causes; and causes always continue actually to exist in their effects. There is a change of the form, but we are compelled to think an identity in the elements of existence:—

"Omnia mutantur; nihil interit."

And we might add,—"Nihil incipit;" for a creative power must always be conceived as pre-existent.

Mutation, Causation, Effectuation, are only the same thought in different respects; they may, therefore, be regarded as virtually terms convertible. Every change is an effect; every effect is a change. An effect is in truth just a change of power into act; every effect being an actualisation of the potential.

But what is now considered as the cause may at another time be viewed as the effect; and *vice versâ*. Thus, we can extract the acid or the alkali, as effect, out of the salt, as principal concause; and the square which, as effect, is made up of two triangles in conjunction, may be viewed as cause when cut into these figures. In opposite views, Addition and Multiplication, Subtraction and Division, may be regarded as causes, or as effects.

Power is an attribute or property of existence, but not coexten-

[a] See above, vol. I. lect. iii. p. 59.—Ed.

sive with it: for we may suppose, (negatively think), things to exist which have no capacity of change, no capacity of appearing.

Creation is the existing subsequently in act of what previously existed in power; annihilation, on the contrary, is the subsequent existence in power of what previously existed in act.

Except the first and last causal agencies, (and these, as Divine operations, are by us incomprehensible), every other is conceived also as an effect; therefore, every event is, in different relations, a power and an act. Considered as a cause, it is a power,—a power to co-operate an effect. Considered as an effect, it is an act,—an act co-operated by causes.

Change, (cause and effect), must be *within existence;* it must be merely of phænomenal existence. Since change can be for us only as it appears to us,—only as it is known by us; and we cannot know, we cannot even think a change either from non-existence to existence, or from existence to non-existence. The change must be from substance to substance; but substances, apart from phænomena, are (positively) inconceivable, as phænomena are (positively) inconceivable apart from substances. For thought requires as its condition the correlatives both of an appearing and of something that appears.

And here I must observe that we are unable to think the Divine Attributes as in themselves they are, we cannot think God without impiety, unless we also implicitly confess our impotence to think Him worthily; and if we should assert that God is as we think or can affirm Him to be, we actually blaspheme. For the Deity is adequately inconceivable, is adequately ineffable; since human thought and human language are equally incompetent to His Infinities.

(*b*.) The Question of Liberty and Necessity as Viewed by the Scottish School.

Written in connection with proposed Memoir of Mr Stewart. On Desk, May 1856; written Autumn 1855.—Ed.)

The Scottish School of Philosophy has much merit in regard to the problem of the Morality of human actions; but its success in the polemic which it has waged in this respect, consists rather in having intrenched the position maintained behind the common sense or natural convictions of mankind, than in having rendered the problem and the thesis adopted intelligible to the philosopher. This, indeed, could not be accomplished. It would, therefore, have been better to show articulately that Liberty and Necessity are both incomprehensible, as both beyond the limits of legitimate thought; but that though the Free-agency of Man cannot be speculatively proved, so neither can it be speculatively disproved; while we may claim for it as a *fact* of real actuality, though of inconceivable possibility, the testimony of consciousness,—that we are morally free, as we are morally accountable for our actions. In this manner, the whole question of free and bond-will is in theory abolished, leaving, however, practically our Liberty, and all the moral interests of man entire.

Mr Stewart seems, indeed, disposed to acknowledge, against Reid, that, in certain respects, the problem is beyond the capacity of human thought, and to admit that all reasoning for, as all reasoning against, our liberty, is on that account invalid. Thus in reference to the arguments against human free-agency, drawn from the prescience of the Deity, he says, "In reviewing the arguments that have been advanced on the opposite sides of this question, I have hitherto taken no notice of those which the Necessitarians have founded on the prescience of the Deity, because I do not think these fairly applicable to the subject; inasmuch as they draw an inference from what is altogether *placed beyond the reach of our faculties*, against a fact for which every man *has the evidence of his own consciousness*."[a]

(c.) Liberty and Necessity.

(Written in connection with proposed Memoir of Mr Stewart. On Desk, May 1856; written Autumn 1855.—Ed.)

The question of Liberty and Necessity may be dealt with in two ways:—

I. The opposing parties may endeavour to show each that his thesis is distinct, intelligible, and consistent, whereas that the

[a] *Active and Moral Powers*, vol. i. *Works*, vol. vi. p. 396.

anti-thesis of his opponent is indistinct, unintelligible, and contradictory.

II. An opposing party may endeavour to show that the thesis of either side is unthinkable, and thus abolish logically the whole problem, as, on both alternatives, beyond the limits of human thought; it being, however, open to him to argue that, though unthinkable, his thesis is not annihilated, there being contradictory opposites, one of which must consequently be held as true, though we be unable to think the possibility of either opposite; whilst he may be able to appeal to a direct or indirect declaration of our conscious nature in favour of the alternative which he maintains.

The former of these modes of arguing has been the one exclusively employed in this controversy. The Libertarian, indeed, has often endeavoured to strengthen his position by calling in a deliverance of consciousness; the Necessitarian, on the contrary, has no such deliverance to appeal to, and he has only attempted, at best, to deprive his adversary of this ground of argumentation by denying the fact or extenuating the authority of the deliverance.

The latter of these lines of argumentation, I may also observe, was, I believe, for the first time employed, or, at least, for the first time legitimately employed, by myself: for Kant could not consistently defer to the authority of Reason in its practical relations, after having shown that Reason in its speculative operations resulted only in a complexus of antilogies. On the contrary, I have endeavoured to show that Reason,—that Consciousness within its legitimate limits, is always veracious,—that in generating its antinomies, Kant's Reason transcended its limits, violated its laws, —that Consciousness, in fact, is never spontaneously false, and that Reason is only self-contradictory when driven beyond its legitimate bounds. We are, therefore, warranted to rely on a deliverance of Consciousness, when that deliverance is *that* a thing is, though we may be unable to think *how* it can be.

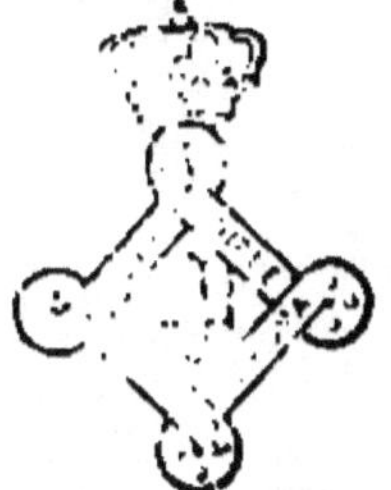

INDEX.

faculty and power, i. 178; quoted in support of the author's doctrine that the mind is never wholly inactive, and that we are never wholly unconscious of its activity, and of sundry other conclusions, 321 *et seq.*; holds that the mind is frequently awake when the senses are asleep, 324; thinks it probable that the mind is always awake, 325; gives induction of facts in support of this conclusion, 325 *et seq.*; gives analysis and explanation of the phenomena adduced, 320 *et seq.*; holds distraction and nondistraction matters of intelligence, 324; applies foregoing analysis to phenomena of sleep, 329; his doctrine illustrated by personal experience, 330 *et seq.*; by experience of those attendant on the sick, 331; by awakening at an appointed hour, 332; his general conclusions, 333 *et seq.*; his theory corroborated by the case of the postman of Halle, 334 *et seq.*; belonged to the Scoto-Gallican School of Philosophy, 392.
Judgment, *see* Elaborative Faculty.
Juvenal, quoted, i. 386; 387; ii. 343.

Kaestner, ii. 416, *see* Feelings; quoted on Descartes' doctrine of pleasure, 401.
Kames, referred to on question of mental latency, i. 363; quoted on utility of Abstraction, ii. 281.
Kant, quoted, i. 39; referred to on definition of philosophy, 49; 58; on the love of unity, 69; his anticipation of the discovery of Uranus, 70; his division of philosophy, 120; 141; admits the fact of the testimony of consciousness in perception, 291; 299; maintains that we are always consciously active, 318; 324; 373; doubts the truth of the testimony of consciousness to our Mental Unity, 373; and to our Mental Identity, 374; a Scotchman by descent, 396; his philosophy originated in a recoil against the scepticism of Hume, *ib.*; 397; his doctrine of space and time, 402; 404; ii. 8; enunciated the law by which Perception and Sensation are governed in their reciprocal relations, 99; divides the senses into two,—*Sensus Vagus* and *Sensus Fixus*, 157; 185, *see* Necessity; quoted on proper application of term *Abstraction*, 282; 416; 428; 472; on Beauty, 516, *see* Feelings; referred to on the Sublime, 513; quoted, 515, *see* Feelings; his analysis of judgments, 526.
Keckermann, distinguished Reflexion from Observation, i. 234-5; ii. 348.
Kepler, i. 76.
Know thyself, i. 26.
Knowledge, discriminated from intellectual cultivation, i. 8; whether knowledge or mental exercise the superior end, considered, 8-13; popular solution of this question,—that knowledge is the higher end,—and its results, 9; knowledge either practical or speculative, *ib.*; the end of practical knowledge, *ib.*, 10; the end of speculative knowledge, 10; the question resolved by philosophers in contradiction to the ordinary opinion, 11; this contradiction even involved in the term *Philosophy*, *ib.*; authorities adduced as to mental exercise being higher than knowledge,—Plato, Prior, Aristotle, Aquinas, Scotus, Malebranche, Lessing, Von Müller, Jean Paul Richter, 12-13; knowledge philosophical, scientific or rational, and empirical or historical discriminated, 53-8; empirical, the knowledge that a thing is,—τὸ ὅτι, 55-6; examples of, 56; this expression how rendered in Latin, *ib.*, *see* Empirical; philosophical, the knowledge why or how a thing is,—τὸ διότι, 58; man's knowledge relative, 61, 137-48; the representation of multitude in unity, 68-9, *see* Unity; faculties of, one grand division of powers of mind, 123; testimonies to relativity of,—Aristotle, Augustin, Melanchthon, elder Scaliger, 138-40; all existence not comprised in what is relative to us, 140; this principle has two branches, 141; the first, 141-45; the second, 145-48; three senses in which knowledge relative, 148; two opposite series of expressions applied to, *ib.*; faculty of, regarded by some philosophers as the fundamental power of mind, 187; distribution of the special faculties of, ii. 1 *et seq.*; the special faculties of, evolved out of consciousness, 10; enumeration of the special faculties of, 10-17, 23-28; *a priori* and *a posteriori*, 26; relation of, to experience, how best expressed, 27; special faculties of, considered in detail, 28 *et seq.*; the distinction of Intuitive or Immediate, and Representative or Mediate Knowledge, 66 *et seq.*, and i. 218-19; the contrasts between these two kinds of, ii. 69-71; this distinction taken by certain of the schoolmen, 71; that the relation of knowledge supposes a similarity, or sameness, between subject and object an influential principle in philosophy, 120-21; the opposite of this principle held by some, 121; refuted, 122 *et seq.*; the essential peculiarities of knowledge, 431 *et seq.*
Knowledges, term used by Bacon and Sergeant, i. 57-8.
Krug, i. 47; on definition of philosophy, 49-50; attacked the Kantian division of the mental phenomena, 187, ii. 421, *see* Feelings; 418; 419.
Kuster, i. 162.

Laboulinière, ii. 161.
Lactantius, his doctrine of mental powers, ii. 6; 35; denied the necessity of visual species, *ib.*
Laertius, Diogenes, i. 45; 114; uses *σύν-εσις* for consciousness, 192.
Language, Does it originate in General

Appellatives or by Proper Names? ii. 319 *et seq.*; this the question of the *Primum Cognitum*, 320; 1. That all terms, as at first employed, are expressive of individual objects, maintained by Vives and others, 320; Vives quoted to this effect, *ib.*; Locke quoted, 321; Adam Smith quoted to same effect, 321-4; 2. An opposite doctrine maintained by many of the schoolmen, 324 *et seq.*; by Campanella, 324; Leibnitz quoted to this effect, 324-5; Turgot cited to same effect, 326; 3. A third or intermediate opinion,—that language at first expresses only the vague and confused, 327 *et seq.*; Perception commences with masses, 327, *see also* 149; the mind in elaborating its knowledge proceeds by analysis from the whole to the parts, 328-34; Degerando quoted to this effect, 329-30; the intermediate opinion maintained by Aristotle, 330-1; and by Julius Cæsar Scaliger, 331; reason of the ambiguity of words denoting objects that lie within the mind, 417-18.

Laromiguière, quoted on hypothesis of Occasional Causes, i. 388 *et seq.*; on Pre-established Harmony, 392 *et seq.*; on Plastic Medium, 391; on Physical Influence, 395 *et seq.*; quoted on Abstraction, ii. 284-6.

Latency, mental, what, and its three degrees, i. 339 *et seq.* *See* Consciousness.

Latin language, expresses syntactical relations by flexion, i. 253.

Laval, Comtesse de, case of, i. 343.

Law, Bishop, his doctrine of substance, i. 155.

Le Clerc, *see* Clerc.

Lee (Dr Henry), referred to on Locke, ii. 159.

Leibnitz, referred to on definition of philosophy, i. 52; 60; 135; first to limit the term *capacity* to passivity of mind, 177; regarded faculty of knowledge as the fundamental power of mind, 187; quoted on veracity of consciousness, 285; 300; held hypothesis of Pre-established Harmony, 300, 303; opposed Locke's doctrine that the mind is not always conscious, 317; but does not precisely answer the question mooted, 318; referred to on *minima* of sense, 351; the first to proclaim the doctrine of mental latency, 361; unfortunate in the terms he employed to designate the latent modifications of mind, 362; referred to on our mental identity, 374; ii. 8; 20; 195, *see* Necessity; 209; 321, *see* Language; 319; 351, *see* Regulative Faculty; 452, *see* Feelings; 528.

Leidenfrost, ii. 156; the first to distinguish the Vital Sense from the Organic Senses, 157.

Leo Hebræus, ii. 34.

Lessing, quoted, i. 13. *See* Knowledge.

Lewd, its etymology, i. 74.

Liberty of Will, ii. 411 *et seq.*; the question of, as viewed by the Scottish school, 541; may be dealt with in two ways, 542-3.

Lichetus, i. 251.

Locke, i. 73; adopted Gassendi's division of philosophy, 120; quoted on power, 174-5; his doctrine of Reflection as a source of knowledge, 235; held that the mind cannot exist at the same moment in two different states, 249; his doctrine on this point refuted by Leibnitz, 250; denied that the mind is always conscious, 314-17; his assumption that consciousness and the recollection of consciousness are convertible, disproved by somnambulism, 319; erroneously attributed the doctrine of latent mental modifications to the Cartesians, 361; on mental identity, 371; his doctrine of Perception, ii. 33; general character of his philosophical style, 55-6; quoted on the doctrine that the secondary qualities of matter are merely mental states, 57-8; his distinction of primary and secondary qualities, 109; did not originate the question regarding plurality of senses under Touch, 150; 177; neglected the Critical Method in philosophy, 194; has his philosophy been misrepresented by Condillac? 193 *et seq.*; Stewart, quoted in vindication of, 196-8; Stewart's vindication of, unsatisfactory, 198; Condillac justified in his simplification of the doctrine of, *ib.*; his Reflection compatible with Sensualism, 199; 201; quoted on Conceptualism, 310; 321, *see* Language; 390, *see* Causality; 396.

Logic, defined, i. 43, 123; as initiative course of philosophy, 11, 128; class of, how to be conducted, 14-16, *see* Philosophy; presupposes a certain knowledge of the operations of the mind, 62; controversy among the ancients regarding its relation to philosophy, 114-15; why usually designated an *art*, 118; a nomological science, 123; Dianoetic best name of, 125; its place in philosophy, and in a course of philosophical instruction, 128.

Lombard, Peter, ii. 71.

Lossius, *Lexicon*, ii. 397; 435; 475.

Lucan, quoted, ii. 492.

Lucretius, quoted, i. 265; 301; ii. 39; 486; on mixed feeling of the sublime, 514.

Lüders, ii. 442.

Luther, i. 67; 69.

Lydus, Priscianus, on unity of knowledge, i. 69; the Platonic doctrine of Perception as expounded by, ii. 38.

Maass, i. 363.

Mackintosh, Sir James, i. 131; his great memory, ii. 226.

Macrobius, referred to, on definition of philosophy, i. 51; 164.

Maine de Biran, ii. 252; 390, *see* Causality.

Major, John, referred to on Intuitive and Abstractive Knowledge, ii. 71.

END OF THE SECOND VOLUME.

www.ingramcontent.com/pod-product-compliance
Lightning Source LLC
Chambersburg PA
CBHW031936290426
44108CB00011B/572

9783742803566